The Best Gun in the World

The

BEST

in the

GUN WORLD

GEORGE WOODWARD MORSE
AND THE SOUTH CAROLINA
STATE MILITARY WORKS

Robert S. Seigler

THE UNIVERSITY OF SOUTH CAROLINA PRESS

Published by the University of South Carolina Press
Columbia, South Carolina 29208

www.sc.edu/uscpress

Manufactured in the United States of America

26 25 24 23 22 21 20 19 18 17
10 9 8 7 6 5 4 3 2 1

Library of Congress Cataloging-in-Publication Data
can be found at http://catalog.loc.gov/.

ISBN: 978-1-61117-792-3 (cloth)
ISBN: 978-1-61117-793-0 (ebook)

CONTENTS

ACKNOWLEDGMENTS

So many people have committed decades of research to the life of George Woodward Morse and the South Carolina State Military Works (State Works) that I feel like little more than an interloper who has collated their efforts into a single volume. That being true, it has been nothing short of a privilege and honor to do so. The Board of Directors of the Museum and Library of Confederate History in Greenville, South Carolina, conceived the idea for this book. Led by Executive Director V. Michael Couch Sr., they desired to accurately preserve the history of the State Works and the people who were intimately involved with it, especially George Woodward Morse. It was only after the board asked me to write the book that I realized I had driven past the site of the State Works hundreds of times and that one of my great-grandfathers had carried a Morse carbine during the latter stages of the war. Curator and board member Webster Jones is the museum's firearms specialist, and he tutored me in the workings of Morse's designs. Another board member, Barton Cox, has spent innumerable hours studying Morse and the State Works over the years and has gathered a large file of information which became the framework for the book. Barton is a tenacious researcher who shared all of his knowledge with me. His analysis and encouragement during the research and writing phases of the project have been invaluable. The primary repository of historical records dealing with the State Works is the South Carolina Department of Archives and History. Patrick McCawley is the supervisor of archival processing there. His professional advice, research assistance, and willingness to introduce me to other researchers who have knowledge of Morse and the State Works proved to be of great benefit.

Several men, both living and deceased, have been instrumental in preserving the firearms and legacy of George Woodward Morse and the State Works. It is their work that provided the basis for my research. Three men once owned large collections of Morse's firearms. George W. Wray Jr. was not only a collector of Morse firearms and related artifacts, but he was a prolific researcher into Morse's personal background. His collection is preserved at the Atlanta History Center. It includes an impressive file of genealogical data on Morse, his family, and his achievements. Dr. Gordon L. Jones, senior military historian and curator at the Atlanta History Center, has been most gracious in allowing me access to Wray's artifacts and files. Dr. Jones's expertise and knowledge of George Morse have been invaluable in writing this book. A contemporary of Wray was Dr. H. Lloyd Sutherland, a veterinarian in Union, South Carolina, who studied Morse and once owned a large collection of his guns. Sutherland presented his findings in an article, "Arms Manufactury in Greenville County," in the early 1970s. Though Sutherland's collection was dispersed after his death, his legacy has been preserved by Dr. Jack Meyer, retired professor of history at the University of South Carolina. Dr. Meyer graciously shared his extensive, long-standing knowledge of Morse firearms and South Carolina history and has

provided helpful guidance for my research. A third man, Dr. John Murphy, accumulated an extensive collection of Civil War firearms and accoutrements, including an impressive number of Morse artifacts, and donated it to the Greensboro (North Carolina) Historical Museum upon his death. Jon A. Zachman, curator of collections, Greensboro Historical Museum, most kindly opened the Murphy collection for in-depth examination.

Barry L. Stiefel, PhD, is in the Historic Preservation and Community Planning Program at the College of Charleston. Stiefel is David Lopez's biographer, and he patiently answered many of my questions about Lopez's life and family. Many others have contributed detailed research which has helped to complete this project, and to them I am grateful.

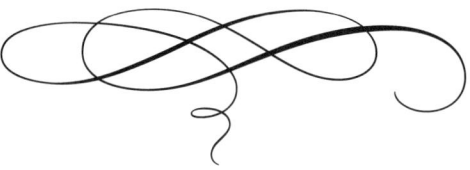

Introduction

About a year after South Carolina seceded from the Union, the state government recognized the acute need to expand its capability of manufacturing weapons. Taking action in early 1862, the Executive Council created an armory called the South Carolina State Military Works (State Works) and assigned to it a mission of manufacturing cannon, shot, and shell, as well as making and repairing small arms and other implements of war. Located briefly at a temporary location on the State House grounds in Columbia, the State Works was soon moved to a permanent site in the small Upcountry town of Greenville, where it operated until April 1865.

South Carolina ex-governor William Henry Gist and a prominent Charleston builder, David Lopez, were responsible for creating, constructing, and managing the State Works

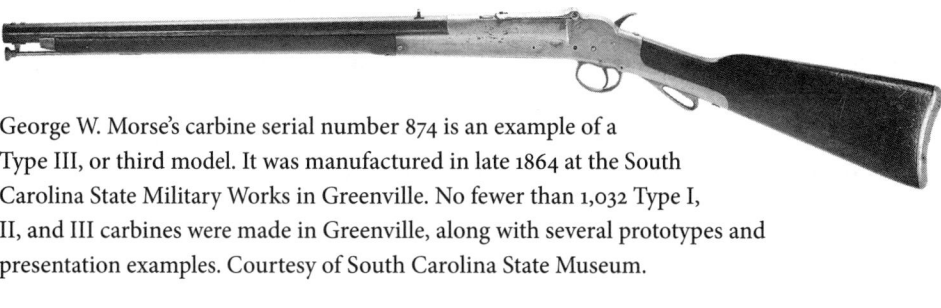

George W. Morse's carbine serial number 874 is an example of a Type III, or third model. It was manufactured in late 1864 at the South Carolina State Military Works in Greenville. No fewer than 1,032 Type I, II, and III carbines were made in Greenville, along with several prototypes and presentation examples. Courtesy of South Carolina State Museum.

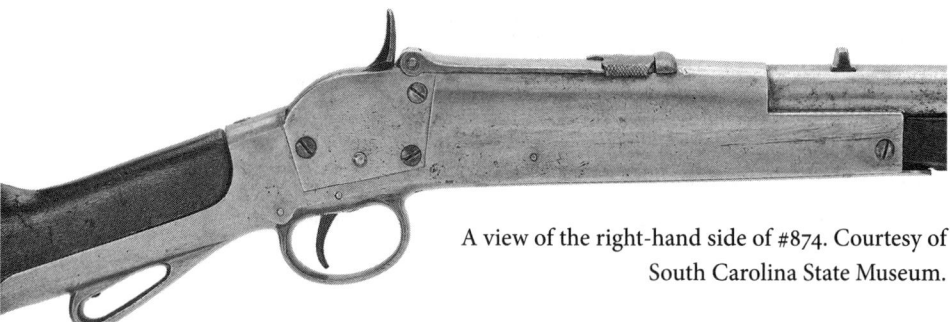

A view of the right-hand side of #874. Courtesy of South Carolina State Museum.

in 1862. An arms inventor, George Woodward Morse also appeared on the scene in 1862 to manufacture his innovative breech-loading carbine at the Greenville workshop, ultimately producing more than one thousand of them. In August 1863, a Charleston machinist, J. Ralph Smith, took over from Lopez as general superintendent, and in February 1865, Governor A. G. Magrath appointed another Charleston machinist, James M. Eason, to supervise Smith in managing the facility.

Amidst the turbulence of war, a complex series of events culminated in Greenville, where an interaction took place between Lopez and Morse, two men of nearly the same age but from very different backgrounds. Both were stellar innovators at their respective crafts, but neither is well remembered by history. Though many men played a role in the management of the State Works, it was primarily Lopez and Morse who defined it.

Both were exceptional men. David Lopez Jr. was a slaveholding descendent of Sephardic Jews and a prominent Charleston building contractor. He was an intelligent, hardworking man, who started out as a carpenter and became an innovator in building large structures. Lopez built Charleston's Institute Hall, Kahal Kadosh Beth Elohim (KKBE) Synagogue, and a large four-story department store as well as numerous smaller structures. He was instrumental in choosing the Greenville location for the State Works and personally supervised the construction of the physical plant there, not realizing the extreme limitations under which it would be forced to operate as the war progressed. Lopez's decisions were not flawless, but his eighteen-month association with the State Works resulted in the establishment of an armory that persevered through financial adversity and managed to remain operational until the very end of the war.

George Woodward Morse was also a hardworking man who rose to the top of his profession, but he was more complicated and worldly than Lopez. He was a New Hampshire–born machinist by trade, and later a surveyor, engineer, Louisiana slaveholding plantation owner, and firearms inventor who rubbed shoulders with governors, congressmen, secretaries of war, and the president of the Confederate States. He lived in England for two years, where he interacted with engineers and statesmen of international stature. Firmly believing that his inventions would literally change the world, Morse aggressively promoted himself and his ideas. His supporters correctly called him an inventive genius. Three years after Morse's death in 1891, a detractor characterized him: "What a brazen imposter! a prevaricator! a falsifier! a consummate, designing, scheming rebel! wickedly energetic and enthusiastic."[1]

In 1856 Morse invented not only an improved breech-loading firearm, which some said was the best in the world at the time, but, more importantly, the first fully functional metallic, center-fire cartridge. It was the combination of the two that was considered by the U.S. Congress in 1876 to be a complete success after centuries of failed attempts to solve the problems of breechloaders.[2] Though Morse held three patents for breech-loading firearms by 1858, his greatest contribution was a cartridge design upon which the modern center-fire cartridge is based. In fact Morse himself wrote that the cartridge, not the firearm, was the very soul of his invention.[3] Yet his name remains relatively unknown, and his contributions to the development of firearms and cartridges have never been completely appreciated.

Morse secured several contracts with the U.S. government to produce his cartridge and pioneering breechloader design in the late 1850s. Though he suffered professional

James J. Mackey (1817–1901), South Carolina gunsmith who, as foreman of the gun-repair shop at the South Carolina State Works, made a one-of-a-kind, presentation cased set of George W. Morse's brass-frame carbine intended as a gift to Governor M. L. Bonham in 1864. It is now owned by the McKissick Museum of the University of South Carolina and is on display at the South Carolina State Museum in Columbia. Photograph courtesy of Upcountry History Museum, Furman University, Greenville, South Carolina.

setbacks between 1856 and 1862, he finally experienced success when the State Works produced more than a thousand copies of his brass-frame carbine between 1863 and 1865.

As with many bright and gifted men, friction arose between Lopez and Morse, but each played an essential role at the State Works. In the end, it was Lopez who created the State Works and kept it running, and it was Morse's carbine that gave it a place in history. Without the State Works, it is very likely that Morse's carbine would never have been mass-produced. Without Morse's carbine, the State Works would likely be remembered as a small munitions and materiel factory. This is the story of the South Carolina State Military Works, David Lopez, and George Woodward Morse, inventor of an innovative breech-loading carbine and the modern, center-fire cartridge.

The pinnacle of achievement for both Lopez's armory and Morse's inventions is characterized by a one-of-a-kind firearm, a presentation cased-set made for South Carolina governor M. L. Bonham which represents both the highest-quality craftsmanship to come out of the State Works and the culmination of Morse's designs. James J. Mackey made the gun for Bonham at the State Works in 1864. Mackey, born in New York City on January 26, 1817, was a firefighter there as a young adult. As early as 1841, he moved to Columbia, South Carolina, where he was a gunsmith until the outbreak of the war. By the summer of 1863 the forty-six-year-old gunsmith was foreman of the musket repair shop at the State Works. (See color plate 1.)[4]

The State Works received a payment of $345 from the "Hon. M. L. Bonham for one model MBL [Morse's Breech Loading] carbine with cartridges, implements, etc. complete," on December 23, 1864, only five days after he left office as governor.[5] This one-of-a-kind Morse firearm is a cased set with three barrels, a 24-gauge shotgun, a .45 caliber rifle, and a .52 caliber carbine.[6] It has a Type III Morse-designed brass frame with no serial number, indicating that it was manufactured as a gift and not as a military production

carbine, which shared an identical brass receiver but had a permanently affixed carbine-length barrel. The fact that Bonham's gun is a Type III frame definitively places its location of manufacture at Greenville and its date of manufacture as no earlier than late 1863 and probably 1864. Each barrel has its own cherry forestock, and each is threaded in such a way that it screws into the brass frame, just like the Morse guns made by Nathan Muzzy in Worcester, Massachusetts, in 1857 and 1858. Each rifled barrel has a cleaning rod and brass rod holder, but the shotgun barrel is missing its cleaning rod. The shoulder stock is walnut and has finely cut checkering on the hand grip, unlike Morse's military carbines of similar design. Accoutrements include three horn powder measures and three brass bullet molds. It also has three metal tools used to load powder and bullets into the cartridge cases and a punch tool for cutting the india rubber wad and center hole for the primer. Several modern-cast .45 and .52 caliber conical balls and .45 caliber round balls are in the case. There are 53 brass shell casings in the set, and most are primed with expended copper percussion caps and are of the Muzzy-style cartridge with an india rubber or gutta percha wad. Two cartridges are open, exposing Morse's distinct "anvil," which offered resistance to the percussion cap when struck by the hammer.[7]

Both Bonham and Morse confirmed the provenance of the cased set. Bonham wrote in 1884 that Mackey had made the "beautiful Morse rifle for me when I was Governor."[8] Morse recalled in 1885 that the carbine was made for Bonham at the State Works.[9] Bonham also wrote that Morse had intended to present the cased set to him as a gift, but he declined to receive it as such, likely fearing that it would be perceived as a bribe. Morse might have intended the gun as an inducement to encourage Bonham to use his influence to continue production of the carbine at the financially struggling State Works, or at least not to sell the State Works, a path Bonham had pursued when he was governor. It is equally likely that Morse intended the cased set as a token of appreciation to Bonham for canceling a public auction of the State Works only days before he left office. Bonham offered to buy it at cost as soon as he left office, which he recalled was $410 and "some cents perhaps" or about $84 dollars in gold. He was correct in recalling that the cost of a Morse carbine at the time was about $400, but State Works records show that Bonham paid $345 for the gun in December 1864.

Mackey, who was employed continuously at the State Works through at least the end of February 1865, was a gunsmith and farmer in Greenville after the war.[10] Bonham sent the broken gun to Mackey for repair in 1884 and wrote, "I prized it very highly but do not know if it can now be utilized. But be pleased to put it in complete working repair for me, and let me know when finished, with the bill."[11] The gun remained with Mackey's family for several decades. Bonham died in 1890; Mackey, in 1901.[12] His descendants found the cased set along with accompanying letters of documentation in their attic in 1929 and returned it to Bonham's family, who donated it to the South Caroliniana Library in 1942. In 1974 it was transferred to the University of South Carolina's McKissick Museum, and today it is on display at the South Carolina State Museum.[13]

CHAPTER 1

George Woodward Morse

George Woodward Morse ca. 1846–1847. Courtesy of Atlanta History Center.

George Woodward Morse was an eighth-generation American of English descent. The American patriarch of the family was Anthony Morse, who immigrated to New England in 1635. Stephen Morse, George's grandfather, was among the first settlers of Haverhill, New Hampshire, where George was born.[1] Bryan Morse, George's father, was a blacksmith, cabinetmaker, staunch abolitionist, and preacher in the Methodist Episcopal Church in Haverhill. He married twice. With his first wife, Susannah Stevens, he had six children, including George, and with his second wife, Eliza D. Torr Repill, there were three offspring.[2] We know very little about the relationship between Bryan and his children, but the four who moved to Louisiana in the 1830s held views of slavery that were diametrically opposed to Bryan's. Reflecting the division within the family concerning slavery, the *New Hampshire Gazette* printed in 1850 that George's older brother, Peabody, had emigrated many years ago to New Orleans, "where he forgot all his father's counsels, who was an early and zealous abolitionist."[3] Though George became a strong advocate for both the personal and public ownership of slaves, he maintained a relationship with his father until the late 1850s and with his brother, Isaac, until well after the end of the Civil War.

Haverhill Academy, Haverhill, New Hampshire, where George W. Morse was educated. By permission of Haverhill Historical Society.

Reverend Bryan Morse (1781–1863), George's father. He was a blacksmith, cabinetmaker, Methodist Episcopal preacher, and zealous abolitionist. Four of his children spent most of their adult lives as slaveholders in Louisiana. Courtesy of Atlanta History Center.

Many writers incorrectly claim a close relationship between George W. Morse and Samuel Finley Breese Morse, coinventor of the Morse code. The error originated in George's obituary and was perpetuated in a variety of journals and publications.[4] One of George's attorneys, James A. Skilton, wrote in 1872 that George was the namesake and relative of S. F. B. Morse, who had died that April. Skilton was probably embellishing the relationship between the two men so that the U.S. Congress would look favorably upon Morse's current appeals for extension of his patents.[5] George, himself, never claimed a close relationship. In all likelihood, the two men's common ancestor was Anthony Morse, George's great-great-great-great-grandfather.

George and his five full siblings, Horace Bassett, Peabody Atkinson, Priscilla Peabody, Isaac Stevens, and Rebecca Carleton, were born in Haverhill. All but Horace played a role in George's adult life. His three half-siblings—Joseph, Mary, and Virginia—were the children of Bryan and Eliza; they were all born in Haverhill and all died at a young age in the 1820s.

The oldest, Horace Bassett, was born in 1804. He was graduated from Dartmouth in 1823 and was the principal of Portsmouth Academy when he died in a drowning accident in Portsmouth Harbor in 1825 at the age of twenty-one.[6] The family was soon divided. George's father; his brother, Isaac; and probably his sisters moved to Lowell, Massachusetts, in 1833. George's father later moved to Groveland, Massachusetts, where he died in 1863.[7] Also in the early 1830s, Peabody began what would become a family migration to

Louisiana, and eventually Priscilla, Rebecca, and George joined him in a settled lifestyle there by 1840.

Though the fifth-born sibling, Isaac Steven, remained in Massachusetts, he played a significant role in George's later life. He attended Dartmouth College and Cambridge Law School, was admitted to the bar in 1840, practiced law in Lowell and Cambridge, Massachusetts, and was district attorney for Middlesex County from 1855 to 1871.[8] Highly respected in his profession, Isaac maintained contact with George well into the 1870s, assisting him in his postwar lawsuits against the federal government.

Peabody Atkinson Morse was the second-born and was seven years older than George. His major influence on the family was that he led three of his siblings, including George, to a life in Louisiana. Peabody was born in 1805 in Haverhill and was graduated from Dartmouth College in the class of 1830.[9] From 1830 to 1833 he lived in Fredericksburg, Virginia, where he studied law under Virginia Court of Appeals justice Francis Taliaferro Brooke.[10] Peabody established a law practice in Natchitoches, Louisiana, about 1833.[11] He married Virginia Sompayrac in 1837 and established the nearby town of Ninock the same year. Peabody served in the Louisiana House of Representatives and rose to the rank of major general in the Louisiana Militia.[12] About 1849, Peabody moved to San Francisco, California, where he served as chief justice of the Superior Court of San Francisco in 1850 and 1851.[13] By August 1855 Peabody moved back to Louisiana, where he remained active in politics and in managing his wife's large estates until his death in 1878.[14] George's two younger sisters, Priscilla and Rebecca, followed Peabody to Natchitoches in the 1830s

Peabody Atkinson Morse (1805–1878), George's brother, photograph taken ca. 1858–59. Peabody was a Louisiana attorney, chief justice of the Superior Court of San Francisco, and major general in the Louisiana militia. Courtesy of Atlanta History Center.

and, in 1840, both married Virginia Sompayrac's brothers, deepening the Morse family roots in northwest Louisiana. Priscilla, born in 1814, married Adolphe Sompayrac, and Rebecca, born in 1820, married Paul Victor Sompayrac. George soon joined his siblings in Louisiana.

George was born in Haverhill, New Hampshire, on May 10, 1812, the third-born child.[15] He was named for his father's good friend and prominent Haverhill attorney and banker, George Woodward (August 20, 1776–December 5, 1836).[16] George Morse was educated at the Haverhill Academy, the extent of his formal education. He displayed mechanical and inventive skills at an early age. His personal letters and professional reports from the 1840s through the 1870s reflect a man of intelligence, education, and creativity. Morse's grammar, vocabulary, and mastery of the complex technical aspects of the disciplines of surveying and civil engineering seem to reflect a college education, though he had none.

Various family traditions and undocumented claims state incorrectly that George moved to Louisiana in the early 1830s when, in fact, he was living in Boston and England until very late in the decade.[17] The correct version of events comes from an affidavit written in 1870 by George's brother and Boston-area attorney, Isaac, and a deposition given by George in 1872.[18] George testified that he worked with Captain Eliphalet Smith in Quincy Point, Massachusetts, in 1829 or 1830, when he was seventeen or eighteen years old. In 1830 or 1831, George moved to Boston to work with Otis Tufts, a machinist who invented a steam pile driver. After only fifteen months of employment, Morse was foreman of Tufts's machine shop making steam engines, printing presses, and running gear for railroad cars. Morse even lived in Tufts's home in 1834 and 1835.[19]

It was during his years spent with Tufts that Morse began to acquire experience with mechanical processes. His obituary states, "as a boy [he] identified with labor-saving inventions and manufactures in Boston."[20] Morse assisted Tufts in the development of a simplified printing press that could print eighteen sheets per minute. When they added steam power to the new machine, it became the first steam-powered printing press in the United States. Tufts introduced the new machine to the public, and Morse set it up in various Boston offices and even learned to print books on it.[21] Because the new press was so promising, a group of capitalists planned to apply for patents in England and delegated Morse as their representative.[22] Accompanying the new invention, Morse sailed for London about 1835 on a two-year journey that exposed him to high-ranking British officials and engineers who would later play a role in the development of his firearms. The printing press proved to be such a labor-saving machine that some workmen intentionally damaged it to prevent its coming into general use. Morse repaired it and subsequently printed three novels by the Irish author Marguerite Gardiner, Countess of Blessington, and a popular annual she edited in 1837 called the *Keepsake*. Building on his success in England, Morse made a contract with noted engineer brothers George and Sir John Rennie of Blackfriars Road, London, to manufacture one hundred presses. While still in London, Morse worked on a new steam engine, which brought him into contact with a noted American inventor, Jacob Perkins; a Scottish diplomat, Lord Napier; and another famous English engineer, I. K. Brunel. During that time the Rennies offered Morse an annual salary of five hundred pounds if he would remain as superintendent of their machine works, but he declined the offer.[23]

Between 1830 and 1837, Morse developed from an eighteen-year-old machinist's apprentice into a twenty-five-year-old young man who had helped develop a new invention with worldwide practical applications and had accompanied the device across the ocean as its sole representative. His contact with Tufts provided him with two critical experiences. First, it allowed him to acquire practical skills and knowledge of manufacturing processes which led to his later inventions.[24] Second, it fostered his inventive spirit and allowed him to participate in a highly successful invention from concept to market. His sojourn in England provided him with invaluable experience in explaining innovative, complicated technology to skeptics, and in becoming comfortable around men of wealth, power, and professional stature. These experiences paid off in the late 1850s when he was trying to get the U.S. government to see the remarkable features of his new breechloader and its accompanying metallic, center-fire cartridge.

After a two-year stay in London, Morse arrived in New York onboard the packet ship *St. James* on December 27, 1837.[25] He soon joined Peabody, Priscilla, and Rebecca in Natchitoches. A gap in our knowledge of Morse's exact whereabouts exists for the years 1838 and 1839, but he was likely living in Louisiana at that time. Isaac recalled after the Civil War that George returned to the United States about 1839 and soon moved to Louisiana.[26] It is possible that George made a second trip to England in 1838 or 1839, but more likely Isaac remembered the date incorrectly. A search of city directories from England between 1836 and 1840 did not reveal Morse's name. We know that George was living in Louisiana in 1840.[27] The U.S. Congress passed an act on September 4, 1841, giving land to the state of Louisiana. The Louisiana secretary of state appointed Morse as one of the agents to locate the land, and Morse accepted the position on May 30, 1842.[28] He served as a deputy U.S. surveyor in the Northwestern District of Louisiana for the remainder of the decade.[29]

Morse married a young widow, Marianne Bludworth Terrett, on June 26, 1848, at Trinity Episcopal Church in Natchitoches.[30] He was a thirty-six-year-old Methodist Episcopal, and she was a twenty-three-year-old Catholic. The marriage would last until George's death forty years later. Marianne was born in Natchitoches on October 11, 1824, and had married a Virginian, Burdett Ashton Terrett, in 1840. Marianne delivered her first-born at Jefferson Barracks, near St. Louis, Missouri, on May 18, 1842.[31] Terrett was captain of Company A, First United States Dragoons, stationed at Fort Scott, Kansas, when, on March 17, 1845, he was killed by the accidental discharge of his pistol while dismounting his horse.[32] Their second child, James Bludworth Terrett, died at the age of one month on March 1, 1845, only two weeks before Terrett's death. The grieving widow and her son, Burdett Ashton Terrett Jr., soon returned to live with her family in Natchitoches, where she married Morse three years later.[33]

Both George and Marianne brought land and slaves into the marriage.[34] After having lived for less than ten years in Louisiana, George owned twelve slaves and had acquired a three-hundred-acre plantation on Fausse River, another parcel of land on Bayou Saline, and a four-hundred-acre tract in Caddo Parish on the Red River.[35] Marianne owned a half-share in a seven-hundred-acre plantation with her brother, Poitevent Bludworth Jr., on Old River, also called Sabine River, in Natchitoches Parish. She also owned a half-share in seventeen slaves and another three outright.[36] Nat, a thirty-year-old slave living on Marianne's plantation, was a carpenter, and was with Morse at the State Works from

George Woodward Morse and his wife, Marianne Bludworth Terrett Morse, are shown in this daguerreotype made about 1853–54. George is holding Lelia in his lap, and Peabody is to the left of his mother, who is pregnant with their third child, Bryan. Courtesy of Atlanta History Center.

1862 until the end of the war. In 1850 George was a surveyor in Natchitoches, and he owned twenty-five slaves and real estate valued at twenty thousand dollars.[37]

Marianne bore six children.[38] Her first husband was the father of two, an unnamed infant and Burdett Terrett Jr., who lived with his mother and stepfather, George, until he went off to the U.S. Military Academy at West Point in 1860. When the war broke out, he refused to take the oath of allegiance to the United States and left the academy on April 22, 1861, to side with the South.[39] Burdett was severely wounded at the Battle of Shiloh but survived the war. He died in Natchitoches in 1874 and was buried at the American Cemetery there. Morse was the father of three children, and we do not know which man fathered the sixth, who probably died in infancy. Peabody Atkinson Morse was born in Natchitoches in 1849, served as a private in Company A, Second Regiment, Louisiana Cavalry, and died in Louisiana in 1872. Lelia Evariste Morse was born in 1851 in Natchitoches and never married. She died in Washington, D.C., on October 2, 1914, and was buried there at Oak Hill Cemetery near her mother and father. Bryant Herbert Morse, probably named for George's father, was born in 1854 in Louisiana and died in Washington, D.C., in 1920.[40]

At some time during the years George lived in Louisiana, he was a colonel in the Fifth Brigade, Louisiana Militia, which his brother, Peabody, commanded, but it is not known

how long he served. Morse only occasionally used the title of "colonel." Workers at the South Carolina State Works referred to him as such, and in April 1874 he presented a fifteen-page memorial to the Forty-third U.S. Congress under the name "Colonel George W. Morse." But when he presented another memorial to the Forty-fourth Congress only two years later, he did not use any military rank. Likewise, in numerous letters and documents signed by Morse, he did not refer to himself as "colonel."[41]

Louisiana governor Joseph M. Walker appointed the forty-year-old Morse to the post of state engineer of Louisiana, or as Morse described it, chief engineer for the state of Louisiana, in 1852.[42] Morse implied in his 1853 annual report that he assumed the office sometime between March 1852 and April 1853 and stated in 1877 that he served as chief engineer under Walker, who left office in January 1853.[43] According to Walker's successor, Paul Octave Hebert, Morse was state engineer for the years 1853 and 1854.[44] Morse's office was in Baton Rouge, and George and Marianne made the 150-mile move from Natchitoches to the state capital probably by the spring of 1853. As state engineer, Morse reported to Louisiana legislators and socialized with his neighbor Governor Hebert, gaining additional experience that would serve him well in future personal dealings with high-ranking men. As state engineer, Morse had widespread responsibilities. He managed the office efficiently and on a tight budget. In 1853 he recommended repeal of a one-year-old law that allowed the state engineer a thousand-dollar contingency fund for travel and office expenses, stating that his office would be run strictly on a voucher system and would always be prepared to account for its expenses. Morse demanded a strict accounting of all disbursements, requiring all orders to be copied into an order book and all letters to be recorded. He insisted that, in his absence from the office, a good clerk be available to handle urgent matters. Morse did plan to travel, not only to make the surveys and examinations of waterways as required by the legislature, but also to oversee and direct the state workforce to make it as useful as possible. In his years as state engineer, Morse developed a distinct managerial style; ten years later in South Carolina he would clash with David Lopez over Lopez's management skills, which Morse considered to be inferior to his own and inadequate to manage the State Works.

Morse's primary responsibility as state engineer was to supervise the removal of obstacles to navigation in the state's waterways using both white and slave labor. Building and maintaining canal locks, levees, bridges, and roads as well as surveying and examining waterways also fell within his purview. Morse advocated expanding the labor force primarily by the use of slave labor. In 1855 he recommended that the state enlarge his department from one hundred slaves to four hundred.[45]

Morse had only one boat at his disposal in April 1853 when he recommended that the State spend ten thousand dollars to build two new snag-boats, which were shallow-draft, steam-powered barges equipped with cranes and thick hawsers used to dislodge waterway obstructions. Displaying his creative mechanical inclinations, he wrote, "In construction of the new boats, I wish to make some improvements in the model and working machinery, which I believe will not only enable them to move from one portion of the State to the other with facility, but to be efficient when at work."[46] In January 1854 he wrote about submerged obstructions that the snag-boat was unable to remove from a river bottom: "I have already a machine planned which I intend to apply to them, that will have sufficient force either to pull them up or pull the boat under water."[47] By 1854 both boats

were under construction. Morse was persistent, and by January 1855 he had a total of four boats either in service or soon to be.[48] Morse also experimented with underwater blasting to clear waterways of floating rafts of debris, submerged stumps, and sunken boats.

Morse made annual reports in 1853, 1854, and 1855. Displaying an air of self-confidence that would continue to manifest itself throughout the rest of his life, he wrote in his January 1854 report, "I am convinced that great good has been accomplished here in a very short time."[49] Though no records exist, it appears that 1855 was Morse's last year as Louisiana state engineer.

Sometime in 1854 or 1855, Morse began to envision the firearm and cartridge that would permanently alter his life and significantly affect the future development of firearms. When he applied for his first patent, in January 1856, Morse was living in Baton Rouge.[50] By July of that year he had a new job as commissioner for the First Swamp Land District, a position he would later describe as chief, or chairman, of the Board of Swamp Land Commissioners, whose function was to secure the reclamation of overflowed lands in the state.[51] He later recalled that, in spite of Governor Hebert's pleas for him to remain in Louisiana, he was determined to have the government adopt his new gun for military service. Though he held the title until mid-1858, Morse effectively gave up the job as swamp commissioner when he went to West Point for gun trials in the summer of 1857.[52]

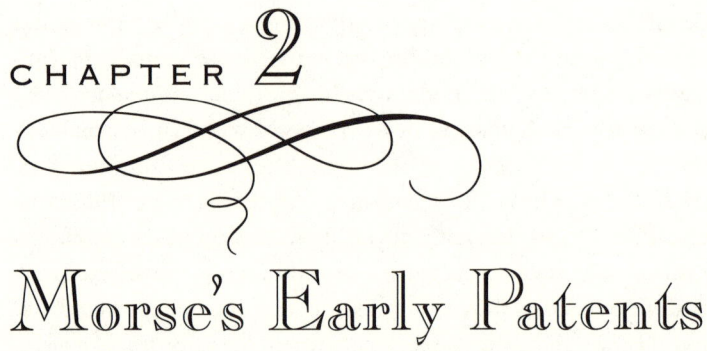

CHAPTER 2

Morse's Early Patents

By the mid-1850s, the concept of a breech-loading firearm was centuries old and that of a self-contained cartridge was decades old. Both ideas had been improved upon with modest success in recent decades by men such as two Americans, John H. Hall and William Montgomery Storm; a Swiss, Jean Samuel Pauly; a Frenchman, Casimir Lefaucheux; and many others.[1] Another Frenchman, Bourcier, had developed a metallic cartridge similar to Morse's, and an Englishman, A. E. L. Bellford, had invented a movable breech block with a center-fire mechanism. Yet another Frenchman, Clement Pottet, patented an improvement of Pauly's center-fire cartridge in 1829. L. N. Flobert's saloon pistol showed some similarities to Morse's design, and both Beranger's and Smith & Wesson's cartridges shared some features with that of Morse. Many American inventors submitted breechloaders for examination by U.S. Army ordnance boards in the decade before Morse began his work. Between 1846 and 1855 they included Jenks, Sharps, Klein, Jennings, Perry, Welch, Hall and Moore, Adams, Maton, Greene, Starr, and Symmes. Numerous other inventors presented breechloaders for evaluation to the 1857 and 1858 Ordnance Boards at West Point. They were A. E. Burnside; Samuel Colt; R. V. DeWitt; W. C. Freeman; S. H. Gibbs; H. Gross; A. V. Hoffer; R. Howe; Merrill; Latrobe & Thomas; George Patton; I. Schenckle; Sharps of Hartford, Sharps & Company of Philadelphia; John F. Shearman; J. C. Symmes; Samuel Wells; Joslyn; Soule; Storm; Maynard; Gilbert; and Schroeder.[2]

Virgil D. Stockbridge, who had twenty years' experience as an examiner specializing in firearms at the U.S. Patent Office and later as a patent attorney, wrote that, before Morse's first two patents—one for a breech-loading firearm and the other for an accompanying center-fire cartridge—were granted in 1856, there were numerous attempts with variable success both in the United States and Europe to design successful breech-loading firearms. Only within the decade preceding the mid-1850s had breechloaders been made that were reasonably free from serious and dangerous defects. According to Stockbridge, among nearly all breechloaders designed prior to Morse's patent there were certain distinct, separate, and excellent features. None, however, combined all of the characteristics of a center-fire, metallic cartridge with a functional breech-loading design in the way Morse did.[3]

Morse possessed both an inventive mind and an interest in firearms from a young age. In his late teens and early twenties, he began to develop creative mechanical skills by working as a machinist, designer, and inventor. He testified in 1870 that he first began experimenting with guns when he was seventeen years old.[4] As a teenager living in Haverhill, Morse learned how to work with gun metals when he made six steel, muzzle-loading pistols and a working small-caliber rifle in a cabinetmaker's shop.[5] The *Hampden Whig* of Springfield, Massachusetts, described Morse's first invention in an 1831 article: "Gun Locks—The New Hampshire papers describe a newly invented percussion gun lock, which is so constructed as to discharge the gun sixty times with one priming. The inventor is George W. Morse, son of Rev. B. Morse, of Haverhill, a lad of 17 years of age. The lock is entirely sealed within the stock, which is of the common form, with the exception of the trigger and guard, the latter divided into two parts, one of which is movable, and drawn back when the gun is cocked, and serves as the hammer by which the percussion is produced."[6] Morse did not patent this design, and it would be more than twenty-five years before he obtained his first patent.

Morse testified in 1870 that as early as 1830 he developed a serious interest in breech-loading firearms and had studied the subject.[7] He offered more detail in an 1872 deposition given as part of his efforts to extend his patents, stating that he was working with Captain Eliphalet Smith at Quincy Point, Massachusetts, in 1829 or 1830. Smith and Morse experimented with various mechanical contrivances and undertook to alter an old United States musket to a breechloader using a system in which a faucet-like device could be rotated so as to reload powder from a reservoir underneath the gun. The primer was external to the gun, but Morse worked out the details of an automatic primer. Because it was not practically functional, Smith and Morse gave up on the idea. Subsequently Morse tried to make another breechloader based on John H. Hall's plan, but that experiment was also a failure because he could not make a tight joint between the breech and barrel. Considering the solution impossible, Morse gave up in despair. Shortly afterward, he did make one more small-caliber rifle which had some type of center-fire arrangement, but he lost interest in the idea of a breechloader and did not revisit it again until the mid-1850s.[8] For the next twenty-five years, Morse was occupied with his employment by Tufts, the extended stay in England, and his move to Louisiana, marriage, and professional activities. In 1850 he was worth at least $20,000 and by 1860, $160,000. Having accumulated enough personal wealth by the mid-1850s to support his inventive nature, an epiphany launched Morse's career as an inventor of a functional breech-loading firearm.

An encounter at a dinner party led to Morse's idea for a solution to the vexing problem of how to design a successful breechloader. Governor Hebert was Morse's neighbor in Baton Rouge in the mid-1850s. In late 1854 or early 1855 he invited Morse and several other gentlemen to a dinner party at his home.[9] One of the guests was Thomas J. Rodman, a West Point graduate who was an artillerist and inventor of a new type of cannon and gunpowder. At the time, he was in command of the U.S. Arsenal at Baton Rouge. Hebert was also interested in arms and ammunition and had recently visited Rodman, who was working with a new bullet invented by a Frenchman, Claude-Étienne Minié, in the late 1840s. The projectile operated on the principle that its base expanded when the gunpowder exploded, allowing the bullet to conform to the rifling of the barrel. Rodman gave a

Minié ball to Hebert as a souvenir, and he took it home and placed it on the mantle in his parlor. As the group of gentlemen gathered in Hebert's parlor before dinner, Morse and Hebert focused their conversation on the Minié ball. Hebert playfully challenged Morse to improve upon it, and Morse replied that he could not do that, but "that he could then (now) make a perfect breech-loading gun by using the gas-tight principle claimed for a ball, applied in [the] rear of the charge."[10]

Morse wrote in October 1855 that he took the idea for his breechloader from the Minié ball itself and intended to use the same principle to develop a breech-loading firearm which, when fired, would be sealed at its breech by the expansion of the cartridge. Morse recalled a more detailed version of the event in 1870: "I was asked to improve this [Minié] ball. I thought a moment, and replied that they were asking too much of me, but that I had discovered the means of making a perfectly tight Jointed Breech-loading gun; and when asked how it was done, I replied that we had only to remove the ordinary breech-screw of a gun, make its seat smooth, then insert the charge of powder and ball, and then insert behind the powder a reversed Minnie [*sic*] ball, and that its expanding butt end behind the powder would close the Joint behind the charge as well as it would before the charge."[11] Morse explained to Hebert in complete detail how he would construct both the gun and its cartridge, to which Hebert exclaimed, "Morse, you have got it."[12] Very soon he came to the conclusion that the cartridge must be self-contained and comprise the powder, ball, and percussion cap, also called the fulminate. Morse wrote in 1855 that he initially planned to make a cartridge using lead and paper, indicating that his focus was on solving the problem of breechloaders and not on simply trying to improve upon existing metallic cartridges. In October 1855, he sent his patent attorney a drawing of his paper cartridge, but it soon gave way to a metallic casing.[13] Therein lay George Morse's major contribution to the development of modern firearms, the metallic, center-fire, pre-primed, reloadable cartridge.

From the beginning, Morse envisioned his new idea as a military firearm. Sporting breechloaders already existed; their tight-fitting breech joints were sealed by contact between the metal of the breech block and the metal of the barrel, an application that was impractical with military firearms. This concept worked well for a sporting gun that would be fired only occasionally, cleaned often, and not exposed to harsh environments for prolonged periods of time. Because military firearms endured much harsher use and conditions, the tight breech joint would become ineffective due to rust, dirt, and powder residue. In 1848 the U.S. Army Ordnance Board observed two problems with breechloaders. "1st. Want of solidity of the parts most exposed to the action of the charge. 2nd. Liability of the moveable parts to become unserviceable by their getting fast from rust or dirt deposited at each discharge, and the escape of the gas through the joints or junction of the different parts."[14] In other words, a breech block joint that was too loose would leak gas when the weapon was fired, and one that was too tight would become unserviceable under strenuous field conditions. Morse's solution was to make a relatively loose-fitting breech joint that became tightly sealed only as the cartridge expanded when the weapon was fired. Morse's idea revolutionized the concept of the center-fire, expanding metallic cartridge. That improvement, more so than the firearm itself, became his legacy. In a single, simple sentence, William F. Whitcher, author of the *History of Haverhill*, summarized the significance of Morse's contributions: "in 1856, he

[Morse] invented the 'metallic cartridge case' which made breech loading small arms a success."[15]

Morse's ideas and patents all revolved around variations of his novel breech-loading firearm and its companion metallic, center-fire cartridge. Though neither breech-loading firearms nor metallic cartridges were new in the mid-1850s, Stockbridge believed that it was Morse who came up with the first practical design for a breechloader and that all subsequent breechloaders followed his design. Stockbridge's 1888 deposition summarized the state of the art of breechloaders in 1856: "The breech loading arms known at that date (1856) were somewhat numerous but each and all of them depended for making a tight joint at the breech upon the contact and fit of the parts which go to make up the gun itself, or upon the fit between the parts of the gun and that device which served to carry the charge. In short, a tight joint or fit as made between the parts at the breech *before* firing" (emphasis added).[16]

Stockbridge continued, "Now [1888], all nations are armed with breech loading rifles and machine guns constructed upon the plan of [an] open breech joint, sealed both as to powder and priming by loosely fitting cartridge cases. Previous to the date of Morse's patents, all were armed with muzzle-loading smooth bored muskets and rifles with few exceptions and those exceptions were tight joint breech loaders."[17] He testified further, "This plan of constructing guns with open breech-joints or without any joint at all at the breech, to be sealed both as to powder and priming by a metallic cartridge case, *at once solved the problem that had vexed inventors for centuries*" (emphasis added).[18] Though his opinion was not shared by everyone, Stockbridge testified in 1888 that Morse's breech block design could be found in all the arms used by all the armies of the world, including the Springfield rifle, an altered Sharps carbine, Spencer's gun, the Gatling gun, and the Hotchkiss machine gun.[19] Stockbridge summarized his thoughts by testifying that Morse was "The first inventor of a device for sealing a breech joint, purposely made open, in a breech loading gun, by the expansion of a yielding metallic cartridge case purposely made smaller than the cartridge chamber."[20]

Soon after the dinner party at Hebert's home in late 1854 or early 1855, Morse made a wooden model from an old cigar box. He and Hebert soon visited Rodman, who felt that the invention was feasible and practical, but warned Morse that it would take a long time before he could expect the firearm to be introduced into military use. Rodman encouraged Morse to make a working model, and Morse soon conducted a successful experiment using an old pistol barrel, and for the base of his cartridge, a reversed Minié ball which had been made thinner and sharper on its rim. He wrote, "Soon becoming satisfied on this point, I at once began the construction of an operating gun. I believed that I had made a great discovery, and I wished to prove it."[21] Morse's new idea permanently altered the trajectory of his life, consuming him for the next thirty-five years.

Morse's earliest designs were crude at best. His first attempt at a cartridge was not metallic, but rather a paper-covered cartridge backed by a "reversed" Minié ball made of lead. Had he merely copied the 1829 center-fire French patent of Clement Pottet, Morse's first drawings would have been much more sophisticated and not based on a paper cartridge. The image below is the first known drawing of Morse's design, and it conveys his early thoughts as to how his new idea might work. The drawing was made in October 1855 and preceded his first patent application by several months. Morse wrote the legend for

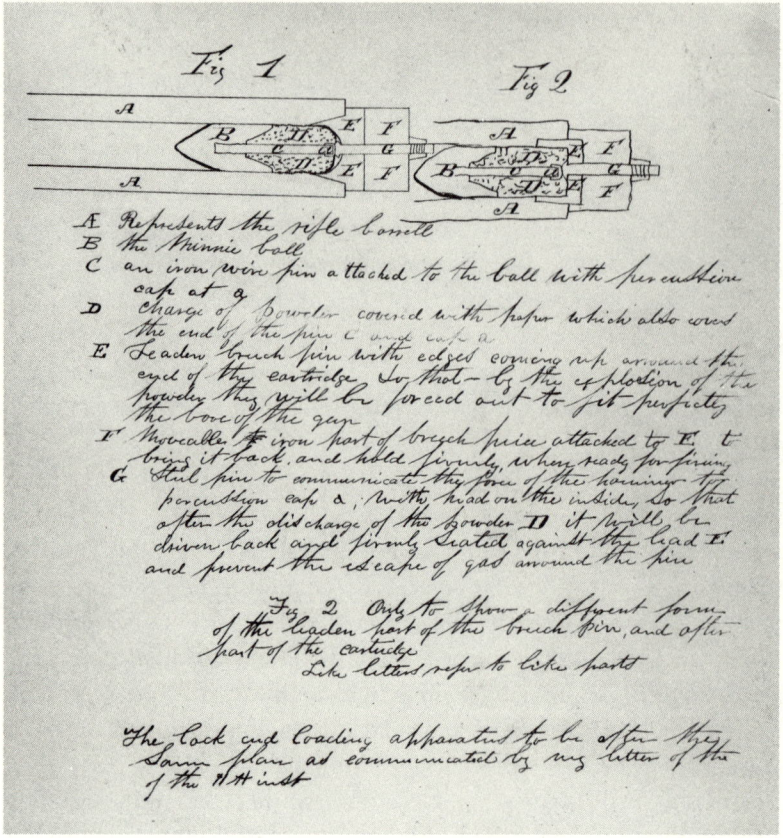

This is the first known drawing of Morse's cartridge, made of paper and lead, taken from a letter he wrote in late October 1855. It predates his first patent drawings of 1856. From National Archives, Record Group 123, Box 449A, Image 530.

the drawings. "A" represents the rifle barrel, and "B" is the Minnie [*sic*] ball. "C" is an iron wire pin attached to the ball with percussion cap at "a." This pin would soon evolve into an anvil-shaped structure made of heavy-gauge wire. "D" is a charge of powder covered with paper which also covered the end of the pin "C" and cap "a." "E" is Morse's "reversed" Minié ball, a leaden breech pin with the edges coming up around the end of the cartridge so that by the explosion of the powder they will be forced out to fit perfectly the bore of the gun. This statement represents the essence of Morse's concept of creating a tight breech joint by the expansion of the cartridge when the gun was fired. Figure 2 is an alternate design of "E" as compared to figure 1, the primary difference being the shape of "E." "F" is a movable iron part of the breech piece attached to "E" to bring it back and hold it firmly when ready for firing. "G" is a steel pin to communicate the force of the hammer to the percussion cap "a" with the head on the inside so that after the discharge of the powder "D" it will be driven back and firmly seated against the lead "E" and prevent the escape of gas around the pin. Morse makes it clear that a single unit comprised "E," "F," and "G" and that "E" was not a part of the cartridge but part of the sliding breech block/firing pin mechanism. He nearly immediately abandoned that concept and changed the design such that "E," the expanding element, became an integral part of the cartridge

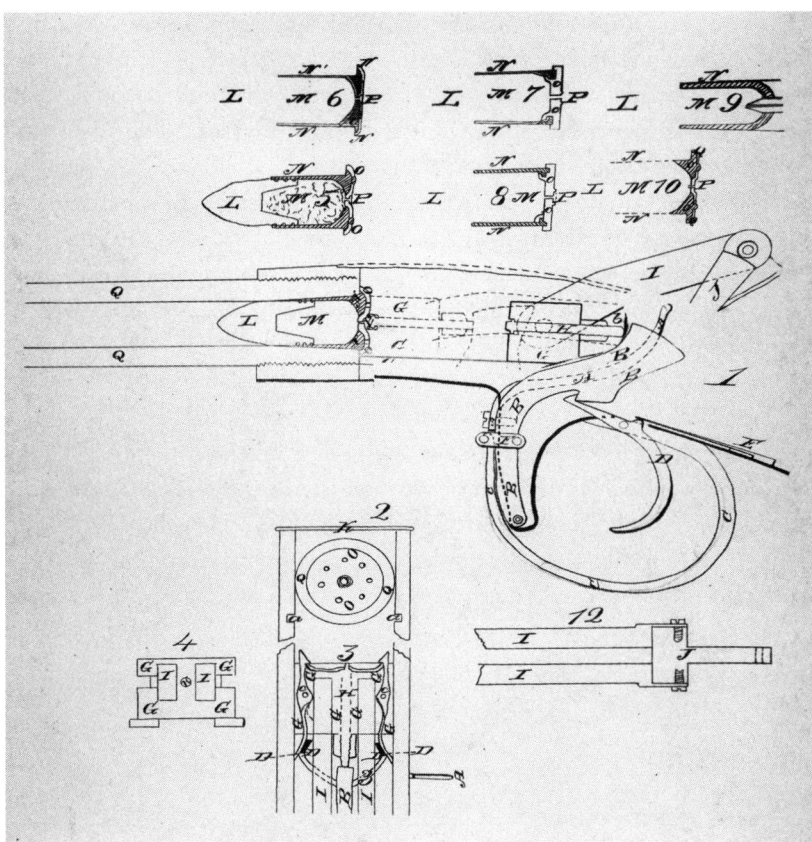

This is the first known drawing of Morse's firearm, from a late October 1855 letter written by Morse and shows his earliest designs, predating his first patent drawings of 1856. It is certainly the gun that Daniel Searles made for Morse, and it shows the rapid evolution of his cartridge. Though this drawing accompanies the previous one, Morse had already abandoned his original paper cartridge and developed a metallic cartridge that contained the expansive rear element as an integral part of the cartridge itself. From National Archives, Record Group 123, Box 449A.

itself. Stockbridge wrote later that Morse's first attempt at a cartridge was actually a reversed Minié ball made of lead, and it tended to stick in the chamber after the powder exploded, forcing him to develop a mechanical means of extracting the spent cartridge. The process of trial and error can be seen in Morse's earliest cartridge design, the failures ultimately producing a more functional product. Morse would soon give up on the idea of a paper cartridge and develop one made of metal.

Morse soon engaged Daniel Searles, an eccentric Baton Rouge gunsmith, to manufacture a prototype. Hebert later recalled that this event took place in the summer of 1855, but it was probably later that year. Morse wrote to his Washington attorney in October 1855 that all of his ideas were still limited to paper and wooden models because a yellow fever outbreak had kept him away from town. But now that the fever had subsided, he was planning to start work that very week and to have the gun made in three or four weeks.[22]

Searles was making guns in Baton Rouge as early as 1827 and, later, Bowie knives, one of which is on display at the Alamo Museum in San Antonio, Texas.[23] He manufactured Morse's experimental gun and cartridges in late 1855 or early 1856, and Hebert recalled successfully testing them in Searles's backyard in Baton Rouge. Morse said that firing the weapon "was a practical and complete success."[24]

Searles made more than one firearm for Morse. We do not know if his original experimental model exists. One possibility is that the experimental model is the same gun as Morse's original patent model, a rifle built by Searles in late 1855 or early 1856 and now owned by the Smithsonian Institution. Another possibility is that a Morse sporting-rifle prototype owned by the Buffalo Bill Center of the West in Cody, Wyoming, was Searles's first Morse firearm. A third possibility is a Searles-built carbine that Morse used in the 1857 West Point trials. It was a .53 caliber carbine with a 23.75-inch barrel, weighing 7.25 pounds, and having five grooves.[25] An extant photograph reveals it to be a different firearm than the patent model or the sporting prototype. It is a cased set with three interchangeable barrels, like a later Nathan Muzzy–built cased set. When shown a Muzzy cased set in 1876, Hebert recalled that the Searles-built experimental firearm was very similar.[26] According to the photograph's caption, the cased set was used by Morse at the

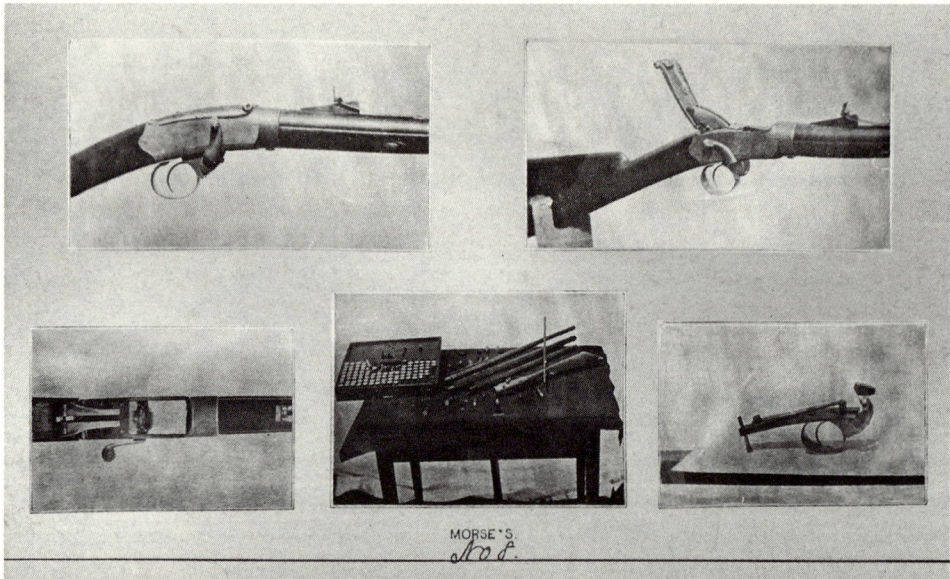

MORSE'S.

This gun was made either by Daniel Searles in 1855–56 or by Nathan Muzzy in late 1857 or 1858. The photographs were sent from the War Department to the US Court of Claims in 1887 as evidence in one of Morse's postwar lawsuits. An accompanying description states that this is the very firearm that Morse used in the West Point trials of 1857 and 1858. Morse used a Searles-built gun in the 1857 trials because the Muzzy-built guns were not yet made, and he probably used a Muzzy-built gun in the 1858 trials. Though this firearm looks very similar to the Muzzy-built guns, there are subtle differences, including the shape of the wooden box and the shape of the lockplate behind the trigger guard. From National Archives, Record Group 123, Box 449A, Image 479.

West Point trials in *both* 1857 and 1858. If this gun was, in fact, used in the 1857 trials, it was made by Searles. If it was used in the 1858 trials, it was probably made by Muzzy. Searles worked for Morse no later than 1856 or 1857. He was a seventy-eight-year-old reclusive widower when he committed suicide in 1860.[27]

Morse was granted seven U.S. patents and one British patent between 1856 and 1887. Six of them were issued in a flurry of activity between 1856 and 1858; three were for improvements on a breech-loading firearm, and three were for improvements on a center-fire, metallic cartridge. Morse's seventh patent, granted in 1886, was for yet another improvement on the cartridge, and his eighth was in 1887 for a handheld reloading tool. Throughout his career as an inventor, Morse habitually made frequent, nearly incessant, design changes, and wrote long letters to those whom he wanted to implement the changes. This pattern hindered the production of his designs in the late 1850s and actually impaired the chances of his guns being produced in larger numbers.

Between 1856 and 1864 Morse made numerous variations of his breech-loading mechanism and metallic cartridge. Some of his improvements show up in U.S. patents, but other major and minor variations are preserved only in drawings and surviving examples. His most notable firearm, the brass-frame carbine made in South Carolina during the Civil War, was never patented with the U.S. Patent Office, though it did utilize the same cartridge and breech-loading concepts of his earlier patents. Likewise, his inside lock muzzle loader was also made in South Carolina during the Civil War. The Confederacy's chief of ordnance, Josiah Gorgas, encouraged gunsmiths to design and perfect new breech-loading firearms and issued twelve such patents between August 1861 and September 1863, but George Morse's name was not one of the patentees.[28] The inside lock was Morse's one new invention that was patentable after the Civil War broke out, but no known Confederate or U.S. patents for it exists. Some postwar court testimony indicated that Morse might have registered his patents with the Confederate States Patent Office, but those records were destroyed in a fire in 1865.

15995—October 28, 1856, Morse's First Patent for a Breech-loading Firearm

After mid-1855 Morse devoted almost all of his time to the development, improvement, and practical application of his patents.[29] In 1872 he wrote that he spent about $6,000 to $7,000 (about $100,000 current value) of his personal income on the constant labor "of both head and hands while experimenting upon, perfecting and seeking to introduce his said inventions, during said four or five years, previous to the year 1861."[30] He testified in 1876 that his occupation prior to 1857 was endeavoring to produce the breechloader he had made and to induce its adoption by the government of the United States.[31]

Morse, then living in East Baton Rouge Parish, made initial contact with the U.S. Patent Office on August 13, 1855, stating that he had made certain improvements in breech-loading guns and cartridges for the same. He requested that the description be filed as a caveat in the confidential archives at the Patent Office, serving notice that he was currently engaged in making experiments to perfect the idea preparatory to his applying for letters patent.[32] The Patent Office received the letter on August 24. Morse and Searles spent the next few months developing the gun and cartridge. By late October 1855

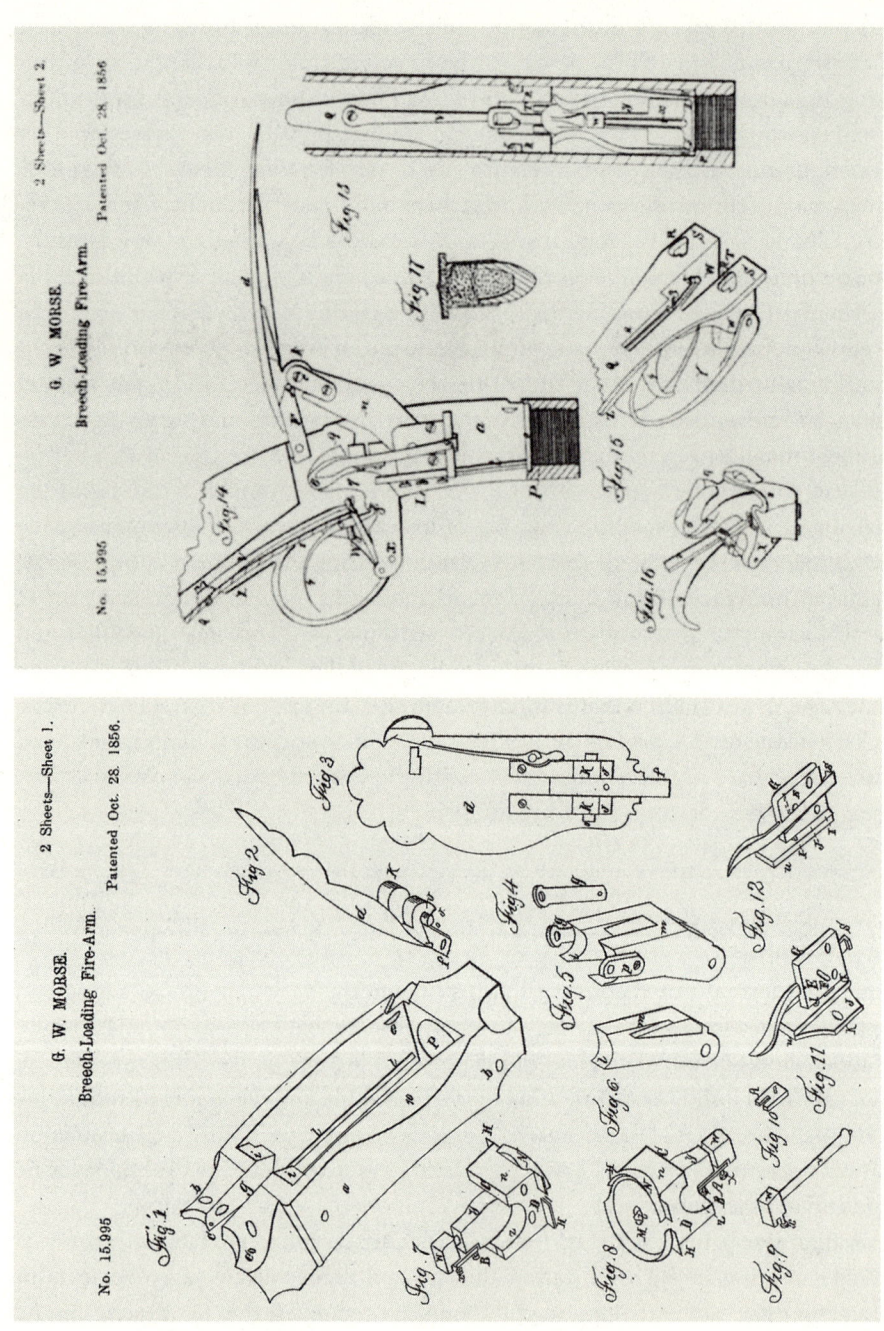

George W. Morse's first firearm patent, 15995, October 28, 1856: G. W. Morse, Breech-Loading Fire-Arm, Letters Patent October 28, 1856, Patent Case File No. 15995; Patented Case Files, 1836–1973; Records of the Patent and Trademark Office, Record Group 241; National Archives at College Park, Md.

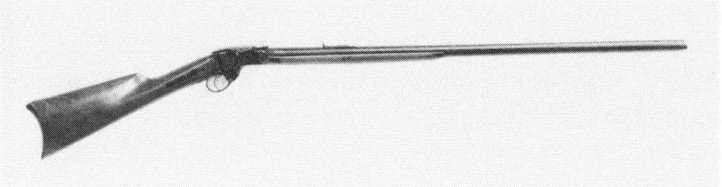

This is the Smithsonian patent model for US Patent 15995, on display at Buffalo Bill Center of the West. It is the earliest known Morse firearm and was certainly made by Daniel Searles of Baton Rouge, Louisiana, in late 1855 or early 1856. Courtesy of Armed Forces History, National Museum of American History, Smithsonian Institution.

Morse was anxious to obtain patents on his firearm and cartridge, but he admitted to his attorney, Thomas G. Clinton, that he was "badly informed upon the subject of obtaining patents" and sought Clinton's advice on how to proceed.[33] Morse applied for what was to become his first patent on January 22, 1856. He gave it the preliminary title of "Breech Loading Guns, etc.," and submitted the required petition, affidavit, specifications, drawings, working model, and a ten-dollar fee with the application. (See color plate 2.)

The patent model is part of the collection at the Smithsonian Institution but is on loan to the Buffalo Bill Center of the West, where it is on display. It was certainly made by Daniel Searles because he was the only gunsmith making Morse firearms in 1856. There are no external markings on the gun except for the number "252571" stamped on the wood above the side tang, or faux "lockplate," on the left side. The Buffalo Bill Center also owns a Morse prototype sporting rifle, which is nearly identical to the Smithsonian patent prototype. The maker of the prototype sporting rifle is unknown, but was probably Daniel Searles because is nearly identical to the Smithsonian model, including screws, buttplate, blued finish, and rear sight. It, like the Smithsonian patent model, has no markings and no serial number. The Smithsonian model has two "ears" used to open the action in the same position as a Muzzy-built Morse breechloader, but they are flatter and have no checkering. The "ears" are not on the Buffalo Bill prototype sporting rifle, but may have been at one time. The prototype sporting rifle barrel is 34½ inches long and measures .583 caliber. Its center-fire action is nearly identical to a standard Muzzy cased set, but there are several external differences between this piece and a Muzzy. The action measures 5-⅛ inches long whereas a standard Muzzy measures 6½ inches. The breech cover has an elaborate scalloped shape, as seen on the original patent drawing for 15995. The standard Muzzy buttplate is iron and measures 4-⅞ inches, whereas the Buffalo Bill prototype sporting rifle has a brass buttplate which measures 4-⅜ inches. The barrel is held on by two screws, unlike the Muzzy production barrels which threaded in position and were easily interchangeable.

Morse's first two patents, both granted on October 28, 1856, were complementary pieces. One, 15995, was for a breech-loading mechanism, and the second, 15996, was for a metallic, center-fire cartridge. Morse's greatest mistake was in seeking a separate patent for each instead of a single patent for the combined system. Patent 15995 was titled "Improvement in Breech Loading Firearms." Declaring that he had invented a new and useful breech-loading gun, Morse described the concept of a relatively loose-fitting breech-joint: "My gun is so constructed that before loading it I can blow through it or run water

through it with facility; but when the charge is in, it becomes comparatively air and water tight, because I use a cartridge-case which seals the breech-joint, both as to powder and the priming."[34] He stated that, since the cartridge was waterproof, placing it underwater would not cause it to malfunction. He even claimed it could be fired underwater.

Morse made three claims in this patent. First, he claimed the use of a cartridge inserted into the chamber of the weapon. Second, he described the use of nippers for the purpose of extracting the cartridge. Third, he claimed that by the single motion of pulling up and back on the cover of the case, "The combination of moveable parts, or their equivalents, whereby I retract or deliver the gun of a cartridge, drop it, open and clear the way for the insertion of another cartridge, whether the previous cartridge was fired or failed to fire, and cock the hammer automatically at one motion, substantially in the manner described."[35] The three novel aspects of the patent were the combination of features that allowed the breech to be opened and closed, the cartridge to be inserted and extracted, and the hammer to be cocked.[36] Considering Morse's "nippers" to be the most novel part of the patent, Stockbridge wrote several years later, "Cartridge retractors had been known and used prior to the granting of this patent, but none to my knowledge having the quality of yielding so as to ride over the flange of the cartridge as in this case. Substantially the same kind of retractor as shown and claimed in this case is now used in combination with nearly all of the modern patterns of Breech Loaders where breech plugs or closers operate reciprocally in a line with the base of the barrel. The retractor in combination with its operating parts may therefore be considered the distinctive important feature of this invention."[37]

Patent 15995 described the basic mechanism upon which most of Morse's future weapons were based, including those built by Nathan Muzzy in Worcester, Massachusetts, in 1857 and 1858, those converted at the Springfield Armory in 1859, and the brass-frame carbine manufactured at the South Carolina State Works between 1862 and 1865. The common feature of all of Morse's breechloader patents was what he termed the "cover of the case," or breech cover, which, when lifted up and back, exposed the open breech. In most of Morse's patents, lifting the case cover would cause "nippers," attached to a sliding breech block through which ran a firing pin, to extract the spent cartridge, thus emptying the breech.[38]

Major variations did appear as Morse refined future patents and built more advanced weapons: (1) In some of Morse's future weapons the act of lifting the cover of the case did not automatically cock the hammer as it did in 15995. (2) Likewise, in some of Morse's future weapons the act of lifting the case cover did not automatically extract the spent cartridge with a single motion as it did in 15995. (3) In 15995 the case cover had a spring-catch to hold it in place when closed, but some future Morse designs did not have this feature. (4) In 15995, the spent cartridge was removed from the top of the receiver, but some of Morse's future weapons ejected the spent cartridge downward through the frame. (5) The hammer style on 15995 was a side-mounted cocking/de-cocking device, but its location in subsequent models varied from side-mounted to a more modern one mounted within the frame, or "guarded," as Morse called it. (6) In 15995, the sliding breech block, or as Morse termed it, "breech piece," was a complex, smaller design, whereas the complexity of the breech piece changed over time, and in one patent it did not slide at all but was a rigid, integral part of the breech cover. And (7) 15995 was designed to accept a threaded

barrel, which could be changed between carbine, rifle, and shotgun to meet the shooter's needs, whereas military versions had a single, permanently affixed barrel. In many ways, Morse's earlier breechloader patents were his most complex, and, as time went by, he responded to criticisms of army ordnance officers by simplifying the design, culminating in a hybrid of the level of complexity used on the brass-frame carbine made at the South Carolina State Works. Morse's new invention was a weapon which foreshadowed the future of the modern breech-loading mechanism.

15996—October 28, 1856, Morse's First Cartridge Patent

Morse first approached the U.S. Patent Office about the new cartridge that complemented his breechloader on November 22, 1855. He submitted all the necessary components, including a model, with his patent application on January 22, 1856. Morse submitted some amendments on March 20, and the next day he wrote to Charles Mason, commissioner of patents, "With a view to have time to take proper steps in regards to European Patents, I request you will place my application for Letters Patents for Improved Cartridge-case, passed yesterday, in the secret archives of your office, so long as the law permits, or so

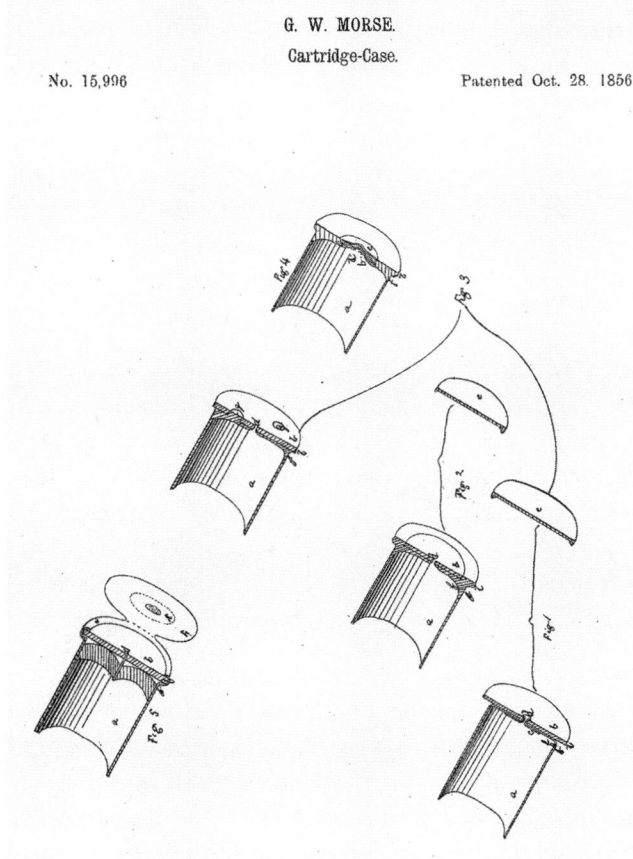

G. W. MORSE.
Cartridge-Case.

No. 15,996

Patented Oct. 28. 1856.

US Patent 15996. Morse's first attempt at a metallic, pre-primed, center-fire cartridge was patented in October 1856. Unlike his first paper cartridge, the expanding rear portion is contained within the cartridge itself. G. W. Morse, Cartridge-Case, Letters Patent October 28, 1856, Patent Case File No. 15996; Patented Case Files, 1836–1973; Records of the Patent and Trademark Office, Record Group 241; National Archives at College Park, Md.

much [of] that time as my interests require."[39] Mason complied with Morse's request but placed a six-month time limit on it.

With time running out, Morse wrote to Mason again on September 7, 1856, asking for an extension. Reflecting the fact that he was rapidly making design changes, Morse doubted that the two patents for his gun and cartridge adequately covered all of the improvements he had made. In fact, he had made several claims on the breech-loading gun that were not approved in the final version of the patent. Still living in Baton Rouge, Morse hoped to be in Washington before the patents were issued and published in October. He appealed to Mason for an extension of the six-month deadline, asking him to keep the relevant papers secret because Morse could not make it to Washington before the deadline expired. Morse's reason was that he had been working constantly on the construction of a new gun and on experiments with different kinds of cartridges until six weeks prior and had been stricken by a malady which had prevented his writing until that day. Mason responded on September 17, extending the deadline to October 15 because of Morse's illness.[40] The patents for both breechloader and cartridge were issued on October 28, 1856.

Morse's intention was that his new cartridge could be easily inserted into the chamber of the firearm, but when the powder was ignited, the rear of the cartridge would expand, thus tightly sealing off the breech joint from escaping gases. He described his earliest attempt at a cartridge on October 24, 1855, as "a cartridge that can be made with facility possessing all the requisites of strength, means of closing the bore of the gun, and an independent priming."[41] In the earlier models of the metallic cartridge, Morse considered using lead, felt, and gutta percha as the wad, or base, to hold the primer, but he settled on india rubber, also called caoutchouc.

In the patent's final version, Morse wrote that he had invented a "new and useful breech and prime-sealing (capable of being reprimed) and automatically retractable cartridge shell or case for breech-loading fire-arms or cannon, light enough for transportation, yet strong enough to resist the blow of the hammer."[42] Morse described his new invention, probably not realizing that it was a rudimentary version of the cartridge of the future:

> The nature of my invention consists in providing a soft-metal cartridge-case with a priming apparatus, whereby I seal the breech and vent of a breech-loading gun at firing as effectually as if the breech were solid metal around and back of the bore, leaving no escape for gas except at the muzzle. My cartridge-case is also capable of being automatically withdrawn, whether fired or not fired, and is so constructed that sufficient resistance is made to the blow of the hammer so as always to effect ignition if the primer does not fail, and if this does fail, it can be readily renewed or replaced. My cartridge-case is also capable of being practically useful for breech-loading cannon as well as small-arms.[43]

Either Morse or his attorney surely did some research between October 1855, when his cartridge was paper-based, and January 1856, when his patent application described a metallic cartridge. In 1856 he enumerated three similar attempts to invent a self-contained metallic cartridge but claimed that his was distinct. First he described a metallic cartridge of elastic metal that had been invented and used for the purpose of preventing

the escape of gas from the joints of breech-loading guns. He also acknowledged that a cartridge had been made that contained powder within the cavity of the projectile, and the powder was covered by a cylindrical cap of metal with a small orifice allowing ignition, but also allowing escape of gases. Finally, he was aware of a percussion cap consisting of powder being placed between a hard metal disk and a soft copper outer shell. Morse claimed that his cartridge, consisting of a metallic case and priming apparatus, effected "the entire exclusion of any and all escape of the gas produced by the combustion of the powder of the cartridge and priming, except by the one channel—the bore of the barrel of the gun—the breech-joints and priming-vent being thereby so effectually sealed and closed that no air can escape at these parts of the gun after the charge is fired until the cartridge case is withdrawn from the bore."[44] The primary features of the new cartridge were: (1) It held its own primer. (2) It was capable of sealing the breech-joint against the escape of gas when it was fired. And (3) it had a flange that was strong enough to allow it to be extracted from the chamber.[45]

Morse summarized his view of his first two patents in one of his postwar lawsuits:

> My invention consisted both in the cartridge and in the arm. The cartridge differed from those in use by being made of metal, primed with a flange at its rear end, for the purpose of extraction, and a cap, which sealed fully the vent and all vent discharge; at the same time the cartridge expanded at its rear end, so as to close the joint between the breech-block, and the end of the barrel. Metallic cartridge[s] had been in existence, but none had been made which closed that particular part of the gun; that is, the joint between a movable breech-block and the stationary gun-barrel; no metallic cartridge had been made which did that thing. These metallic cartridges of mine protected the whole charge, both the powder and the fulminate, held the projectile in the axis of the gun at the moment of discharge and guided it thence into the groove of the barrel; so that a breech-loading gun using my cartridge would make more perfect shooting and make a better target than could ever be done by a muzzle-loading gun.[46]

1357—May 13, 1857, Morse's Second Patent for a Breech-loading Firearm

Morse's plea to Mason for extra time to develop his interests in European patents in March 1856 culminated in British Patent 1357 for "An Improved Breech-loading Fire-arm" being issued to Morse on May 13, 1857.[47] This weapon was made on the same basic principle as his first U.S. firearm patent, 15995, but it was more complicated and was probably his most complex design. Just like its U.S. counterpart, the British patent had a metal case cover, which, when lifted up and back, caused the nippers of a sliding breech block to extract the spent cartridge, thus exposing the open breech and automatically cocking the hammer in the process. Morse described how to operate the gun: "In order to charge and prepare the gun to be fired, first unlatch the clasp lever E and pull it upward and backward. This will draw back the breech slide and cock the hammer, so as to enable the cartridge to be inserted in the gun; having inserted it therein, the lever clasp E should be moved back again to its place, the gun then being ready to be fired. If desirable the hammer may be uncocked and again cocked as occasion may require."[48]

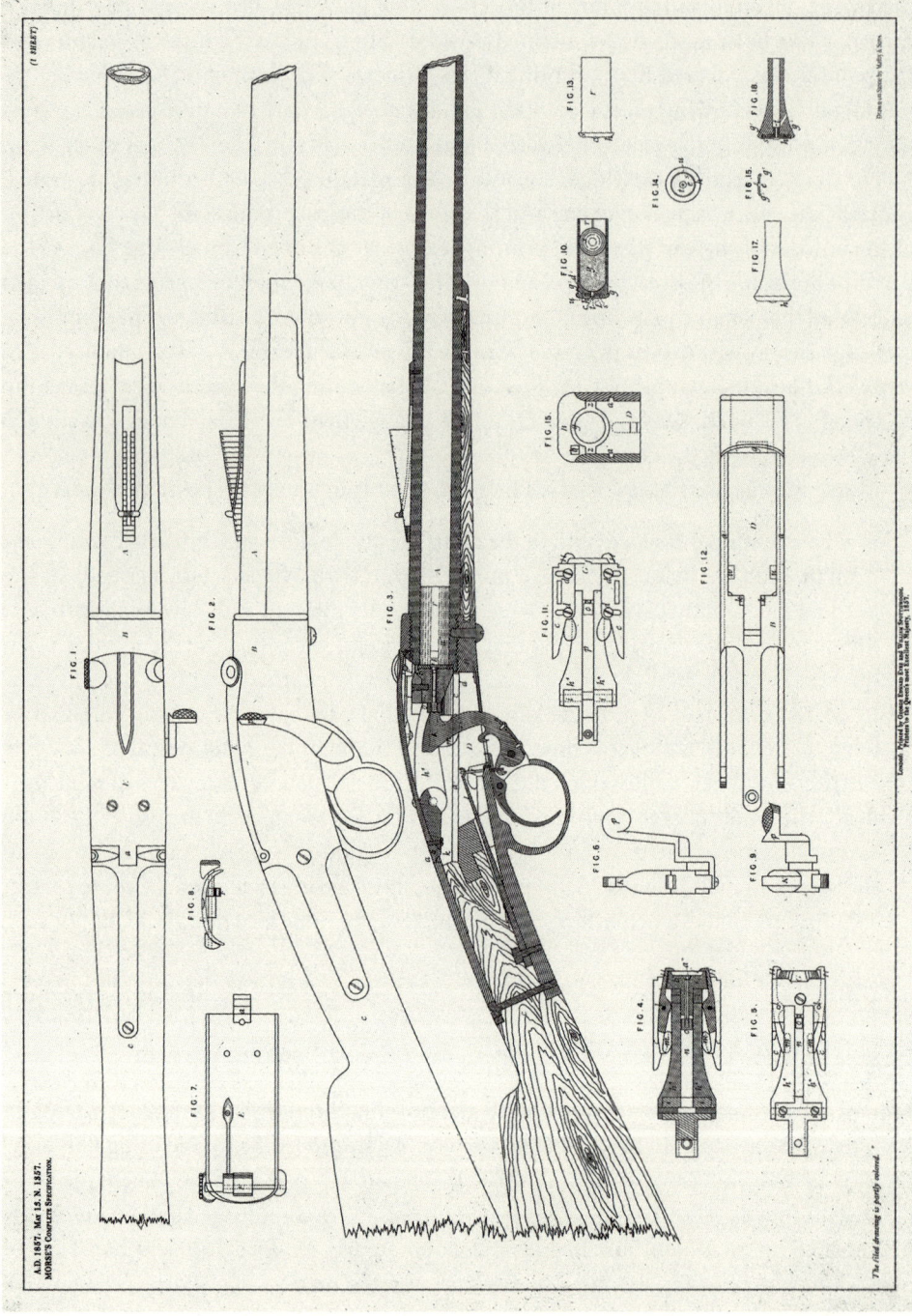

The original drawings from Morse's British Patent 1357, granted in 1857. From the author's collection.

Morse wrote that he placed the hinge at the rear of the breech cover because the upward and backward motion of opening the breech was the most natural and convenient, especially as it carried the shooter's hand by the same motion towards his cartridge box.[49] He implied that he considered placing the hinge at the front of the breech cover, as seen in the Erskine Allin–designed "trapdoor" mechanisms made at the Springfield Armory in the 1860s and 1870s, which Morse considered to be an infringement on his patents. Other similarities between Morse's British Patent 1357 and U.S. Patent 15995 were that the case cover had a spring bolt to hold it in place when closed, and the cocking/de-cocking lever was side-mounted, though it appears somewhat larger than 15995. The sliding breech block of the British design was a complex, smaller design, and it also had a threaded, interchangeable, barrel. Unlike the U.S. patent, the British gun utilized the force of the main spring via the percussion pin to eject the cartridge.

Morse made the following claims for the British patent relative to its extraction mechanism: (1) the function of the nippers extracting the cartridge as the breech slide is withdrawn, (2) The opening in the bottom of the casing such that the spent cartridge would fall out of the bottom of the gun after it was extracted, (3) the automatic ability of the mainspring to expel the spent cartridge from the bottom of the breech when the nipper's grasp was released, (4) automatic cocking of the trigger by the operation of the breech slide, (5) a mechanism whereby, when the weapon is discharged, the explosion would force the firing pin backward upon a valve that would close on its seat, thus preventing the escape of any gases into the pin passage or lock, and (6) the cartridge as described in the text.[50] The cartridge in the British patent was essentially the same as that of U.S. Patent 15996. Morse wrote that his invention could be applied to double-barrel shotguns as well as to heavy ordnance, and he included a drawing of a proposed cartridge for cannon. Morse wrote in 1870 that he sold part of his European patent for $6,000, but he did not say to whom. He also wrote that the $6,000 sale included "part of one [patent] subsequently made, which grew out of it," not being clear if this was U.S. Patent 20503, 20214, or 20727.[51] Lord Francis Napier observed several demonstrations of Morse's gun and, acting on behalf of the British government, ordered some of his models. Because Morse had become so preoccupied with trying to convince the U.S. government to develop his guns in the late 1850s, he completely neglected to fill the order for the British government.[52] Elements of Morse's British patent and U.S. Patent 15995 were incorporated into a hundred firearms produced by Nathan Muzzy of Worcester, Massachusetts, in 1857 and 1858.

CHAPTER 3

Nathan M. Muzzy

Morse lost no time in manufacturing and marketing his inventions, even as he continued to perfect the design (see color plate 3). His younger brother, Isaac, was a respected Massachusetts attorney when he wrote in 1870 that Morse deemed his new invention of great value and "immediately started North to try and get aid in its manufacture and introduction" in early 1857.[1] Only a year after his first patent was granted in October 1856, Morse was making the cartridge and preparing to produce the firearm.

Both the U.S. Congress and European powers began to recognize the potential military advantages of breech-loading firearms in the 1850s. Acknowledging the shortcomings of thousands of smoothbore muskets and obsolete rifles in the U.S. inventory, Congress hoped to convert some of them to breechloaders, thus salvaging some of its older firearms for ongoing use.[2] Congress appropriated $90,000 on August 5, 1854, "for the purchase in the opinion of the Secretary of War, [of] the best breech loading rifles for the use of the U.S. Army: Provided that the Secretary of War, after a fair practical test thereof shall deem the purchase advisable and proper."[3] The purpose of the measure was to stimulate inventors to develop effective models of breechloaders, and it worked. Brigadier General Stephen Vincent Benet wrote, "it is here properly that the era of breechloaders in this country begins."[4] A board of officers met at the Washington Arsenal in December 1854 to consider purchasing the most effective existing breechloader designs for further field-testing, but Secretary of War Jefferson Davis ordered the board to consider only carbines for use by the cavalry. Morse did not present a firearm for this trial because he had not yet made his first prototype.

Morse wrote in September 1870 that he spoke with Secretary Davis in late 1856 or early 1857. Morse showed Davis his recently completed firearm and cartridges, certainly a Searles-built gun, and requested that Davis order its testing for military use. Davis, "like all military officers" in Morse's opinion, was opposed to breech-loading guns and offered him no encouragement.[5] Davis favored muzzle loaders over existing breechloaders for infantry use and expended very little of the designated funds for breechloader development during his tenure. When Davis resigned from the War Office in March 1857, $82,143.50 of the original $90,000 fund remained unspent. Morse was confident in believing that his gun was a world-changing invention. He did not allow the disappointing meeting with

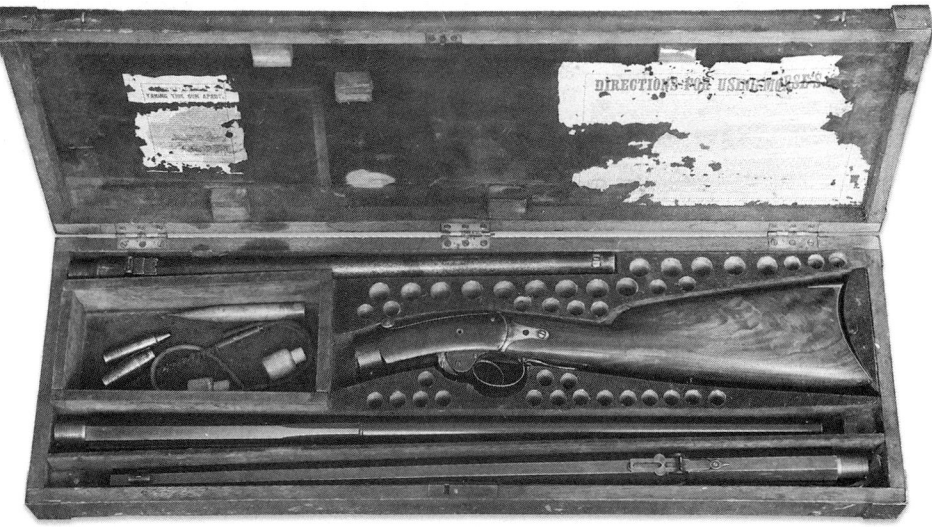

This firearm is serial number 60 of 100, part of a cased set made by Nathan M. Muzzy of Worcester, Massachusetts, in 1857 or 1858. The breech mechanism is a combination of Morse's US Patent 15995 and British Patent 1357. Courtesy of Museum and Library of Confederate History, Greenville, S.C.

Davis to deter him, and he decided to attack the problem along two simultaneous tracks. One was to persist in his pursuit of a military contract with the new secretary of war, John B. Floyd, who officially took over from Davis on March 6, 1857. The other path was to produce a sporting version of his firearm and sell it on the civilian market in hopes that he could attract general acceptance of the design.

Floyd, a graduate of South Carolina College and former governor of Virginia, held strong Southern sympathies and would become a Confederate brigadier general. He favored research involving breechloaders and began an active program to examine various

A close-up view of the engravings seen on Nathan Muzzy's cased sets, Muzzy & Co. Worcester. Courtesy of Robert M. Holter.

(BELOW) A close-up view of the rear sight and threaded barrel on Nathan Muzzy's cased sets. Courtesy of Robert M. Holter.

breech-loading designs. Floyd became so supportive of Morse's work in the late 1850s that the U.S. Court of Claims referred to Floyd as Morse's "special friend" in one of Morse's postwar lawsuits against the United States.[6]

Within days of Floyd taking office, Morse presented a Searles-built firearm to the Ordnance Office and met with success. Morse recalled many years later, incorrectly, that it was one of Nathan Muzzy's best firearms that he took to Floyd's office in March. Muzzy, a gunsmith working in Worcester, Massachusetts, would begin manufacturing one hundred of Morse's firearms by year's end but as of early 1857 had not yet begun to do so. Floyd was highly impressed with Morse's gun and probably instructed the chief of ordnance, Colonel Henry Knox Craig, to assess it. Craig complied when he wrote to William H. Bell, major of ordnance and commanding officer at the U.S. Arsenal at Washington, D.C., on March 5, "Mr. Morse, of Louisiana, will present at the arsenal a breech-loading rifle, which please cause to be tried, and report the result to this office, with your opinion as to the fitness of the arm for military service."[7]

Morse presented a .54 caliber rifle with a thirty-inch, threaded barrel for testing, and he testified in 1877 that it was the Searles-built rifle that was used for this test.[8] Bell test-fired forty rounds from Morse's firearm on March 6 and issued a favorable critique the same day.[9] He described the gun in detail and reported that it performed exceedingly well.[10] Bell wrote, "This arm, though complicated in its machinery, is amazingly ingenious and very durable in its construction. It is worthy of the consideration of the department, and were it not for its complicated machinery, it would probably be as well suited to military as it certainly is to civil purposes."[11]

Captain D. N. Ingraham, chief of the Bureau of Ordnance and Hydrography at the Washington Navy Yard, also ordered an evaluation of Morse's invention. Commander John A. Dahlgren test-fired a Searles-built firearm with both a rifle barrel and a carbine barrel on March 14, 1857, at the Navy Yard, paying special attention to exposure to water.[12] Morse testified in 1877 that it was the same Searles-built rifle that Bell had tested a week earlier, confirming that the Searles-built gun had interchangeable, threaded barrels.[13] In one of Dahlgren's tests, an assistant poured water over the lock and breech as Morse fired the weapon, and there were no failures. In another test, cartridges were coated with shellac and grease, then submerged in water for forty-four hours, after which they were test-fired. Seven of twenty-five cartridges failed to fire. The powder in the cartridges was found to be dry, and the problem was determined to be from a lack of powder in the caps. In a third experiment, the loaded firearm was completely immersed in water and successfully fired on all five attempts. Dahlgren's favorable report to Ingraham issued on March 17 stated that "the apparatus performs its functions perfectly and without fail," and he recommended that it was worthy of a trial in service by a group of picked men.[14] In 1876 the U.S. Congress considered Dahlgren's report to be an argument that Morse was due some remuneration for his services to the country.[15] And in 1883 the chief of ordnance, Brigadier General S. V. Benet, acknowledged that this was the first evidence of such a center-fire, metallic cartridge in the U.S. Ordnance Department.[16]

Floyd was impressed with both reports but did not immediately offer Morse a contract. Floyd, who was already a supporter of breechloader research and would soon be an advocate of Morse's designs, wrote prophetically: "I think it may be fairly asserted now, that the highest efficiency of a body of men with fire-arms can only be secured by putting in their hands the best breech-loading arm. The long habit of using muzzle-loading arms will resist what seems to be so great an innovation, and ignorance may condemn; but as certainly as the percussion cap has superseded the flint and steel, so surely will the breech-loading gun drive out of use those that load at the muzzle. For cavalry, the revolver and breech-loader will supersede the saber."[17]

After gaining a favorable assessment but no contract from Floyd in early 1857, Morse approached a Worcester, Massachusetts, gunsmith named Nathan Mills Muzzy to help him develop a civilian sporting version of his invention as a practical demonstration of its utility. Muzzy, born on June 14, 1822, in Worcester, was a blacksmith and armorer there from the mid-1840s into the 1850s. Dexter Hitchcock, born about 1826, was a rifle maker in Worcester by 1851. About 1855, he and Muzzy employed sixteen workers and formed a company called "Hitchcock and Muzzy" in Worcester.[18] The company manufactured guns and ammunition in the Merrifield Building, which burned in 1854 but was rebuilt by 1856.

Morse testified in 1877 that he left Louisiana to attend the breechloader trials at West Point, which were held in August 1857, and that he used a Searles-built gun in the trials.[19] Morse also made a visit or two to Worcester in the summer of 1857 and signed papers for his patent attorney there on both July 14 and September 7.[20] Certainly he was there to discuss business arrangements with Muzzy for the manufacture of his new breechloader and cartridges. They agreed that Muzzy would produce one hundred three-barrel cased sets of Morse's firearm to be sold to the public.[21]

Other than those made by Searles in Baton Rouge, Muzzy and Company produced

the earliest Morse cartridges. Most likely, Muzzy made cartridges for Morse to use with the Searles-built gun in the 1857 West Point trials before he started making the firearm itself.[22] Robert M. Holter described the early production of Morse's cartridge case: "At first such cases were cast of bronze or copper and among the early military shipments were cartridge cases and molds for casting them. Inspection of the cartridge cases in my old oak gun box suggests that at least some were cast. At an unspecified date, but probably as early as 1858, Morse cartridge cases of drawn brass began to appear."[23]

Hitchcock, Muzzy & Co. ceased to exist sometime in 1857, but Muzzy started his own company about that time called "Messrs. Muzzy and Co. Worcester, Massachusetts." Morse wrote in 1870 that he paid Muzzy a large sum of money, "the larger part of $15,000" to manufacture one hundred of his sporting arms. The retail price was probably $125, that being the amount the U.S. government reimbursed Morse for the use of one such gun on the Utah Expedition in April 1858.

Though we do not know exactly when Muzzy built his first Morse firearm, we do know that Morse used a Searles-built gun for the military trials held at West Point in August and September 1857, and he did show the board a new version of his gun—possibly built by Muzzy—on September 17, as well as an improved cartridge, but it was too late to enter the new gun in the competition.[24] We also know that Morse showed a Muzzy-built version to Floyd in November 1857. Muzzy was still producing Morse's design at the same time that Morse was attempting to interest the U.S. government in utilizing his inventions for military purposes in 1858. Hicks writes, "As soon as the license was granted [two thousand alterations for five dollars each on September 9, 1858] Morse went to Worcester, Mass., and had the altered arms that were before the Board at West Point sent to him."[25] Thus it appears that Muzzy began to manufacture Morse's firearm in the latter half of 1857 and continued well into 1858.

The Muzzy venture was a financial failure. Morse wrote, "When the first hundred arms were turned out, they did not find ready sale, and the company soon failed, and I lost all of my money except for $4,000, besides all my time."[26] Morse wrote in 1870 that it took about a year for Muzzy to manufacture a hundred firearms, "and it was found when they were completed, that they were so inferior in workmanship, that parties who had given me orders for them would not accept them, and in point of fact, I did not wish them to go out, fearing that their failure would injure the value of my patents."[27] Morse also wrote in 1870 that, after Muzzy's company failed, he (Morse) put his own money into creating a new company called the Muzzy Rifle Barrel Manufacturing Company, also known as the Muzzy Rifle and Gun Manufacturing Company, and "Muzzy & Co, Rifle Barrels etc."[28] The company existed until at least 1860, about which time it also failed, and Muzzy and Morse parted ways.

The Muzzy-built Morse firearms incorporated elements of both Morse's U.S. Patent 15995 and his British patent. The Muzzy was sold as a cased-set sporting arm, not a military weapon. Survivors include both complete cased sets and remnants. Each set consisted of one buttstock/receiver and three threaded, interchangeable barrels—a carbine, a rifle, and a shotgun. Reloadable cartridge cases, loading tools, and printed instructions were also included. The tools that accompanied the cased set were bullet molds for the rifle and carbine, two punches to make the base wads—one for making the wad itself and the other, which fit inside the first, for punching a hole in the wad for the percussion cap,

and a bullet seater for placing the projectile into the cartridge. The instructions reveal how Morse perceived the use of his new design:

DIRECTIONS FOR USING MORSE'S GUN

Grasping, with the left hand, the stock, loosen the catches by pressure of the right forefinger and thumb, and bring the gun to full cock by lifting the cap piece. Attach the barrel by screwing it until the shoulder is brought up to the stock so that when the cap piece is shut down, the hooks will fall into the notches cut into the end of the barrel. The set screw on the underside must then, with a screwdriver, be tightened up; so that it will enter the side of the barrel and, thus, hold it in firm position. Open the small hinged door under the frame of the stock; so that the cock can come forward to the firing pin; and the gun is ready for use.

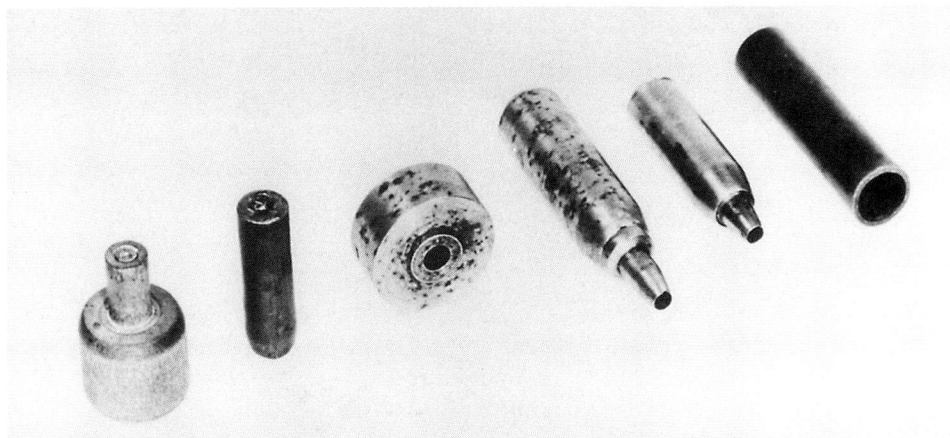

Tools that accompany Morse-designed, Nathan Muzzy–built cased sets. From Wray-Morse File 047. Courtesy of Atlanta History Center.

Muzzy tools, including .55 caliber bullet mold with iron collar, funnel-sprue-cutter, and swedge. From Wray-Morse File 047. Courtesy of Atlanta History Center.

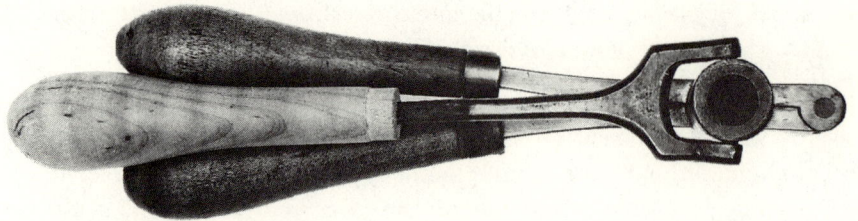

Top view of the Muzzy mold with the sprue cutter in place ready to receive molten lead. Courtesy of Robert M. Holter.

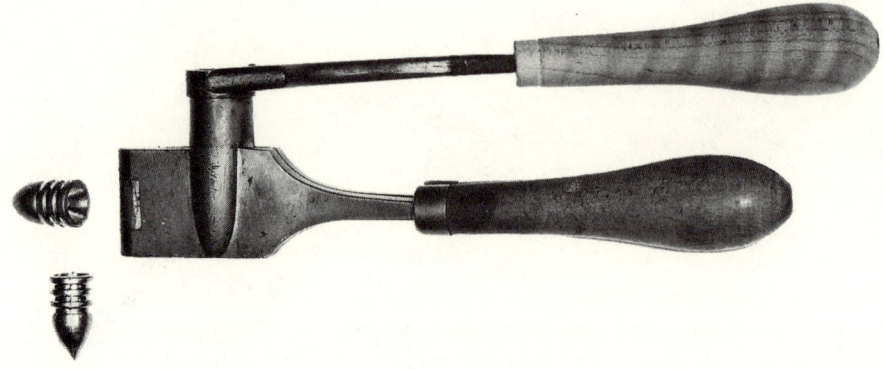

Side view of the Muzzy mold. Courtesy of Robert M. Holter.

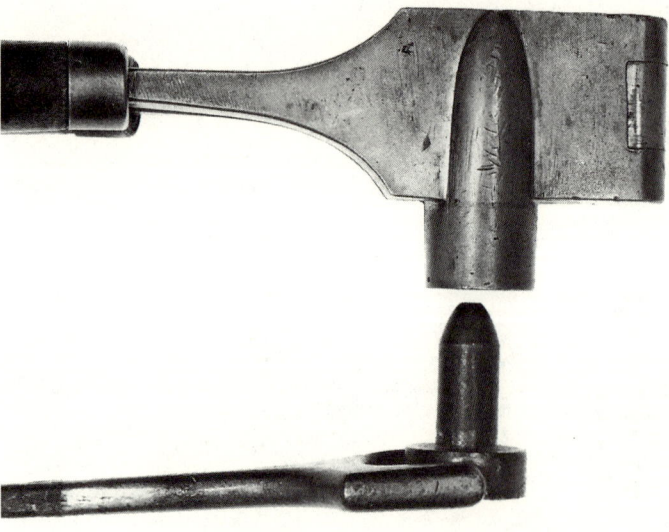

View of the Muzzy mold with the sprue device positioned to insert into the mold. Courtesy of Robert M. Holter.

Muzzy bullet seater. Courtesy of Robert M. Holter.

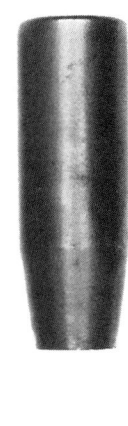

(RIGHT) These two punches were used to make base wads for Morse cartridges. "B" was used first to cut out the wad from a sheet of india rubber; then "A" was inserted into "B" and used to punch a center hole in the wad for the primer. Courtesy of Robert M. Holter.

A   B

Charging the Cases with Balls

The balls being properly swedged insert the open cartridge case up to the head of the loading tool, at the opposite end of which, insert the ball, which is to be forced by the rammer, deep enough to retain it firmly while it is being greased then remove the cartridge from the loading tool, and, with hard tallow, or a candle, run lengthwise the ball until its circular rings are filled with the tallow. The required number of balls having been thus prepared, must be put back, one after another, into the loading tool, and should all be driven down by a small hammer or mallet into the cases, a little beyond the foremost ring of the ball, except when an unusual charge of powder is to be used; for which due allowance must be made.

Charging the Cases with Powder

Coarse common powder is preferable to rifle powder and must be accurately weighed or measured, and the charges may consist, at the discretion and experience of the shooter, of from thirty to sixty grains. It has been ascertained, however, by careful experiment, that forty grains of United States musket powder are enough for the carbine as well as for the rifle; forty-five or fifty grains may however be better for the rifle. The charge having been determined on, the powder is to be poured into the case at its larger end, and is to be pressed down by an india rubber wad, until the percussion cap in its centre so rests on the wires in the case so as to prevent any powder getting between the end of the wires and the fulminate in the cap. When this is properly done, the surface of the wad will be about one sixteenth of an inch below the rim of the case.[29]

To charge with shot, the armed rubber wad is, first, inserted; the powder, forty-five or fifty grains, is the poured into the opposite end and kept down by a wad; the shot are then added; and a wad covering the whole. On this outer wad should be marked the number of shot within, so that the hunter can promptly select the charge most proper for the game he is in quest of.

Percussion Caps

Eley's double waterproof central fire percussion caps are best adapted to Morse's gun; but almost any cap will answer, if it is exactly fitted to the hole in the india rubber wad; and in all cases unvarnished caps are preferable. Hicks's musket caps are very suitable; but care must be taken that the india rubber selected for wads should be rigid enough to prevent the cap from splitting too much.

Wadding

India rubber is the best material, and of the kind used for packing steam joints. Rubber belting and rubber cloth are suitable, but unmixed soft rubber ought never to be used unless sustained by a corresponding piece of hard leather. By this mode old india rubber shoes have been made available for hunting purposes.

The entire wad, whether of packing or belt rubber, or of combined rubber and leather, ought to be about a quarter of an inch in thickness. In a wad composed of rubber and leather, pasteboard, &etc., the percussion cap should be thrust through the central hole *from the rubber side,* and it must be placed in the cartridge case with the rubber side touching the powder.

Loading and Firing

The gun being in the left hand, the knobs on the cap piece are grasped with the right thumb and finger, and, at the same time, the cap piece is lifted and pulled back until the piece is at full cock. The cartridge is then inserted, and the cap piece is forced down to its place. If the gun, thus loaded, is not to be immediately fired, the hammer can be let down in the usual way; but if otherwise, the shooter will bring the gun to his shoulder, and, as he grasps the handle of the stock, the second finger of his right hand rests behind the guard and upon the safety catch, which he will move involuntarily, as he touches the trigger.

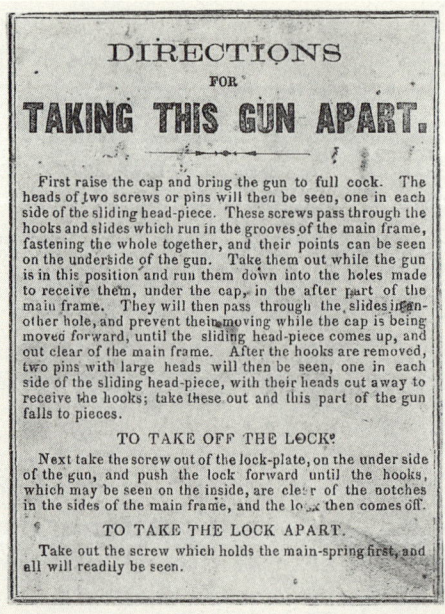

"Directions for Taking This Gun Apart."
Courtesy of Atlanta History Center.

DIRECTIONS FOR USING MORSE'S GUN.

Grasping, with the left hand, the stock, loosen the catches by pressure of the right forefinger and thumb, and bring the gun to full cock by lifting the cap piece. Attach the barrel by screwing it up till the shoulder is brought up to the stock. So that, when the cap piece is shut down, the hooks will fall into the notches cut into the end of the barrel. The set screw on the under side must then, with a screwdriver, be turned up; so that it will enter the side of the barrel and, thus, hold it firmly in position. Open the small hinged door under the frame of the stock; so that the cock can come forward to the firing pin: and the gun is ready for use.

Charging the Cases with Balls.

The balls being properly swaged, insert the open cartridge case up to the head of the loading tool, at the opposite end of which, insert the ball, which is to be forced by the rammer, deep enough to retain it firmly while it is being greased; then remove the cartridge from the loading tool and, with hard tallow, or a camphene, and lengthwise of the ball, till its circular rings are filled with the tallow. The required number of balls having been thus prepared must be put back, one after another, into the loading tool, and should all be driven down, by a small hammer or mallet, into the cases, a little beyond the foremost ring of the ball; except when an unusual charge of powder is to be used, for which due allowance must be made.

Charging the Cases with Powder.

Coarse common powder is preferable to rifle powder and must be accurately weighed or measured, and the charges may consist, at the discretion and experience of the shooter, of from thirty to sixty grains. It has been ascertained, however, by faithful experiment, that forty grains of United States musket powder are enough for the carbine as well as for the rifle; forty-five or fifty grains may, however, be better for the rifle. The charge having been determined on, the powder is to be poured into the case at its larger end, and is to be pressed down by an india rubber wad, until the percussion cap in its centre so rests on the wires in the case as to prevent any powder getting between the end of the wires and the fulminate in the cap. When this is properly done, the surface of the wad will be about one sixteenth of an inch below the rim of the case.

To charge with shot, the armed rubber wad is, first, inserted; the powder, forty-five or fifty grains, is then poured into the opposite end and kept down by a wad; the shot are then added, and a wad covers the whole. On this outer wad should be marked the number of shot within, so that the hunter can promptly select the charge most proper for the game he is in quest of.

Percussion Caps.

Eley's double waterproof central fire percussion caps are best adapted to Morse's gun; but almost any cap will answer, if it is exactly fitted to the hole in the india rubber wad; and in all cases unvarnished caps are preferable. Hicks's musket caps are very suitable; but care must be taken that the india rubber selected for wads should be rigid enough to prevent the cap from splitting too much.

Wadding.

India rubber is the best material, and of the kind used for packing steam joints. Rubber belting and rubber cloth are suitable, but unmixed soft rubber ought never to be used unless sustained by a corresponding piece of hard leather. By this mode old india rubber shoes have been made available for hunting purposes.

The entire wad, whether of packing or belt rubber, or of combined rubber and leather, ought to be about a quarter of an inch in thickness. In a wad composed of rubber and leather, pasteboard, &c., the percussion cap should be thrust through the central hole from the rubber side, and is must be placed in the cartridge case with the rubber side touching the powder.

Loading and Firing.

The gun being in the left hand, the knobs on the cap piece are pressed with the right thumb and finger, and, at the same time, the cap piece is lifted and pulled back until the piece is at full cock. The cartridge is then inserted, and the cap piece is forced down to its place. If the gun, thus loaded, is not to be immediately fired, the hammer can be let down in the usual way; but if otherwise, the shooter will bring the gun to his shoulder, and, as he grasps the handle of the stock, the second finger of his right hand rests behind the guard, and upon the safety catch, which he will move, involuntarily, as he touches the trigger.

These Guns are manufactured by Messrs. Muzzy & Co., Worcester, Massachusetts, who will, on application by mail, supply all the tools, implements, ammunition, &c., necessary for the shooter, whose address is accessible by means of the various express companies.

These Guns are manufactured by Messrs. Muzzy & Co., Worcester, Massachusetts, who will, on application by mail, supply all the tools, implements, ammunition necessary for the shooter, whose address is accessible by means of the various express companies.[30]

Though the basic design of Muzzy-built receivers is identical, significant variations exist among surviving cased sets. Some have elaborate engraving on the side of the receiver, and others do not. The threads of one privately owned Muzzy barrel do not fit a receiver made for another Muzzy cased set.[31] The caliber of the various barrels varies widely. The Atlanta History Center has two Muzzy cased sets, an early serial number, 7, and a late serial number, 94. Both have a .69 caliber (16-gauge) shotgun, a .48 caliber rifle, and a .52 caliber carbine. Another survivor, serial number 46, has a .69 caliber (16-gauge) shotgun, a .48 caliber rifle, and a .54 caliber carbine.[32] This weapon has a trigger guard, buttplate and sights made of iron, and a walnut stock. Lewis described variations among the Muzzy-built Morse firearms, a habit that was typical of Morse, who was constantly altering and improving his designs but not always seeking patents to cover the changes. Lewis wrote that the first of the cased sets came with a .54 caliber carbine, .50 caliber rifle, and 16 gauge (.69 caliber) shotgun. His article shows a photograph of a cased set with a .55 caliber carbine, a .55 caliber rifle, and a 12-gauge shotgun.[33] Serial number 60, on display at the Museum and Library of Confederate History in Greenville, South Carolina, has a 16-gauge shotgun and .54 caliber rifle and carbine barrels. Lewis further described a variation in the cartridges that came with the Muzzy cased sets. He wrote that a different-size rubber washer was used for each caliber, though all had the same rim diameter to fit the extractor. Also, Lewis wrote that the 16-gauge rim was too weak to withstand the extraction process and was reinforced by the addition of a steel ring, which then fit the .54 caliber–size rubber washer. Lewis did not cite his references, and his statement appears to disagree with Muzzy's "Directions for Using Morse's Gun," which enclosed a single punch for making all three sizes of india rubber wads.

CHAPTER 4

War Department
Evaluations, 1857–1858

Having ample funds left over from the 1854 appropriation for breechloader evaluations, Floyd issued an order on August 12, 1857, directing a board of officers to assemble at West Point on August 17 "for the purpose of making trials of breech-loading rifles" to determine which was best suited for military service.[1] Board members were Lieutenant Colonel B. S. Beall from the dragoons; Major Henry Hill with the pay department; Captain Henry Heath, an infantry officer; Captain J. G. Benton from ordnance; and First Lieutenant John Gibbon, an artillerist.[2]

Morse was in Worcester on August 3 having discussions with Muzzy when he notified the Ordnance Department that he would present his gun and cartridge at the trials.[3] He later wrote that he had been too busy to make an updated gun, and the one he submitted to the trial in the summer of 1857 was the same Searles-built gun that Bell had tested in March. He later confirmed this in an 1877 deposition.[4] Morse did, however, show the board a new version of his gun—possibly made by Muzzy—on September 17, as well as an improved cartridge, but it was too late to enter it in the competition.[5]

Morse's first patent was less than a year old and, perhaps remembering Rodman's admonition that it would take a long time before the new breech-loading design would see military service, Morse took an aggressive approach. He was a natural lobbyist, a trait he displayed many times throughout his adult life, and he spoke with several of the board members prior to their convening on August 17. Morse also wrote a long letter to the board commending his inventions, including the favorable reports from March 6 and 17. The letter reviewed the two major drawbacks to muzzle loaders: (1) they took a long time to reload compared to breechloaders, and (2) they often failed to fire due to moisture or an occluded nipple. Morse then discussed the major problem of breechloaders, that is, the inability to manufacture a tight breech joint and maintain a serviceable weapon over time. He also pointed out specific problems he had observed with all breechloaders except his. Acknowledging that there were other successful breechloaders that used a metallic cartridge, Morse made a strong argument for his firearm when he summarized its

favorable qualities: (1) A tight seal at the breech joint was guaranteed with and renewable with each firing of the weapon by the action of the expanding cartridge, not by a tight breech joint *per se*. (2) Because there was no nipple to admit moisture into the powder chamber, rain and water would not affect reliable firing of the weapon. Morse reiterated his claim that the gun could be immersed in water when loaded and still be discharged. (3) In a single motion, the cartridge could be withdrawn, dropped on the ground, and the weapon cocked. In the case of a misfire, the unspent cartridge could be cleared in four or five seconds, much quicker than in a muzzle loader. (4) There were great advantages resulting from the simplicity of the motions required in his gun's operations. (5) His gun could accept a wide variety of barrels threaded into a single receiver. (6) His gun met all the conditions for accuracy. (7) The parts of the gun used in loading and discharging were not subject to damage by heat from the firing process. (8) The main working parts of the gun were internal, protecting them from outside forces.[6]

Craig mailed a circular establishing the ground rules for the trials. He stated that no special permission was required to participate and that all firearms would be evaluated by the same method.[7] The circular was initially mailed to the representatives of eight entrants, though there would eventually be eighteen. The entrants represented a mix of carbines and rifles which varied in caliber from .45 to .56 and included a wide variety of designs. The entrants were: Burnside, Colt, Gibbs, Gilbert Smith, Greene, Gross, Howe, Joslyn, Maynard, Merrill, Morse, Schenckle, Schroeder, Sharps, Shearman, Soule, Storm, and Symmes.[8] Shearman withdrew his firearm from the competition when he realized that Morse's was superior to his. Competition was stiff, especially considering the fact that some manufacturers like Colt, Sharps, and Burnside were already producing their guns in well-established factories. Morse's main competition, however, came from William Montgomery Storm of New York and future Union general Ambrose E. Burnside of Rhode Island.[9]

Only the Morse, Schenckle, and Schroeder entries contained a primer within the cartridge. Only the Morse, Burnside, and Maynard firearms used a brass cartridge to prevent gas from escaping at the breech joint, and Smith and Greene used a wad made of india rubber or felt to solve the same problem. All others tried to solve the problem by allowing movement or expansion of metal around the breech to seal it during firing. Almost all the guns used either a percussion cap or Maynard's tape primer. Schenckle's and Schroeder's used a needle, or pin-fire, mechanism, and Sharps used his own primer. Morse's gun stood out because it was the only one that was self-priming with the cap set inside the cartridge.[10] In one test, Morse fired eighteen shots in one minute, forty-five seconds; only Maynard and Greene were faster.

The board's analysis of Morse's carbine was: "[It] is fired by means of a percussion cap enclosed in the cartridge. Three of the 100 shots missed fire, and twelve missed the target. The barrel was found to be much fouled from lead and the residuum of the burnt powder. The packing of the joint by the brass cartridge case, seemed effectually to stop the escape of gas, and no difficulty was experienced in working the machinery."[11] The board considered the most important features of a military breechloader to be simplicity, solidity, strength, ease and rapidity of loading, and certainty of fire. It felt that accuracy and force of fire were dependent on the projectile, powder charge, and barrel, and not on the design of the individual arms.[12] In spite of Morse's best efforts, the board decided unanimously

Mr Morse's Carbine   Range 100 yards.

| Nº | Distance from Centre in Inches. | | | | Miss. | Remarks. Aug 21. |
	Above.	Below.	Right.	Left.		
1		11.	8.			Firstly Capt. Duncan
2		10.5	4.			"
3		6.5		2.		"
4		5.		3.		"
5	2.			2.5		"
6	3.			.5		Capt. Heath
7	8.			1.		"
8	3.			3.25		"
9		5.		2.		"
10	1.			1.		"
11	4.25		5.2			Lieut Gibbon
12	6.			1.5		"
13			1.	2.		"
14	1.		6.5			"
15	1.5		6.5			"
16		4.	5.			"
17		0	9.			"
18	3.		3.5			"
19	1.		4.5			"
20		1.5	4.5			"

Gun:
Diameter of bore: .53
Length of barrel: 23.75
Weight: 7.25 lb
Grooves Nº of: 5
do Depth of: .01
do Twist of: ... to muzzle
Ammunition.
Ball, weight of: 445 grs
Powder do: 40
Wind: Left

One-hundred-yard data card from the West Point trials for Morse's Searles-built carbine, August 21, 1857. From National Archives, Record Group 123, Box 449B.

in favor Burnside's breechloader, which, like Morse's, used a metallic cartridge and effectively sealed the breech upon discharge.[13]

Though it chose the best available breechloader, the board did not look favorably upon the design in general, and wrote in its report on October 6, "In submitting this opinion the board feel it their duty to state that they have seen nothing in these trials to lead them to think that a breech-loading arm has yet been invented which is suited to replace the muzzle-loading gun for foot troops. On the contrary, they have seen much to impress them with an opinion unfavorable to the use of a breech-loading arm for *general* military purposes."[14] As a result of the West Point trials, Floyd ordered Burnside carbines into production on November 9, but he subsequently cancelled the order and ended up spending the 1854 appropriation on Joslyn carbines and Colt, Maynard, and Sharps rifles. Morse retained some hope for his design when the board attached to its report the opinions of Erskine Allin, master armorer at Springfield Armory, and James Stillman, an arms inspector, whom the board had requested to attend the trials and render their opinion. They felt that Sharps's entry was the best breechloader for use in the cavalry service. Their primary objection to Morse's gun was that, if a cartridge missed fire, the entire cartridge had to be removed and replaced by another. Stillman and Allin, who would cross paths

with Morse again, wrote of Morse's gun, "It is not suitable for military purposes, although very ingeniously made."[15]

Craig wrote to Floyd on October 24 to suggest that the lengthy process of evaluating new inventions via preliminary trials be stopped and replaced by field evaluations so that a weapon or weapons could be chosen for use in the service. Craig reminded Floyd that Green and Sharps's carbines and Colt's rifle were already in service, and that Symmes's carbine was at the Arsenal ready for issue, and Burnside's would be soon.

At some time in late 1857, both Morse and an armorer were injured in separate incidents when the breech cover rose up unexpectedly as they test-fired the weapon. Morse was firing the gun one day when it blew up, and the breech block struck him in the face, forcing gunpowder up his nose. During his three-week recuperation, he came up with the idea of milling out the barrels of old muskets and installing a new breech-loading receiver instead of making a new receiver and threading the old barrels into them. Thus was born the idea of converting old muzzle loaders to more modern breechloaders without building a completely new gun.

Throughout the summer and fall of 1857, Floyd ordered Craig to report on the available number of muskets and other small arms that had been converted from flintlock to percussion and to render his opinion about selling some or all them. As part of that discussion, the possibility of converting some of those muzzle loaders to breechloaders using Morse's design came up about the same time Floyd cancelled the order for Burnside's carbines. In early November 1857 Morse presented the new three-barrel Muzzy-built firearm to Floyd in an effort to convince him that it was the best gun in the world for military purposes.[16] As Morse was demonstrating the shotgun barrel to Floyd in his office, he accidentally pulled the trigger, and the loaded weapon spewed shot all over the office. Morse was mortified, but Floyd was impressed with the breechloader. He told Morse that it was the best gun he had ever seen and that he was surprised that the West Point Board had not adopted it. A discussion ensued in which Morse indicated to Floyd that he designed the firearm primarily as a military weapon. Morse readily accepted Floyd's offer to test it further. Floyd then ordered his clerk to tell Bell that he thought this was the best firearm he had ever seen and to cooperate fully with Morse in testing it.[17]

Floyd officially ordered further testing of Morse's design on November 4, 1857, with a view to altering existing muzzle loaders to Morse's design, and on November 9 and 10 the tests took place at the Washington Arsenal under Bell's supervision and by Ingraham at the Navy Yard in Washington.[18] Morse recalled in 1877 that Bell felt the testing was unnecessary because, in his opinion, breechloaders would not make good military guns, but he obeyed Floyd's order to assist Morse. Morse fired the gun 105 times with positive results. The tests went on for two weeks, after which Bell agreed that Morse's was the best gun he had ever seen.[19] The testing officers felt that it was an admirable breechloader because it exhibited moderate recoil, the ammunition would fire after being submerged in water, the bore was not fouled up with lead, and the machinery of the firearm performed perfectly well under testing conditions.[20] Conducting his second evaluation of Morse's gun that year, Ingraham personally supervised the testing on November 9, and made his report the next day. Ingraham's assistant fired twenty rounds in 122.5 seconds, a rate of nearly ten rounds per minute. Ingraham gave the weight of the gun as 8.98 pounds, much heavier than the Searles-built gun used in the trials earlier that summer. Ingraham noted

four positive characteristics of Morse's gun: (1) The breech joint was sealed and did not leak gas when fired. (2) The recoil was moderate. (3) The gun could be immersed in water, removed, and fired; a trait important to the Navy. And (4) the machinery performed perfectly well under severe tests. He wrote, "The machinery of the piece is complicated; but it is at the same time compact and works without strain or apparent liability to derangement."[21] He felt that the gun was unbalanced due to a heavy barrel, a problem that could be easily fixed. Ingraham concluded that Morse's gun was "an admirable breech-loading firearm."[22]

With positive assessments from Bell and Ingraham, Morse surely felt that he had a second opportunity at a large military contract. Referring to the recent experiments, he wrote to Floyd on November 12, "Major Bell is of the opinion that my arm is the most suitable and efficient arm for cavalry service he has yet seen." He wrote further that he had "been able to convince all parties who have witnessed my operations of the great superiority of my weapon over all others." Morse concluded that he confidently expected to prove to Floyd his assertion "that mine is the best gun in the world."[23]

Contrary to the opinion of the West Point Board, Floyd now deemed Morse's breech-loading invention to be the best available. He asked Morse if he could apply his design to alter old muzzle loaders, and, after considering the request for a few days, Morse agreed. Floyd then directed a more complete evaluation of Morse's gun at the Washington Arsenal, issuing the following directive to the Ordnance Bureau of the War Department on December 9, 1857: "The colonel of ordnance will direct Major William H. Bell of the Washington Arsenal to have a model gun constructed on the principle of either of the guns of Mr. Morse which may be considered by Mr. Bell best adapted to Cavalry service. The gun to be of proper caliber, size, weight, form, etc., Mr. Morse furnishing the barrel, [and the] gun to be made in accordance with the directions of the Ordnance officers and assisting in the construction."[24] It would be the first gun of Morse's design to be constructed by the U.S. government. When he wrote "either of the guns," Floyd was referring to one design based on Morse's U.S. Patent 15995 and his British Patent 1357, and a second design based on what would eventually be designated U.S. Patent 20503, which was not issued until June 8, 1858, but was available to Floyd for his examination as early as December 1857. Morse explained in an 1876 deposition that in late 1857 he had been working on a new, simplified mechanism whereby he removed the toggle joints and sliding breech and replaced it with a solid breech block and an extraction lever on the underside of the gun. This is the design that became U.S. Patent 20503, and Floyd allowed Bell to make a gun based on his choice of either 15995/1357 or 20503.[25] Over the next two years, the U.S. government would utilize these patents both to make new firearms and to alter old muzzle loaders to breechloaders.

20503—June 8, 1858, Morse's Third Breech-loading Firearm Patent

Morse was awarded U.S. Patent 20503 for an "Improvement in Breech-Loading Fire-Arms" on June 8, 1858, though he had been working on it since late 1857. This design was substantially different from his first two firearms patents. In this patent, the "cover of the case," or breech cover, functioned like Morse's previous patents in that, when lifted up

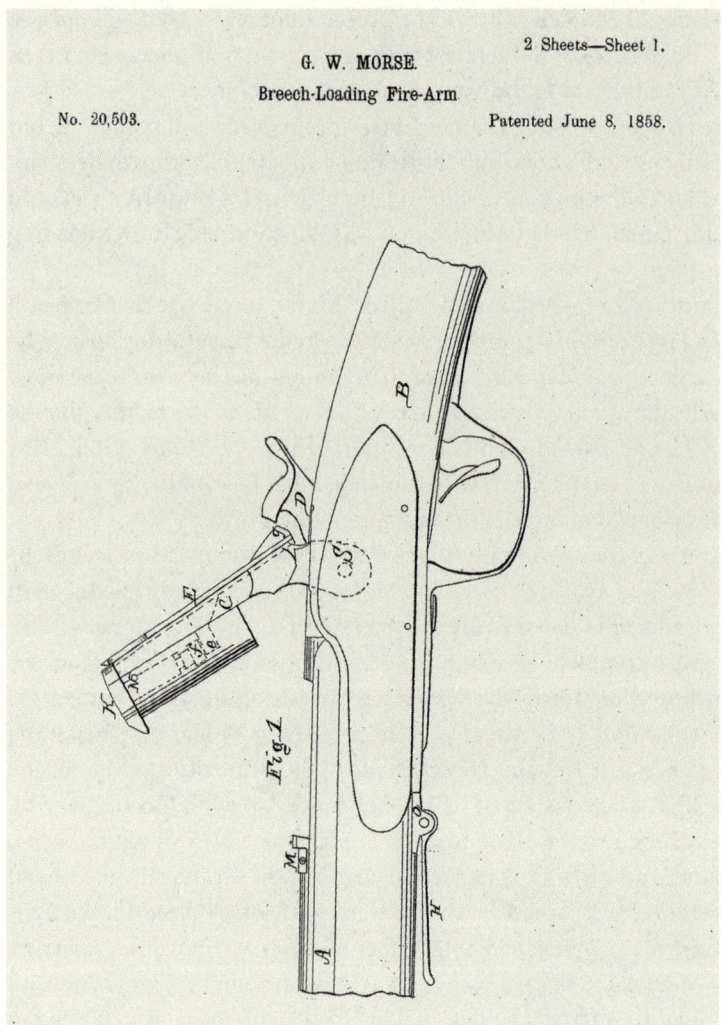

2 Sheets—Sheet 1.

G. W. MORSE.
Breech-Loading Fire-Arm

No. 20,503.   Patented June 8, 1858.

Morse's US patent 20503 was his third firearm patent and was designed to be a military weapon. It differed from his earlier designs by deleting the sliding breech piece and by including a manual lever-operated cartridge extractor on the underside of the gun. G. W. Morse, Breech Loader, Letters Patent June 8, 1858, Patent Case File No. 20503; Patented Case Files, 1836–1973; Records of the Patent and Trademark Office, Record Group 241; National Archives at College Park, Md.

and back, it exposed the open breech. In his two previous patents, the breech block slid back and forth on a toggle as the cover was raised or lowered. In contrast, in the design of 20503 the breech block and breech cover were actually a single piece.

The breech cover in 20503 did not have a spring-catch to hold it in place when closed, but it did have a locking mechanism when it was fired. In contrast to earlier patents, as the breech cover of 20503 was closed, the breech block's front surface, which was "globular," or "in the form of an arc of a circle," dropped down to form a snug fit with the breech opening. When the weapon was fired, the release of the hammer caused two parts contained within the breech cover/block to move forward; a slide on top of the cover

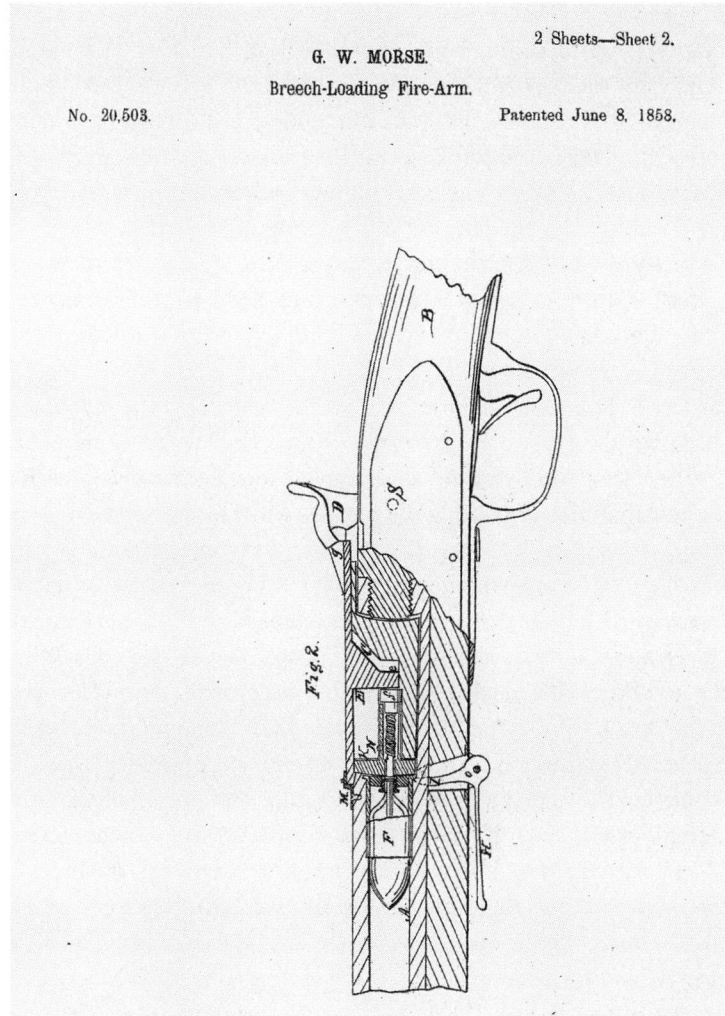

G. W. MORSE.

Breech-Loading Fire-Arm.

2 Sheets—Sheet 2.

No. 20,503.

Patented June 8, 1858.

Fig. 2.

A cutaway view of Morse's US Patent 20503 with a cartridge inserted and the breech closed. G. W. Morse, Breech Loader, Letters Patent June 8, 1858, Patent Case File No. 20503; Patented Case Files, 1836–1973; Records of the Patent and Trademark Office, Record Group 241; National Archives at College Park, Md.

was forced forward and locked the breech in place, and an internal firing pin mechanism moved forward, striking the cartridge primer and discharging the powder load.[26] This patent represented a variation of Morse's combination of a snug-fitting breech block and expanding cartridge that prevented gas from escaping at the breech joint when the weapon was fired.

Patent 20503 was Morse's first design which did not use nippers to extract a cartridge; instead, a hand-operated lever on the bottom of the gun ejected the cartridge upward. Unlike Morse's previous weapons, in which the act of lifting the cover of the case automatically cocked the hammer, in 20503 the hammer was manually cocked in the standard manner.

Morse made three claims with this patent: "1. The percussion-rod in a moveable breech piece, in combination with the sliding bolt, when so arranged that the lock in the act of firing shall both make fast the breech piece and fire the charge. 2. The construction and use of the globular surface on the front end of the moveable breech-piece, in combination with the end of the cylindrical cartridge case, for the purpose of more effectually preventing the escape of gas at the joint. 3. The construction and use of the lever when arranged substantially as described, for the purpose of retracting the cartridge case."[27]

Patent 20503 was designed to be used in the conversion of existing muzzle loaders to breechloaders and therefore was not designed to accept a threaded barrel. Even before the patent was issued, Floyd became interested in it and planned to use it to convert a number of muzzle loaders to breechloaders.

Floyd changed his orders to Bell on December 14, 1857, and instructed the manufacture of four complete guns according to Morse's pattern—a rifle, a musket, a carbine and a pistol—all in .58 caliber. These four guns were not alterations of muzzle loader to breechloaders, but were rather new manufactures using old barrels. Most likely he planned to use the design from Patents 15995 and 1357. Morse lost no time in complying with Floyd's decision; by December 19, he wrote to Floyd requesting that a list of barrels necessary to complete the order be forwarded to Bell at the Washington Arsenal.[28] From Springfield Armory he requested three barrels—one .58 caliber "rifle pistol" barrel, one .58 caliber rifle musket barrel, and a .58 caliber rifle musket barrel with three grooves. From Harpers Ferry Armory he requested five barrels—one .58 caliber rifle barrel, one .58 caliber carbine barrel 24 inches in length, and three U.S. model 1842 smooth-bore barrels.[29]

Morse immediately pursued the manufacture of the new guns and personally guided the process from late December 1857 to early March 1858. Unfortunately, he encountered a serious technical problem due to a design flaw. The center of resistance in the breech connection was not directly in line with the center of the rearward-directed force when the gun was discharged. The forces that were created when the gun was fired naturally tended to separate the barrel and the breech, acting as a lever on the breech cover and, after a number of firings, forcing it upward. The problem was so serious that by March 1, 1858, Morse abandoned the plans to make the four guns Floyd had ordered and returned the eight barrels to the storekeeper.[30] It must have been about this time that he abandoned the thin breech cover of Patent 15995 as seen in the Muzzy-style firearms. None of Morse's firearms after this time used the thin Muzzy breech cover, and, though some had a sliding breech piece and others did not, all incorporated a heavy breech block as part of the cover. Meanwhile, Morse and the armorers tested the new concept of altering an old muzzle loader by firing it successfully 1,200 times.

Craig asked Morse on January 27, 1858, to send one of his breech-loading rifles, along with a hundred cartridges, to Lieutenant General Winfield Scott for him to examine. According to Morse's 1876 deposition, Scott's response was, "If we could only keep this thing from getting to foreign countries, we could master the whole world."[31] Craig asked Morse to ship the items to the Ordnance Office in Washington, D.C., as soon as possible so that they could be forwarded to New York in time to be placed on a steamer which was scheduled to sail on February 5 for Aspinwall, the former name for Colon, a town at the Atlantic terminus of the Panama railroad.[32] Likely, Craig was sending one of Morse's firearms to California, but the reason remains unknown.

Morse wrote to the "board of officers for the examination of altered United States arms" on February 26, urging the use of brass cartridges, and stated that "I do not depend upon the gun for a tight joint at the breech, but upon the cartridge, which seals perfectly, and is renewable after each discharge, so that it can never wear out."[33]

Meanwhile, alterations on muzzle loaders began. Floyd issued new orders on February 28, 1858, that a .69 caliber rifled musket be altered to Morse's design, and on March 13 he ordered the same alteration for a .69 caliber smoothbore musket and a .54 caliber Harpers Ferry rifle. The alterations were to take place at the Washington Arsenal under Bell's direction. It is not entirely clear if Bell used Morse's Patent 15995 or 20503, but it was probably 20503. There is no evidence that Morse ever again used the unmodified 15995 for a military gun. Though 20503 was not granted until June 1858, its development was in progress in February. In the trials conducted in July 1858 to determine the best method to alter old muzzle loaders to breechloaders, Morse used the three guns that Bell altered in February 1858 and reviewed the mechanics of Patent 20503 in detail for the ordnance board, providing further evidence that Bell used 20503 to make these three guns earlier that year.[34]

Floyd was interested both in economy by avoiding the contemplated sale of old muzzle loaders at a loss and in efficiency by producing a more perfect arm than had yet been made. He ordered Bell to conduct the experimental alterations with the least possible change to the old weapons and at the least possible expense. Bell altered the three firearms using Morse's principle, but there were variations between the three. In general, the alterations involved taking an existing muzzle loader, machining out a section of the top of the breech, and installing Morse's breech lock in the empty space. The front section of the old hammer and the nipple were cut off, and the hammer now served as a locking device for the breech block. At half-cock, the breech cover and its attached breech block could be manually raised upward and backward, opening the breech and allowing a new cartridge to be inserted. Squeezing the trigger dropped the hammer, which drove both the locking bolt and firing pin forward, igniting the cartridge. After altering the three weapons, Bell tested each one. He fired the .69 caliber rifled musket 1,300 times and experienced no problems with the breech cover rising as Morse had a few weeks earlier. Bell found its accuracy and range to be the same as the muzzle loader, and its penetrating power was slightly better than the muzzle loader. He estimated the cost of alteration to be between $2.50 and $3.50 per gun. Bell fired the smoothbore 425 times and also had no problems with the breech cover rising up. While this gun was being tested, Floyd visited the armory and examined the process. In doing so, he agreed that a sliding breech block, as found in Morse's Patent 15995 but not in 20503, was unnecessary and only made the firearm more complicated. Finally, Bell concluded his experiments when he fired the .54 caliber rifle 74 times on May 9 and, once again, had no issues with the breech cover rising up. Floyd personally fired this gun on May 3, but it had a crooked barrel. After the defect was corrected, Bell found that the .54 caliber rifle alteration was as accurate and had as much penetrating power as a .58 caliber rifled muzzle loader and its cost of alteration was about the same as the .69 caliber rifled musket. At the end of the experiments, Floyd ordered that a .69 caliber smoothbore musket be altered to incorporate all of the lessons learned, and work on that gun had started when Bell made his report to Craig and Floyd on May 12, 1858.

As the War Department was going through the process of evaluating the best available breechloader and the best method of converting existing muzzle loaders to breechloaders, Morse was simultaneously negotiating a contract to provide newly manufactured carbines to the War Department in early 1858. It is clear that he was working on a design similar to, if not identical to, those built by Muzzy. He had not yet developed the design for the brass-frame carbine, which would eventually be made in Greenville, South Carolina. The earliest evidence that Morse, who was living in Washington at the time, was working on the carbine at Springfield and that he agreed to produce them for the government comes from a directive that Floyd issued on March 5, 1858, for one hundred of the "Morse guns" to be purchased at forty dollars apiece and to be paid for in lots of twenty-five each as soon as they were inspected and delivered.[35] The War Department placed the order on March 16, calling for Messrs. Muzzy of Worcester, Massachusetts, to do the manufacturing work.[36] This is clearly a separate order from the two thousand muzzle-loader conversions at five dollars each issued in September 1858.

Ironically Morse now had his first military order, but he had already lost faith in the Muzzy design using Patent 15995. Nevertheless, Morse informed Craig on March 17 that he accepted the order for one hundred guns and would supply them as soon as possible. Possibly he was considering using a new design, but there is no documentation for that. Morse said the guns would be supplied with one hundred cartridge cases each and with as many bullet molds as were usually furnished with that number of arms in the U.S. service. Morse agreed that Muzzy would manufacture the guns, and that he would send Craig the first completed gun for his inspection. Though Morse accepted the order, he ultimately provided only a single sample model.[37] Morse returned to Baton Rouge no later than May 1858.[38]

In a successful effort to resolve a conflict known as the Utah War, President James Buchanan sent two peace commissioners to Utah in April 1858 to negotiate a nonviolent solution with Brigham Young, leader of the Mormon Church. One was Benjamin McCulloch, U.S. marshal for the Eastern District of Texas, and the other was former Kentucky governor and U.S. senator-elect Lazarus W. Powell. Born in 1811, McCulloch had been a soldier in the Texas Revolution and Mexican-American War, a Texas Ranger, surveyor, and legislator.[39] Because of his exploits as an officer commanding a unit of Texas Ranger scouts in the Mexican-American War, his name was recognized throughout the nation. His lifelong goal had been to obtain an appointment as colonel of a regiment of U.S. cavalry on the Texas frontier. Though not formally educated, he studied assiduously and was well recognized as having a deep knowledge of military strategy and tactics and as being an intrepid fighter. McCulloch finally achieved the military rank he deserved when Jefferson Davis appointed him colonel in the Confederate Army on February 14, 1861, and advanced him to brigadier general on May 11. McCulloch was in his late forties when he first came into contact with a Morse breechloader, and he was killed at the battle of Elkhorn Tavern on March 7, 1862, with one in his hands.[40]

Secretary of War Floyd gave an order on April 5, 1858, to issue McCulloch and Powell one Morse breechloader and one Colt's "gun" as well as two Colt's Navy pistols to take with them on the overland expedition to Utah.[41] Undoubtedly Floyd was anxious to see how Morse's carbine functioned in the field. The fifteen-man expedition left Fort Leavenworth on April 25 and reached Camp Scott, Utah, on May 27 after traveling more than a

Benjamin McCulloch said that Morse's carbine was "the best arm ever invented for mounted troops" (Cutter, *Ben McCulloch and the Frontier Military Tradition,* 1993). He was a nationally known Texan who had been a soldier in the Texas Revolution and Mexican-American War, a Texas Ranger, and a US Marshal. He became a strong supporter of Morse after carrying one of his breechloaders to Utah in 1858. Confederate Brigadier General McCulloch was killed in action at Elkhorn Tavern in March 1862 while carrying a Morse breechloader, probably one of the five carbines made at Springfield Armory in 1860. Photograph of McCulloch, 1862, di-02298, Dolph Briscoe Center for American History, University of Texas at Austin.

thousand miles.[42] After two days of discussions, McCulloch and Powell reached an agreement with Mormon leaders on June 12, averting armed conflict.

Morse was reimbursed $125 for the Muzzy-built carbine that McCulloch carried with him on the trip. McCulloch wrote to Morse on August 17, 1858, that he carried six U.S. rifles and six Sharps's carbines on the expedition but that none was equal to Morse's in accuracy, range, or certainty of fire under all conditions. McCulloch's party experienced much bad weather, in which the other weapons repeatedly failed, but Morse's misfired only twice, and both of those events were attributed to absence of powder in the primers. McCulloch wrote, "I do not hesitate to say that I consider your primed metallic cartridge,

for all practical use, the best that has yet been invented."[43] McCulloch became a strong supporter of Morse's breechloaders over the next few years, and because of his national reputation as a fighter and his numerous Washington contacts, his endorsement became invaluable to the promotion of Morse's inventions. Powell also wrote to Morse on August 17, 1858, "I fully concur with Major McCulloch in what he said in the above note concerning your breech-loading carbine. I frequently used this carbine, and consider it decidedly the best gun I ever used."[44] Such glowing praise from men of the South had a positive effect on Floyd and would later play a significant role in Morse's decision to side with the South in the coming conflict.

Throughout the spring and summer of 1858, Floyd pursued efforts in the U.S. Congress to gain appropriations for the alteration of old muzzle loaders to Morse's breechloader design. Both the House of Representatives and the Senate considered the issue, and Jefferson Davis, by then a senator from Mississippi and still a firm advocate of muzzle loaders, opposed the measure.[45] Morse had a conversation with Floyd in his office about this time and recalled that Floyd was determined to adopt Morse's system for the U.S. Army in spite of the opinion of the Board of Ordnance officers.[46] Floyd told Morse that he would ask Congress for an appropriation for that purpose, and he prepared a letter on May 15 addressed to J. A. Quitman, chairman of the Committee on Military Affairs in the House of Representatives, offering a clear view of his feelings about Morse's design.[47] Floyd wrote, "Valuable improvements have been made in breech-loading arms which promise great advantages to the government, as by them old muskets and rifles which have become unfit for issue to the troops . . . may, for a moderate sum, be altered so as to equal the best breech-loading arms."[48]

Floyd submitted a request to Congress for an appropriation of $100,000 to be used in making the alterations. Simultaneously, Morse's friends in the House and Senate worked for the passage of the bill.[49] One of Morse's strongest supporters from Louisiana, Representative John M. Sandidge, told him that the appropriation was intended for his reimbursement.[50] The Senate requested Floyd to submit his opinion on altering muzzle loaders on May 20, discussing the advantages to the government, costs involved, and the necessary appropriation. Floyd responded the next day by submitting the same letter he sent to Quitman and asking for the same appropriation.

Bell submitted his findings to Floyd on May 12, and Floyd directed him to forward them to the Military Committee of the Senate. On May 26, a week before Morse was issued Patent 20503, Bell delivered his report to Henry Wilson, a member of the Military Committee of the Senate from Massachusetts and future vice-president of the United States, rendering his opinion concerning the alteration of arms using Morse's design. Specifically, Bell reported on his analysis of alterations to the .69 caliber rifle musket, the .69 caliber smooth-bore musket, and the .54 caliber rifle he had recently test-fired.

In all areas of evaluation, Bell was heavily in favor of Morse's design. His report comprised ten points:

1. The change to each firearm, when made by machinery, was very small, involving only part of the lock and the lower part of the barrel. Bell estimated it would cost only $2.50 to $3.50 per conversion and saw this as a general advantage for Morse's design.

2. Another general advantage for Morse's invention was, according to Bell, that the alteration did not affect the overall strength of the firearm. One weapon was converted and underwent 1,350 firings of a service charge without injury to the gun.

3. Because the Morse conversion was a breechloader, the barrel could always be kept level and directed to the front during operation, thus allowing for both more rapid and more accurate fire in any position.

4. The conversion was superior in accuracy when compared to a muzzle loader. Bell theorized the reason was that the cartridge and its chamber in the barrel were in better alignment than that of a muzzle loader at the time of discharge.

5. Bell found that the breechloader penetrated a board 1.0–1.5 inches deeper than did a muzzle loader with the same powder charge at thirty yards, and the projectile from the Morse conversion traveled farther as well. He attributed this to the fact that the breechloader lost no gases through a vent as did the muzzle loaders.

6. Bell wrote that the Morse conversion never misfired with a good cap, and its certainty of fire was greater than any other breech- or muzzle-loading firearm. (This finding was similar to that of the South Carolina legislature when it compared Morse's carbine to Burnside's in late 1862.)

7. Morse's conversion was a much cleaner weapon after multiple discharges because some of the debris from firing was withdrawn with each spent cartridge, all leading to a more accurate firearm.

8. Bell addressed the concern of unplanned discharge due to the priming cap and cartridge being in constant contact prior to being loaded into the chamber. He anticipated that soldiers would load the cartridges with powder and projectile in the field and carry them in wood-bottom cartridge boxes, thus minimizing the chance of an unwanted detonation.

9. Again anticipating that the cartridges would be loaded with powder and lead in the field, Bell foresaw a cost savings by no longer having to manufacture and transport paper cartridges for muzzle loaders.

10. Because Morse's cartridge was waterproof, it could remain functional for extended periods of time in rainy weather when compared to existing muzzle-loader paper cartridges.[51]

In 1876 the Forty-fourth U.S. Congress used Bell's opinion in part to justify its support of Morse's patent appeal. At that time, Congress felt that Bell considered the self-primed metallic cartridge to be the main feature of the new system.[52] Apparently the public press was not so impressed with the concept of breechloaders in 1858. Morse replied to such criticism in an article printed in the *Washington Union* on June 8, 1858: "If this measure should be carried, it will save the government millions of dollars. The opposition to breech-loading is short lived. The day is near at hand when ramrods will be as obsolete as matchlocks."[53] His prediction would soon come true, but he would not receive the credit for making it happen.

In response to a query from the House, Floyd responded on May 27, stating that the Ordnance Bureau showed an inventory of old guns—558,532 muskets and 62,238 rifles— each worth about $2.50 at auction. Floyd's opinion was that, if they were altered to breechloaders, they would be equal to currently manufactured muzzle loaders, each worth

$13.00 to $15.00. Even counting the cost of $2.50 to $4.00 to alter each gun, the government would enjoy a significant cost savings. Floyd acknowledged that there were other methods of converting old arms to breechloaders that might be simple and efficient, but it was Morse's design that he submitted to Congress.[54] Floyd also resubmitted a copy of Bell's analysis in his May 27 response. On the morning of June 7, Floyd sent a rifle and a musket that had been altered to Morse's plan to the Capitol. These must have been either the alterations he had ordered in February–March of that year or the one he had ordered in May. All of his efforts soon paid off.

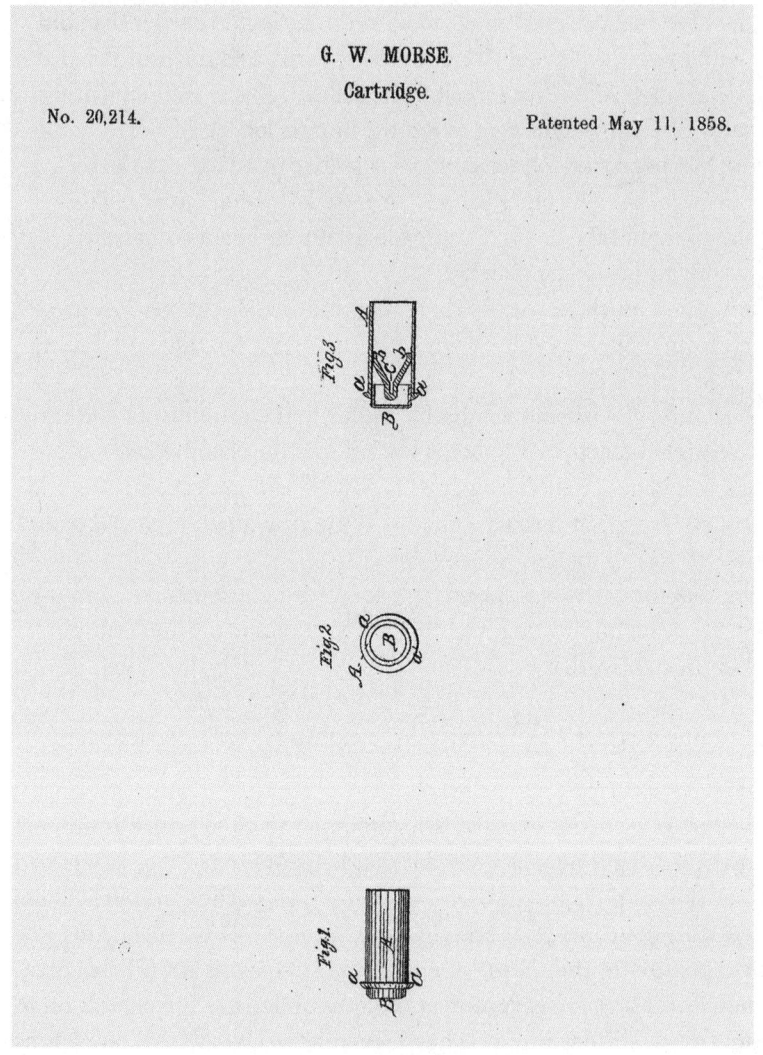

US Patent 20214. As Morse's cartridges rapidly evolved, he patented this version in May 1858. It incorporated his first use of an internal tige, or V-shaped heavy-gauge wire commonly called an anvil, which held the primer and offered resistence when the cup-shaped head of the cartridge was struck by the firing pin. G. W. Morse, Cartridge, Letters Patent May 11, 1858, Patent Case File No. 20214; Patented Case Files, 1836–1973; Records of the Patent and Trademark Office, Record Group 241; National Archives at College Park, Md.

20214—May 11, 1858, Morse's Second Cartridge Patent

Morse applied for a new patent on July 24, 1857, forwarding the required thirty-dollar fee, necessary application, and a working model. It was rejected on September 27 for "want of novelty in the devices," but, after Morse made acceptable amendments, the patent was granted on May 11, 1858, as U.S. Patent 20214 for an "Improved Cartridge Case."[55] Morse's invention was another stepping stone toward the design of the modern center-fire cartridge. It was a tubular cartridge open at both ends with a shoulder around the rear end. A cup-shaped piece of thin metal, which he called the "head, or closer," slid into the back of the tube, thus closing it off. He used a heavy-gauge wire, which he referred to as a tige, in the approximate shape of the letter "V" and soldered each end of the V to the inside of the cartridge. To prepare a cartridge for firing, a percussion cap was mounted on the point of the tige; then the cup, or head, was inserted into the tube, powder and ball were added, and the cartridge was ready for insertion into the open breech. The hammer of the gun would cause the firing pin to strike the base of the cup-shaped head, igniting the percussion cap and, in turn, the powder charge. Upon discharge, both the head and tubular cartridge would expand, preventing gas from escaping from the breech. Morse acknowledged similarities with other patented cartridges, but claimed that his invention differed in that the expanding head fit inside, not outside, the tubular cartridge, that the tubular cartridge had a shoulder that prevented it from being forced out the muzzle of the gun, and that the combination of all the parts ensured an explosion of the percussion cap.[56] Initially, the patent was to be kept in the secret archives of the U.S. Patent Office.

20727—June 29, 1858, Morse's Third Cartridge Patent

Morse received a patent for yet another "Improvement in Cartridges" on June 29, 1858, only six weeks after his previous cartridge-related patent, 20214. As with his previous patents, the application process was time consuming, and Morse probably put this design in use even before the patent was issued. With this most recent improvement, Morse established the essential features of the modern center-fire cartridge, that is, an expandable, reloadable metallic case, a primer preset into the head of the cartridge which sealed the breech opening, and a bullet that was crimped into place.

In Patent 20727 Morse described three styles of a tige, or heavy-gauge bent steel wire, that was soldered to the interior of the cartridge. The tige served two functions; one was to hold a percussion cap in place, and the other was to offer resistance to the blow of the firing pin against the cap. For those reasons, Morse's tige is typically referred to as an anvil. In this one patent alone, Morse offered three designs for the tige, and in practice he continued to experiment with the best design, some having a two-legged anvil and others with four legs. In most designs the flame from the exploding percussion cap traveled around the legs of the anvil, but in at least one design the anvil contained a flash hole through which the flame would travel.

This latest version of the metallic, center-fire cartridge involved a disc made of un-treated natural rubber called caoutchouc, or some other similar elastic material, which

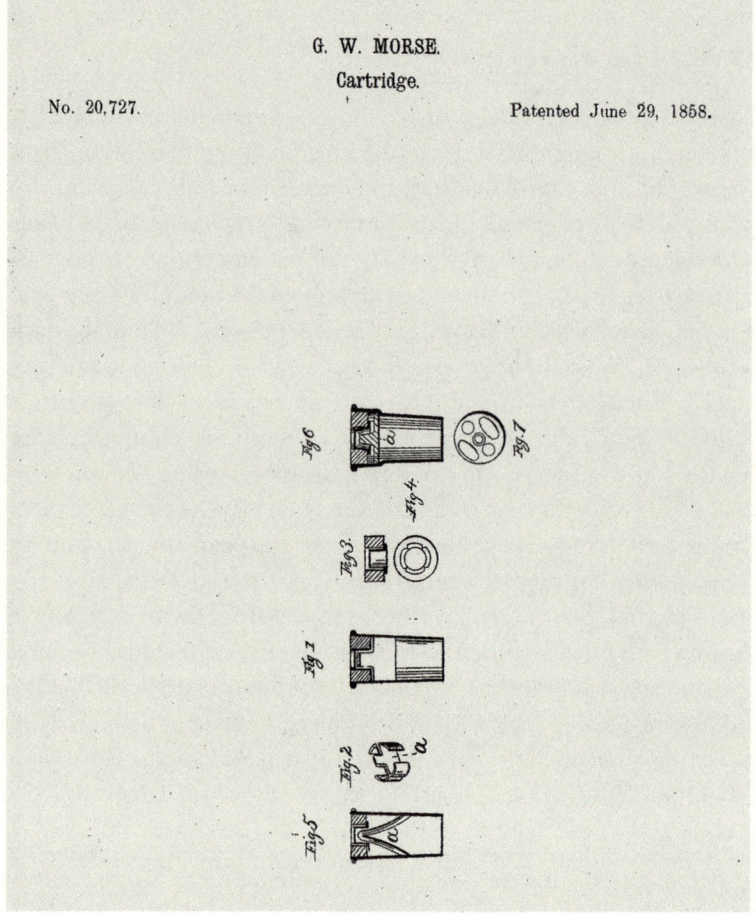

US Patent 20727, issued in June 1858, was Morse's third cartridge patent and represented the cartridge that was used in all of his firearms from 1858 to 1865. G. W. Morse, Cartridge, Letters Patent June 29, 1858, Patent Case File No. 20727; Patented Case Files, 1836–1973; Records of the Patent and Trademark Office, Record Group 241; National Archives at College Park, Md.

was perforated in its center to accept a percussion cap. By sealing the rear end of the cartridge, the combination of caoutchouc disc and percussion cap prevented moisture from seeping into the powder load. The percussion cap was imbedded in the disc such that it was not flush with the rear of the disc but indented about one-sixteenth inch, thus offering some protection against accidental discharge during transportation, a major objection made by military planners to the new self-primed cartridges which Morse specifically addressed in this patent. The disc/percussion cap design also provided a means for dependable discharge only when it was desired. The cartridge itself had a flange on the rear end to facilitate extraction of the spent casing.

　　Morse described three versions of the cartridge in which the shape of the tige was variable. In one, the V-shaped tige was virtually identical to his Patent 20214, issued only six weeks earlier. In the other two styles, the tige was more of a squared off U-shape. Unfortunately, Morse was immediately forced to disclaim the first of his two claims in this

patent. The first claim was for the tige and its variations secured in the cartridge case.[57] The disclaimer, also dated June 29, was made because of an existing patent of M. Chaudron dated March 9, 1855. Morse's second claim was for the combination and arrangement of the percussion cap and perforated disc.[58] Patent 20727 was Morse's last cartridge patent issued before the Civil War and was the basis for the cartridge used in his prewar muzzle-loader alterations and wartime brass-frame carbine.

Morse's cartridges from 1855 to 1865 came in a variety of calibers and lengths and were made from brass, tinned brass, bronze, and copper.[59] More common Morse calibers included .44, .50, .54, .58, .69, and a variety of shotgun gauges, but he experimented with other calibers as well. The wad was made from a variety of materials including leather, gutta percha, and india rubber, or caoutchouc. B. R. Lewis, a cartridge expert, designated three categories of Morse cartridges as Types 1, 2, and 3. Offering a detailed discussion of Morse's different cartridge types, his description seems to be based on extant examples and not Morse's patents. Lewis did not offer a numerical designation for the first

The breech ends of three Morse shotgun cases. The case on the right has an anvil in place. The center case has a base wad and percussion cap properly placed. The left-hand case has only a base wad installed. Courtesy of Robert M. Holter.

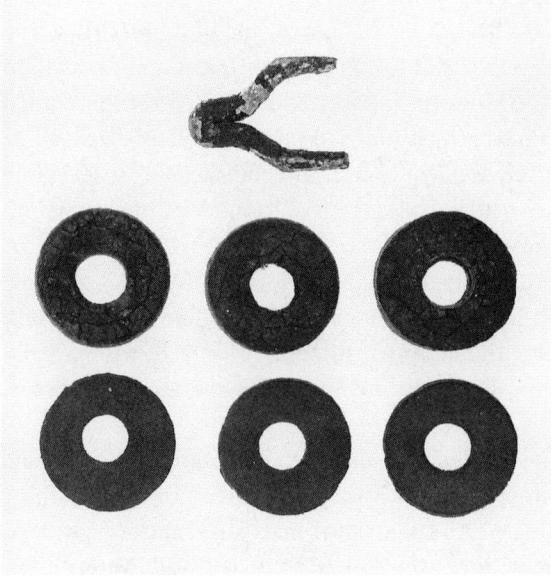

A Morse anvil and six base wads. Morse experimented with and patented several different anvil styles, but the one here was the most commonly used. Courtesy of Robert M. Holter.

(LEFT TO RIGHT) Two .54 caliber cartridges made by the Frankford Arsenal, a .50 caliber cartridge made by the South Carolina State Military works, a .56 caliber cartridge made by Nathan Muzzy, and a .58 caliber cartridge probably made by the Frankford Arsenal. Courtesy of Atlanta History Center.

two patents of Morse's early cartridge designs, 15996 and 20214, nor does he discuss the pre-patent paper cartridge. What he calls Type 1 is contained within Morse's third cartridge patent, 20727. Lewis's Type 2 appears similar to the drawings in Patent 20727, but it has enough dissimilarities that it might have been one of Morse's designs from about 1860 that he never patented. What Lewis referred to as Type 3 is covered in Morse's fourth and last cartridge patent, 345,165, which was granted in 1886.[60]

Congress passed an act on June 12, 1858, which contained clauses providing for two separate appropriations that would affect Morse. The total was $50,000, not the $100,000 that Floyd had requested. Morse felt that it was Jefferson Davis, chairman of the Senate Military Committee, who was instrumental in lowering the number from the original $100,000.[61] One clause appropriated $25,000 for "the purchase of breech-loading carbines of the best model to be selected by a board of ordnance officers." The other provided $25,000 "for the alteration of old arms so as to make them breech-loading arms, upon a model to be selected and approved by a board of ordnance officers."[62] Ultimately, $14,200 went toward the alteration of muzzle loaders to Morse's design. The entire "carbine" fund was awarded to Burnside.

Each clause gave rise to a board of ordnance officers to which Morse would present his new firearms, and on June 14 Floyd ordered one board to assemble at the Washington Arsenal, which they did the next day, to begin the selection of the best breech-loading carbine. Needing more time to issue a public call for potential entrants and desiring to have adequate space to field-test the carbines on horseback, the board recommended adjourning and reconvening at West Point, which they did on July 12. Public advertisements were sent out inviting anyone to submit a breech-loading carbine, at least one hundred rounds of ammunition, and a written explanation of its mode of operation and advantages. In addition, the inventor was required to leave the carbine with the board for examination and trial.[63] The advertisements appeared in newspapers in New York, Boston, Hartford, Baltimore, Philadelphia, Richmond, Cincinnati, Abingdon, Virginia, and Knoxville, Tennessee. Final entrants for the carbine competition were Burnside, Colt, Gibbes, Joslyn, Maynard, Merrill, Morse, Poultney and Smith, Sharps, Storm, Symmes, and Wells.

The board tasked to consider the best breech-loading carbine began testing the carbines on July 13 and 14.[64] Morse probably submitted a Muzzy-built firearm with its carbine barrel. He submitted a .54 caliber carbine with a twenty-two-inch barrel and five grooves, and it weighed seven pounds, ten ounces, dimensions more consistent with Muzzy's gun

than that of Searles. Each presenter fired 40 rounds from his weapon to determine its safety, general working of its parts, and rapidity of fire. Following that, each gun was fired twenty times at 100, 300, and 600 yards to determine its accuracy. The board reserved the right to add additional tests as they felt necessary. Morse fired 40 rounds in three minutes, forty-eight seconds—a rate of 10.5 rounds per minute—with no failures. Only the Sharps and Maynard carbines fired faster. Because Storm and Wells's carbines were not tested, only ten carbines were assessed for accuracy. Morse's carbine ranked eighth in

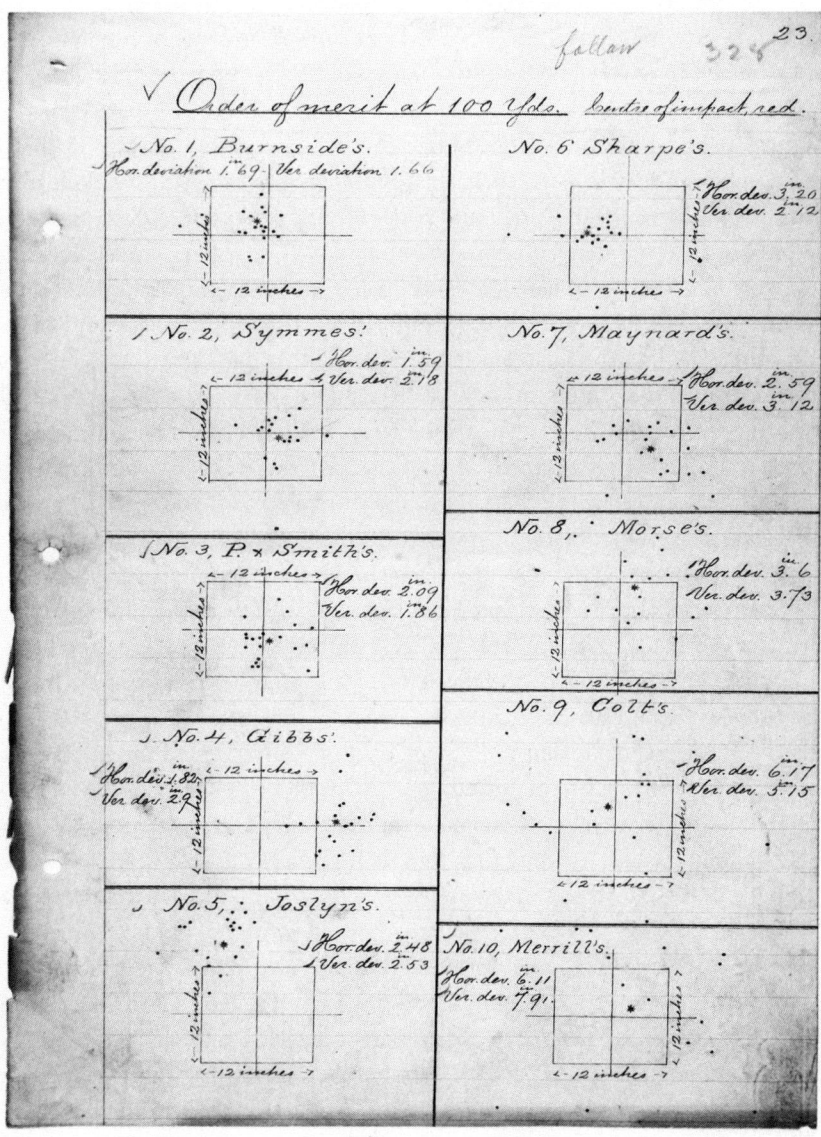

Composite results of ten breech-loading carbines fired at one hundred yards during the trial at West Point in July 1858 for the purpose of determining the best available breech-loading carbine for cavalry service. Morse's carbine came in eighth in this test. From National Archives, Record Group 123, Box 449.

the 100-yard accuracy test, sixth at 300 yards, eighth in the 600-yard range, and seventh overall. All carbines were subjected to rust tests, and Morse's fared better than most. All carbines presented some difficulties while being tested on horseback. The board found two faults with Morse's carbine when used by mounted troopers. First, they were liable to drop the cartridge in the act of loading it, and, second, the carbine failed to extract the spent cartridge on two occasions.

Consistent with his style, Morse wrote a lengthy letter to the board on July 22, in the midst of the testing: "That you may clearly understand what I have sought to accomplish by the construction of my arm, it seems necessary that I should point out what I have conceived to be the essential requisites of a good one as well as some of the defects of those already made."[65] In essence he simply summarized Bell's report from earlier that year.

As testing was nearing its conclusion, Morse shrewdly suggested on July 27 that a simulated rain test be conducted in which water would be poured over the weapons from a watering pot as they were being loaded and fired. He knew from the tests conducted at the Navy Yard in early 1857 that his weapon would do well in such circumstances and others would likely fail. The board assented, and during the following trial, Morse's gun misfired only once in twelve attempts. Colt's arm fired five times in ten tries, and all the others failed to fire. The board concluded its work on July 31 and issued its report the same day. It said in part that "none of these arms are free from serious defects which would render them unsuitable for use when subjected to the ordinary accidents and exposure of military service." The board felt that all the carbines lacked strength, solidity, simplicity of construction, certainty of action of the moveable arts, convenience of use, ease of loading, and security from accidental derangement necessary in a military firearm. Consequently the board was unable to give any of the arms an unqualified approval. Having been ordered to find the best available breech-loading carbine, however, the board decided that Burnside's carbine was "the least objectionable for use in the hands of the mounted troops."[66] An order was then issued in September for the full $25,000 appropriation to be spent on Burnside's carbines.

An interesting but unrelated event took place on July 1 when Floyd ordered one of Morse's guns and one of Sharps's to be boxed up and shipped by express to the Duke of Mecklenburg-Schwerin in care of the Consul General of Prussia in New York.[67] Morse's gun was probably a Muzzy because only four muzzle loaders had been altered to his design, and it is unlikely that Floyd would have allowed any of those to leave the country at that time. The reason for this transfer and the fate of the weapons are unknown.

To consider the second $25,000 appropriation, Floyd issued an order on July 8 for a different board of officers to meet at West Point on July 12 for the purpose of considering all methods for altering old muzzle loaders to breechloaders. Floyd also ordered the same day that the arms already altered to Morse's plan be transferred from the Washington Arsenal to West Point for examination by the board. Four such arms existed, but only three were shipped to West Point. Colonel Craig notified Morse, who was already at West Point, of the plan on July 12.[68] Similar letters went out the same day to other inventors, including T. Poultney of Baltimore; Dr. E. Maynard of Washington, D.C.; and J. H. Merrill of Baltimore.[69]

The three-officer board on "alteration of old arms" met at West Point on July 22 to get organized. They issued a call to competitors and adjourned until July 27 when they reconvened to evaluate firearms submitted by representatives of six entrants—Morse, Joslyn, Montgomery Storm, Maynard, Merrill, and Sharps.[70] Each arm was test-fired and evaluated for accuracy, endurance, rapidity of fire, and penetrating power over the following few days. Morse was present during the trials, and, like all the entrants, he submitted a written description of his invention.[71] In addition, Morse lobbied strongly throughout the process, writing to the board on July 26, 27, and 28. In his communication of July 26, he wrote that he was presenting for trial a .69 caliber smoothbore musket, a .69 caliber rifled musket, and a .54 caliber Harpers Ferry rifle. The .69 caliber rifle had already been test-fired fifteen hundred times. In the same letter, Morse reviewed the mechanics of his patent, 20503, and the advantages of his metallic cartridge, which he called "the best feature of my plans."[72]

Morse stressed the minimal cost of using his alteration, the strength of the stock after being altered, and the certainty, accuracy, and force of fire as well as the durability of the weapon.[73] In the letter of July 27, he stressed the advantages of priming the cartridge, not the gun. In the communication of July 28, Morse attributed an unusually large number of misfires with his weapon earlier that day to a small adjustment he had recently made in the cartridge. He concluded by writing, "The construction and use of my form of cartridge present eminent advantages that are indisputable."[74] In all three documents, Morse went to some length to reiterate many of the same points both he and Bell had made over the past year, and enclosed a copy of Bell's report from May 26. Morse told the board that he was sure of success if his cartridge were maturely considered and properly compared with any other system of ammunition.[75]

Morse's entry came in first place in the competition for penetration, but it placed last in the rapidity test. It misfired on twenty-seven of eighty rounds during the endurance test and was not the most accurate at a hundred yards. In spite of that, Morse won the overall competition, which prepared the way for him to win a new contract with the U.S. government. The board made their final report to Craig on August 3, and he forwarded it to Floyd on August 7. The board felt that they could not approve any model for alteration until it had actually been field-tested but that all models submitted evinced highly creditable mechanical ingenuity in solving the problem. They wrote, "The limited trials . . . do not give sufficiently decisive results of the superiority of one arm over the others to authorize the Board to pronounce that any old arms altered to breechloaders—nor indeed that any breech-loading arms—will be as efficient and suitable, for troops generally, as our U.S. muskets and rifles."[76] Morse attributed this attitude to a pervasive prejudice against all breechloaders among army officers in the late 1850s. The board selected Morse's model because "it differs from the others by including the new and untried principle of a primed metallic cartridge, which may on actual trial be found of advantage."[77] They recommended that Floyd appropriate funds for the alteration of old muzzle loaders to breechloaders using Morse's model. There is some evidence that the board also found favor with the design submitted by William Montgomery Storm based on the fact that both Morse and Storm were soon awarded contracts for the government to produce their firearms.[78]

The hundred-yard results of a .69 caliber rifled musket that had been altered to Morse's design and tested during the West Point trial in July and August 1858. Morse's gun was not the most accurate at one hundred yards, but it did win the overall competition and thus a government contract for Morse. From National Archives, Record Group 123, Box 449.

CHAPTER 5

Springfield Armory

By early August 1858, Morse had lost the competition for the best carbine and had won the contest for the best altered muzzle loader. Merrill challenged the results of the altered-arms competition based on the multiple misfires of Morse's gun during the endurance test and asked for a new trial. Knowing that Floyd favored his design, Morse responded to him on August 19, writing that the misfires did not represent a problem with the gun's design but were caused by a defect in the construction of the cartridge, which had already been corrected. Morse opposed a new trial on the grounds that the original one was fair and that he won because his was the best design. Ironically, in the same letter, Morse challenged the Board of Officers' decision in favor of Burnside's carbine. He brought forth one of the major issues for both boards, that is, whether it was better to place the primer within the cartridge or in the traditional location on a nipple outside of the gun. The "carbine" board chose in favor of priming the gun (Burnside's design), and the "altered gun" board decided in favor of priming the cartridge (Morse's design). Lobbying for the salvation of his carbine, Morse pointed out to Floyd that the "carbine" board acknowledged it was more troublesome to prime the gun than the cartridge while on horseback, but, in spite of that, chose Burnside's carbine. Morse wrote, "The manner in which my carbine withstood all the tests which the carbine board could apply to it, should have given me their award, and any attentive reader of their report will perceive that they could not find any such serious objections to my arm, as they did to the one of their selection."[1] Morse then changed his focus to money, telling Floyd that he was nearly "broken up" in trying to carry out the project of developing his breechloader for the public service, and he hoped that Floyd would see it as his duty as a government officer to come to his relief. "It is necessary that my altered arms should be put into the service immediately," he wrote, while proposing that the work be done in one of the government arsenals under his personal supervision. After laying out the finances in which he said the government would profit $120,000 by altering five thousand muzzle loaders, Morse made a plea to Floyd for "present relief," an entreaty that he would repeat many times over the following thirty years.[2]

The altered Morse firearms used in the West Point trials were returned to the Washington Arsenal on September 6. Morse wrote to Floyd three days later, proposing that the government pay him $12,500 to alter 2,500 old arms in any of the government armories under his supervision. He based this price on conversations he had with Floyd related to the current value of the old arms compared to their worth after being altered.[3] This left $7,500 in the appropriation to pay for labor and materials. Morse even offered that Floyd could retain some of the $12,500 until he was satisfied that he had adequate funds to cover the costs of labor and materials.[4] On September 11, Floyd countered with an offer to alter 2,000 arms, including the privilege to use all of Morse's patents in the process, and he agreed to pay Morse $10,000 out of the special appropriation.[5] Floyd stipulated that, after signing the contract, Morse would be issued three U.S. flintlock muskets of the latest model and three U.S. percussion rifles with seven grooves to be used as patterns for the alterations. Floyd also required Morse to designate one of the two existing U.S. armories, either Springfield or Harpers Ferry, where the completed patterns would be used to alter 2,000 muskets of the same model. The arms were to be rifled and sighted at the armory, and Morse was accorded the privilege of superintending the process and was required to state that the work was done in a satisfactory manner.[6]

Morse signed the contract on September 13. It said, "In consideration of the sum of $10,000, paid to me by warrant on the Treasury of the United States, I hereby grant, sell, and convey to the said United States the right and privilege to alter 2,000 of the muskets now belonging to the United States" according to this pattern.[7] Only three years after beginning the design of his novel breechloader and metallic cartridge, Morse had a second contract with the U.S. government to manufacture his firearm, this time on a large scale. According to Moller, this was the largest arms-related royalty payment made by the federal government up to that time.[8] Morse would later cite this contract when he sued the U.S. government in 1874.[9]

Morse chose to have at least one of the pattern firearms made by Muzzy in Worcester, Massachusetts, certainly because that firm was already familiar with Morse's designs. Morse had Muzzy make a new model similar to the one he had submitted to the West Point Board.[10] Morse wrote to Floyd on September 23 requesting that all the altered arms and tools he used at West Point in July be forwarded to him at Worcester, and by September 29 they were on the way.[11] Logically, Muzzy also manufactured ammunition for Morse's firearms. Craig informed Floyd on November 4 that the first pattern model was ready and that Morse recommended the Springfield Armory for the site of altering the two thousand firearms.[12] Because the pattern model differed slightly from the one tested by the board at West Point, Craig recommended that the Ordnance Bureau ask for an examination by Ramsay and Maynadier, two of the officers who had participated in the West Point trials. Floyd agreed, and the two officers approved the modifications on November 5.[13] Orders were issued on November 6 and 9 for the armory to begin work on the Morse conversions, and Floyd gave specific orders on November 17 not to rifle the muskets.[14] Work at Springfield began immediately under Morse's personal supervision and went on for several months. In a letter dated February 24, 1859, James S. Whitney, superintendent of Springfield Armory, told Craig that he had commenced work on altering model 1822 muzzle loaders, an improved variant of the M1816, to Morse's plan in November 1858 and had spent $671.56 on labor by February 10, 1859. Most of this must

have been start-up costs.[15] The original projection was for the government to spend a few dollars per gun in the alteration process, but only a few guns had been completed by early 1859.

Prior to ramping up the production line, at least two muzzle loaders were converted at Springfield. Theodore Lewis, military storekeeper at the Springfield Armory, provided one of Morse's pattern models to Major Ingersoll in December 1858. Ten years later, in 1868, this model was issued to the commanding officer of the Springfield Museum.[16]

About the same time, one thousand model 1816 muskets that already been converted to percussion were drawn from the Springfield Armory stores to be altered using Morse's patents.[17] The mechanics of the alteration are described on the Springfield Armory web site and indicate he was using the design in Patent 20503 with a solid, not sliding, breech block:

> The alteration required that the top portion of the barrel near the breech and the nip-ple be removed and a new breechblock hinged just forward of the tang be installed. The front of the hammer was cut away and what remained served as a locking bolt lever. This was attached internally to the locking bolt that entered the action from the rear. When the trigger was pulled and the hammer dropped, it forced the locking bolt forward and it, in turn, drove the firing pin forward through an opening in the front of the block's body and into contact with the cartridge's primer. When the hammer was at half-cocked position, the locking bolt was withdrawn from engagement and the action was free to be opened. Pulling the knurled finger latch at the front of the breechblock rearward and up opened the action.[18]

Sixty muzzle loaders were altered to Morse's design at Springfield Armory between late 1858 and November 12, 1859.[19] This number included three model 1841 rifles and 57 model 1816/1822 muskets. A fourth M1841 was altered at Springfield in 1860. Morse's de-sign was the first breech-loading firearm and the first to use a center-fire cartridge to be manufactured at the Springfield Amory.[20]

After work had begun on a planned run of six hundred alterations in late 1858 and early 1859, Morse became dissatisfied with the current design, Patent 20503, and modi-fied it by making yet another major design change that resulted in an improved firearm but a slowdown in productivity. He employed a man whom he later recalled was named Munk, or Munck, to make the changes. This was probably C. H. Munck, a forty-year-old German-born gunsmith living in Washington, D.C., at the time.[21] Munck made a new model that included a "head-block to be moved longitudinally in the cartridge receiver which should force the cartridge all the way into the chamber and retract it wholly out of the chamber."[22] With this change, Morse retained the rear part of the heavy breech block of 20503 and reintroduced a sliding breech block similar to that found in Patent 15995. By doing so, he created a hybrid design which he would use on all of his future breechloaders but for which he never sought a patent. Morse reported the changes to Floyd, who was surprised and annoyed, but at Morse's request called together the members of the West Point Board to examine Munck's modification. The board found the new model favorable and incorporated it into the ongoing work.[23]

Morse stated in 1877 that all of the Springfield alterations, meaning the fifty-four M1816/1822 alterations, were manufactured using the model that Munck had made. We

This alteration of an M1841 Harpers Ferry rifle combines features of Morse's patent 15995 and 20503 and was first made by a Washington gunsmith named Munck. Its hybrid design was used on all Morse breechloaders made from 1859 onward and was virtually identical to that of his new carbine, which was also being developed at the time. Courtesy of Atlanta History Center.

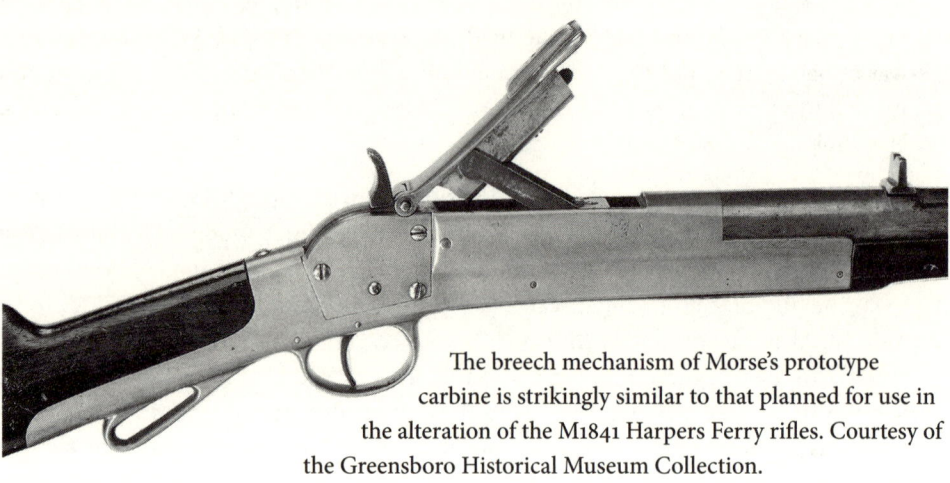

The breech mechanism of Morse's prototype carbine is strikingly similar to that planned for use in the alteration of the M1841 Harpers Ferry rifles. Courtesy of the Greensboro Historical Museum Collection.

know that at least some of the six muzzle loaders—three M1816/1822 and three M1841—used as patterns for the Springfield Armory production run were made according to the "solid-block" 20503 and some were based on Munck's hybrid design. Surviving examples of the Springfield conversions confirm that not all were made exactly alike. Muzzle-loader alterations for each design exist at the Springfield Armory Museum. Three surviving model 1816/1822 muzzle loaders at Springfield Armory Museum were altered using the pattern of Patent 20503 without a sliding breech block and with an ejector lever on the bottom of the breech. It is likely that these were the only three made using this design. Three other survivors at Springfield Armory Museum, one M1841 and two M1816/1822 models, have Munck's hybrid design with a sliding breech block. Most, if not all, others were made using Munck's hybrid design. The Atlanta History Center has two survivors, an M1841 and M1816/1822, both with Munck's hybrid design.

Morse wrote to Whitney on February 7, 1859, to inform him that he had just shipped the pattern musket to be used for making the new alterations. Morse clarified the modifications he had made, including one to the ejection lever, and introduced Whitney to the basic functions of dismantling and operating the gun. Morse was in Washington at the time but planned to be in Springfield within a few days.[24] It is not clear exactly which modification he was referring to, but it was probably the one recently made by Munck. Munck's gun did not have an ejection lever because Morse had gone back to using a sliding breech block with an extraction mechanism built into it. It is likely that the modification that Morse described to Whitney was one involving deletion of the lever ejection mechanism.

On February 8, Floyd instructed Colonel Craig to ask Morse to ship two examples and a hundred rounds of ammunition for each of three conversions to the War Office in Washington as soon as practicable.[25] Craig conveyed the message to Morse at Springfield the same day.[26] The three conversions were the .69 caliber musket and the .54 and .58 caliber rifles. By February 22 the Washington Arsenal had spent $1,253.92 on "altering four service arms to Morse's pattern and work done on his model guns."[27]

Morse was back in Baton Rouge in mid-March when he wrote to Whitney that he regretted the necessity of his being in Louisiana because he had hoped to personally supervise the alterations going on at Springfield. Anticipating quick work, Morse wrote that he expected some or all of the alterations to be completed by the time he could arrive there. Morse made a gross miscalculation because two thousand alterations had been ordered; but a total of only about sixty were ever made, and probably only a fraction of those at the time he wrote the letter. Expecting to be in Springfield soon, Morse asked Whitney not to ship the altered arms to Washington until he could personally test them.[28] Whitney shipped two of the altered model 1822s to Craig on April 22, along with cartridges, powder, shot, and tools for casting the cartridge cases, all having been prepared according to Morse's directions. Whitney hoped to forward two altered rifles, .54 and .58 caliber, by May 15.[29]

Morse's constant changes began to stress the budget. Whitney told Craig that because of recent modifications to the alteration plans, specifically a change in the model, he was forced to fabricate a considerable number of new tools, and he anticipated a cost of about $1,200 to prepare the new tools and models. For that reason, Whitney told Craig that he would not be able to make the specified two thousand alterations at a cost of less than $5 each.[30]

On April 25, Morse requested that the War Department order the commanding officer at the Washington Arsenal to assign an officer to assist him in making trials of the new modification, and Floyd approved the request.[31] Craig wrote to Floyd on April 26, calling his attention to the fact that the funds appropriated for the task of altering two thousand muzzle loaders to Morse's breechloaders were insufficient. Craig summarized that the original appropriation of June 12, 1858, was $20,000. Morse had already been paid $10,000 for the right to alter two thousand firearms according to his plan at Springfield Armory, and William Montgomery Storm had been paid $2,500 for two thousand alterations at Harpers Ferry Armory. This left only $7,500 to cover the costs of materials and labor at both armories. Whitney estimated that it would cost $11,200 to alter two thousand arms, a figure that was $3,700 more than the budget for both armories, placing

the plan for Morse alterations over budget.[32] Floyd responded on May 6 by ordering a reduction in the number of alterations to correspond with available funds. As a result, he ordered that only six hundred guns would be altered to Morse's plan at Springfield Armory, and the number of arms altered using Storm's plan at Harpers Ferry Armory would be dictated by the amount of the appropriation left over.[33] Craig notified the two superintendents accordingly. He limited Whitney at Springfield Armory to spending $1,200 on new tools and $3,000 on altering six hundred muzzle loaders, and he capped A. M. Barbour, superintendent at Harpers Ferry, at $3,300 for all previous and future work on Storm's design.

Morse returned from Louisiana by mid-May. He was in Washington when he wrote to Whitney on May 20, saying that he had learned only three or four days earlier about the new order limiting production to six hundred firearms. The nature of his business in Louisiana that spring is unknown, but it took him away from Washington and Springfield at a critical juncture in the manufacture of his inventions. Morse's letter indicates that he completely disregarded or misunderstood budgetary limitations and continued to suggest new improvements faster than Whitney could keep up. Morse discussed making the cartridge cases lighter. Recently they had been shortened to 1½ inches, requiring counterboring the rifle barrels, and Morse talked about making them from thinner metal like the ones previously produced at Waterbury. Morse also told Whitney that he was planning to ask Floyd to allow him to alter and rifle two hundred muskets and to alter four hundred rifles of either caliber Floyd might choose. He also told Whitney that he wanted to advance the india rubber wad 1/16 of an inch farther into the cartridge case in an effort to prevent accidental discharge. Making these changes would require yet another alteration in which the forward part of the sliding breech piece would now need to project farther into the cartridge. It was evident that Morse had been out of touch with the overall process for some time when he asked Whitney if he had completed the original six alterations and if he had begun altering additional arms.[34]

Morse wrote to Floyd on June 24, suggesting further additional modifications to the muzzle-loader conversions: "The arms which have been altered in accordance with my plans work well, but still some slight modifications are necessary to their perfection."[35] Morse told Floyd of his desire to alter four hundred rifles, anticipating that they would be placed in service and field-tested against other rifled, breechloader conversions—probably Storm's. He also discussed rifling the smoothbore alterations. The additional modifications he recommended involved changing the positions of the inside and outside cocks and lengthening the firing pin mechanism. Erskine S. Allin, master armorer at Springfield, responded to Craig on June 28, to say, in essence, that ongoing experimentation would do nothing but increase the cost of the project, which was already operating on a very limited budget. Seeking Craig's permission to proceed, Allin wrote, "The expense cannot be great, but your authority is deemed necessary before we can follow the repeated suggestions of Mr. Morse in perfecting his invention."[36] Craig wrote to Floyd on July 1 endorsing the changes as being an improvement and recommended that Floyd authorize them with the condition that all work on Morse's alterations be ceased as soon as the appropriation was expended.[37] Morse wrote to Whitney on July 4, expressing regret that he had not made the recommended changes sooner and discussing his optimism that Floyd would approve Craig's endorsement. Morse planned to pass through

Springfield with his family within a few days on his way to spend the summer at Lowell, Massachusetts, probably with his father. Floyd was out of town, and acting secretary of war William R. Drinkard approved the changes recommended by Craig on July 12.[38]

When the changes suggested by Morse on June 28 were approved, he agreed that all work would stop when the expenditure reached $4,200 instead of the previously agreed upon $10,000 regardless of the number of alterations completed by that time. By August 6, the Springfield Armory had reached the $4,200 limit and the appropriation of June 12, 1858, for the alteration of old arms into breechloaders was completely exhausted.[39] From the original $25,000 fund, Morse had been paid $10,000, Springfield Armory had been paid $4,200 for making the Morse alterations, $5,800 had been spent on Storm's design, and $5,000 went to a new method of priming.

Benjamin McCulloch resigned as U.S. marshal on April 1, 1859, and, though he had a myriad of job offers from which to choose, he developed a formal business relationship with Morse no later than July. McCulloch, a nationally recognized arms expert, was enamored with Morse's breechloaders and, because of his extensive contacts both in Washington, D.C., and Texas, he became an excellent field representative for Morse. McCulloch entered into a partnership with Morse, purchasing a share of the rights to Morse's patents. The new venture required all of his attention and financial resources throughout the fall and winter of 1859.[40] McCulloch wrote to his mother about his new business venture, "much money can be made out of it if well managed," and he even predicted that he would be promoting the new Morse firearms in Europe by the end of the year.[41] Like Morse, McCulloch also knew John Floyd, and, recognizing that Floyd was a strong supporter of Morse's designs, he was convinced that Floyd's influence would result in Morse's arms being sold to both the U.S. government and state militias. And if that went well, he envisioned good sales in foreign markets as well. McCulloch wrote, "One of two things will be the consequence of my connection with it. I will either be rich or flat broke in one or two years."[42] By late October, however, McCulloch learned that there were other shareholders and that both his potential profits and control over marketing were thusly diminished. He wrote on October 23: "Matters about the gun are more complicated than I first thought."[43] The European trip was cancelled, but McCulloch's relationship with Morse remained intact. By December, circumstances appeared to be a bit brighter. McCulloch had become increasingly fearful of the collapse of the Union and urged the Texas legislature and Governor Sam Houston to purchase arms so that the state could defend itself if it were to secede. He wrote on January 1, 1860, that he had "a little better prospect of my doing something with the gun now than of late."[44]

In spite of having no budget to continue altering old guns to Morse's design, Floyd planned to do exactly that. He wrote to Craig on December 7, 1859: "You will please cause the musket with bayonet, and the four rifles with sword bayonets which were altered at Springfield Armory to Morse's plan and are now at the Washington Arsenal, to be properly boxed and sent to Mr. Morse, together with the wad cutters, wads, moulds, and cartridge cases belonging to the guns. These guns are to be delivered for experimental purposes to Mr. Morse who will be permitted to obtain from the Arsenal on payment for therefore, the necessary powder and balls for filling the cartridge cases. The arms and appendages are to be returned to the Arsenal when called for."[45]

Floyd also planned to place Morse's breechloaders in service. He had promised Captain Theodore Claiborne that his Company B, Mounted Rifles, would have a share of Morse's arms when they were field-tested. Claiborne wrote Floyd on December 12 from his post at Fort Stanton, New Mexico, saying that he had heard rumors that four companies were to be armed with Morse's rifle. The rumor was premature, and Floyd's reply, if any, is not preserved.[46]

Craig wrote to Morse in Washington on December 28, 1859, telling him that the muskets which had been altered at Springfield had been ordered to be rifled and sighted. One of them was to be sent to Muzzy Rifle Barrel and Gun Manufacturing Company as soon as possible, and the others were to be shipped to the Washington Arsenal. Further, Craig instructed Morse to make twelve thousand cartridge cases, to be paid for by the Ordnance Department.[47] Morse hired Augustus Brown to make the cartridges, and by September 25, 1860, Brown had finished six thousand and nearly completed the remainder of the order.[48]

By the end of 1859, a clear picture of the production of Morse's firearms emerged. Whitney reported that for the fiscal year ending June 30, 1859, Springfield Armory had altered 6 arms using Morse's breechloader pattern and that 698 Morse breechloaders were "in progress."[49] Whitney reported on November 12, 1859, that $4,200 had been expended and that 60 arms had been completed, signaling the end of work on Morse's arms at Springfield Armory.[50] This number included the 6 model arms completed before June 30 and 54 muskets, which were completed after June 30 and were ready to be issued. He also reported that components for an additional 540 were in progress, and the estimated cost of completing them was $4 each.[51] Craig sought Floyd's opinion concerning the disposition of the 54 arms on November 26.

The number of muzzle loaders altered to Morse's breech-loading designs at Springfield Armory between late 1858 and 1860 was 61. (See appendix 5.) However, exact numbers provided in different official documents and personal recollections do not all agree. Floyd *ordered* the following Morse designs to be produced at Springfield: 3 M1841 rifles and 3 M1816/1822 muzzle loaders as patterns followed by 2,000 M1816/1822 muzzle loaders as the production run. But we know that the actual number manufactured was only 61, falling far short of the original goal. One pattern model was made by Nathan Muzzy and another by Munck, neither counting toward the number manufactured at Springfield. The sum total production of Morse muzzle loader alterations at Springfield was 57 M1816/1822s in late 1858 and 1859, 3 M1841s in late 1858 and 1859, and 1 M1841 in 1860.

An extant ledger of all arms manufactured at the Springfield Armory is difficult to interpret because it is based on fiscal years, not calendar years. At first glance, it *appears* to show that the armory had not converted any weapons to the Morse breech-loading mechanism as of December 31, 1859. Further, the report *appears* to show that between January 1, 1860, and December 31, 1860, the Springfield Armory converted 54 model 1822 rifled muskets and 1 percussion rifle to Morse's breech-loading design.[52] The same ledger shows that the Springfield Armory made no Morse conversions after December 31, 1860.[53] The explanation for the apparent discrepancy in dates of manufacture is that the arms were actually altered during the calendar year 1859 but also during the fiscal year that ended June 30, 1860, and they were therefore included on the ledger as being manufactured in 1860, not 1859. This ledger shows neither the 3 M1841 pattern models nor the

Fifty-seven M1816 muzzleloaders, seen here, and its variant M1822, were altered at Springfield Armory using Morse's designs. Courtesy of Buffalo Bill Center of the West, Cody, Wyoming. Gift of Olin Corporation, Winchester Arms Collection, 1988.8.1508.

3 M1816/1822 pattern models, which were probably made in the fiscal year ending June 30, 1859.

Another document providing ample evidence of the Morse alterations that took place at Springfield in 1859 is titled, "Statement of Receipt and Issues of Morse's Breech Loading Arms from December, 1858, to February 7th, 1878."[54] The statement gives an accounting of all work performed at Springfield using Morse's patents. This document correctly states that the majority of Springfield alterations to Morse's design utilized model 1816/1822 muzzle loaders. The statement records that 1,000 model 1822 muskets were taken from the Ordnance Store Keeper in the fiscal year 1859 with the intention of altering them to Morse's breech-loading design. Four hundred were never altered and were returned to the Ordnance Store Keeper in 1859. Of the remaining 600, only 56 were completed. Superintendent Wright wrote to Morse on November 22, 1860, that 544 "parts in progress" were halfway through the alteration process.[55] They were never completed and were carried on the Master Armorer's Inventory until they were sold at public auction between 1868 and 1871. Alterations on the remaining 56 were completed at Springfield, and 2 of those were sent directly to the Ordnance Office in Washington, D.C., on April 22, 1859. Fifty-four were turned over to the Ordnance Store Keeper at Springfield, 1 of which was sent to the Muzzy Rifle and Gun Manufacturing Company in Worcester, and the other 53 issued to Major George D. Ramsey at the Washington Arsenal in February 1860. According to the Springfield Armory web site, a contract was given to Muzzy to make 10,000 .69 caliber Morse cartridges and a single altered musket was sent to Muzzy to test the ammunition.[56] It is possible that Whitney shipped another altered musket to Muzzy on January 10, 1860.[57]

The "Statement of Receipt and Issues of Morse's Breech Loading Arms from December, 1858, to February 7th, 1878," also shows that three Harpers Ferry model 1841 rifles were altered to Morse's design at the Springfield Armory in 1859, and that four "Morse Breech Loading Carbines" were fabricated in part. These seven firearms remained on the master armorer's inventory at Springfield until 1877, when they were all transferred to "Miscellaneous Arms" at the Springfield Museum. Along with the musket alteration sent there in 1868, the total at the museum rose to eight. Of these eight Morse examples, one was sent to General Stewart L. Woodford on June 13, 1877. Six were sent to the Ordnance Office in Washington, D.C., on January 25, 1878, and another was sent there about a week later on February 7.[58] On October 26, 1880, six Morse firearms remained at the War Department in Washington, D.C.[59]

In addition, a fourth model 1841 was altered at Springfield and sent to Harpers Ferry. We know that a rifle was altered to Morse's design and issued to the Harpers Ferry

Armory in July 1860, probably to be used as a pattern model. The Springfield Armory Museum currently has one model 1841 alteration and five model 1816 alterations, as well as a Morse inside-lock muzzle loader and a Morse brass-frame carbine, both of which were manufactured at Greenville, South Carolina. In the Wray Collection at the Atlanta History Center there are two muzzle loaders altered to Morse's design—one model 1841 and one model 1816. Another model 1841 Morse alteration is held in a private collection. The fourth model 1841 alteration made in July 1860 was probably destroyed in the April 1861 fire at Harpers Ferry.

In his annual report to Congress in December 1859, Floyd, who would serve as secretary of war for another year, expressed his ongoing enthusiasm for the introduction of breechloaders into service:

> Under the appropriations heretofore made by Congress to encourage experiments in breech-loading arms, very important results have been arrived at. The ingenuity and invention displayed upon the subject are truly surprising, and it is risking little to say that the arm has been nearly, if not entirely, perfected by several of these plans. These arms commend themselves very strongly for their great range and accuracy of fire at long distances; for the rapidity with which they can be fired; and their exemption from injury by exposure to long continued rain. With the best breech-loading arms, one skillful man would be equal to two, probably three, armed with the ordinary muzzle-loading gun. True policy requires that steps should be taken to introduce these arms gradually into our service, and to this end preparations ought to be made for their manufacture in the public arsenals.[60]

CHAPTER 6

Morse's New Carbine, 1860

Most examples of George Morse's best-known firearm, the brass-frame carbine, were made in Greenville, South Carolina, between 1862 and 1865. Two previously unanswered questions relative to the carbine are "When did he design it?" and "When were the first prototypes made?" We know that the carbine itself was never patented by the U.S. government. We also know that some prototypes were made in late 1862, and that the mechanism was essentially the same as the Munck-built hybrid combination of Patent 20503 with the addition of a sliding breech piece, which existed by late 1858 or early 1859. Evidence presented here strongly suggests that Morse began working on the design of the new carbine in 1858, and five examples of it were made at the Springfield Armory in 1859 or 1860.

Morse agreed on March 16, 1858, to manufacture one hundred carbines for forty dollars each. Twenty-one months later, on November 4, 1859, he was able to present only a single sample carbine to Colonel Craig, most likely of his new design. If this carbine were either a Muzzy-built model or a muzzle-loader alteration, why would it have taken Morse more than one and a half years to produce a single sample? The carbine that Morse presented to Craig had a "ring" for attaching a cavalry sling, but Craig instructed Morse to add a swivel attached under the handle of the stock similar to Sharps's carbine. Arguing for it being a Muzzy-built receiver is that one with a side-mounted sling bar and standard cavalry saddle ring is known to exist. More likely, however, the example that Morse presented to Craig in November 1859 was a prototype of his new carbine, which would eventually be built with a brass frame in Greenville, South Carolina, between 1862 and 1865. The brass-frame carbine was designed to be a cavalry weapon, but it does not have a saddle ring *per se*. It does, however, have a finger rest for the third, fourth, and fifth fingers of the trigger hand. We know that slings were issued with Morse's carbines during the war; Percival's Company was issued seventy-five carbines and seventy-five slings on November 16, 1864, and the only place on the Morse carbine for a sling to attach was the finger rest behind the trigger. Perhaps Morse designed it with a dual purpose—both

An unusual example of a Muzzy-built Morse-designed receiver. It is the only one known to have two sling bars and a standard cavalry saddle ring. From author's collection.

finger rest and sling attachment. Though the finger rest is not in the typical form of a saddle ring, it is possible that it was the "ring" disliked by Craig.

After Morse contracted with Munck in early 1859 to make new modifications, which included a sliding head-block, he determined that this improvement represented the perfection of his breech-loading design. Floyd agreed that Munck's model was better than any of Morse's previous models and proposed to build any new arms using this model. It was the Munck gun that served as a basis not only for the alteration of muzzle loaders at Springfield Armory but also for the firearm that would become the brass-frame carbine made in South Carolina. We do not know for certain if Morse was working on a new design for a carbine when he agreed to produce one hundred in March 1858, but we do know that by early 1859 he intended to use the Munck modifications on all of his future firearms. This could explain why it took him twenty-one months to produce a single carbine, which might not have originally been made of brass but was probably of a similar design to the carbine that was made of brass at the South Carolina State Military Works.

Craig also suggested in November 1859 that Morse provide one two-bullet mold for every ten carbines or one single-bullet mold for every five, as well as a screwdriver and wiping brush or thong with every carbine.[1] Such a bullet mold would have been unnecessary if the new carbine was an alteration to the standard .69 or .54 caliber, as all other alterations had been. Further supporting the supposition that Morse had developed a completely new design for the carbine is that he circulated a batch of letters on December

19, 1859, which prompted a memo recognizing that he had been working on a new carbine since March 1858, but that *it had never been tested by a board of ordnance officers.* All of his earlier designs, both the Muzzy-built firearms and the altered muzzle loaders, had been thoroughly tested by ordnance officers.[2]

On January 26, 1860, Floyd ordered that six of each kind of breech-loading arm purchased by the United States be sent immediately to the Washington Arsenal.[3] Craig issued the order the next day, and all affected weapons were carbines. From the St. Louis Arsenal came six breech-loading carbines each of Burnside, Joslyn, Maynard, Schroeder, and Symmes's designs. From the New York Arsenal, Craig ordered six breech-loading carbines each of Sharps, Merrill, and Morse's designs, but he added a note that Merrill and Morse's carbines would not be sent because none had ever been received from the contractors.[4] By February 1860, nearly two years after the original order, Morse had furnished the government with only the single sample carbine.

After Floyd acknowledged that Munck's model was Morse's best design and that it should be used as a pattern for all of Morse's future firearms, Morse and a man named A. Anderson proposed on February 6, 1860, to alter the two-year-old contract for carbines.[5] Floyd accepted it on February 9, rescinding the previous order for one hundred carbines at forty dollars each. The new agreement stated that Morse had supplied none of the original hundred carbines which he had agreed to manufacture for the U.S. government. Under the new contract, the government purchased the patent rights from Morse and agreed to manufacture one thousand carbines and to pay Morse and Anderson three dollars for each one. Morse agreed to the new contract on February 11, and two days later the U.S. Treasury paid Morse and Anderson three thousand dollars from a fund appropriated for "arming and equipping the militia."[6] The agreement specified that the carbines would be made generally according to Morse's inventions for breech-loading arms (Patent 20503) and for his cartridge (Patent 20727), and specifically according to the plan which had been approved by the secretary of war for the alteration of the U.S. rifle, certainly referring to the Munck modification of Patent 20503. No known examples of a Morse carbine were made using an unmodified version of Patent 20503. It must be kept in mind that the Munck modification of 20503 was never patented *per se,* and that Patent 20503 was the most recent Morse firearm patent to which the government could refer in the contract.

Benjamin McCulloch was a business partner with Morse, Anderson, and possibly others at this time. McCulloch's biographer wrote that Morse and his partners received an order from the secretary of war for one thousand carbines at thirty-five dollars each. McCulloch wrote to his brother, Henry, "This makes the neat sum of $35,000."[7] Morse's most recent contract was for three thousand, not thirty-five thousand dollars. It is possible that another contract was agreed to and never signed, but it is more likely that McCulloch simply got the numbers wrong. McCulloch wrote that he had an influence over the sale, but that he would not receive any direct financial benefit from it. All the profits would go to the other owners, presumably Morse and Anderson, who according to McCulloch, would sign over all of the patent rights to McCulloch. He wrote, "The Secretary says they shall do it or he shall never give them another order of any kind." In a testament to his belief in the usefulness of the carbine, McCulloch was convinced that he could "do something handsome with this gun" once he controlled the patent.[8] He had

high hopes that the Southern states' need for arms would create a lucrative opportunity for him. There appears to be a missing link in this scenario, however; perhaps McCulloch misunderstood the details of the contract, or maybe the deal fell through. In all of the thousands of pages of testimony in Morse's postwar lawsuits, McCulloch's name never came up, indicating that the patent rights to Morse's inventions were never transferred to McCulloch. In spite of all the confusion, McCulloch continued to represent Morse's firearms throughout 1860. He attempted to convince the governor of Alabama to join the state of Mississippi in placing an order for ten thousand Morse rifles, a number sufficient to justify their manufacture by a private firm. McCulloch also traveled across the South during that time demonstrating Morse's firearms and advocating secession. Georgia's *Milledgeville Herald* reported in November that McCulloch had recently demonstrated a short-barrel, breech-loading rifle of a "new patent." He placed thirty or forty shots within one foot of the target at four hundred yards. The article did not specifically state that it was Morse's design, but it most certainly was.[9] Back in Austin, Texas, by the winter of 1860–61, he held a shooting demonstration of Morse's rifle on the banks of the Colorado River every morning at 10:00 o'clock.[10]

During its development, there was much disagreement about what caliber to use in manufacturing the new carbines. Dahlgren wanted it to be .69 caliber for its stopping power. Captain Maynadier wanted .50 caliber, and others wanted .54. Morse preferred .45 caliber for its flatter trajectory and longer range.[11] Floyd resolved the issue when he wrote on February 28, 1860, that preparatory to the government beginning to manufacture 1,000 Morse carbines, the superintendent of the Springfield Armory was to manufacture four of the carbines as models under Morse's supervision.[12] Two were to be in the .54 caliber of the Harpers Ferry model 1841 rifle, and two in the .44 caliber of the Colt's army pistol.[13] Though the production of one thousand carbines stipulated in the contract of February 1860 never took place at either Springfield or Harpers Ferry, five "Morse Breech Loading Carbines" were fabricated at Springfield Armory in 1860.[14] Some evidence exists, albeit inaccurate personal recollection twenty years after the events, that the carbines were never made at Springfield in 1860. One of Morse's postwar lawsuits states inaccurately that no guns were made or altered on Morse's plan under the contract of February 11, 1860.[15] Additional evidence presented below supports the fact that five carbines were actually completed at Springfield by late 1860.

Meanwhile, Floyd was planning to pursue the production of both the altered M1841 rifles and the new carbines at Harpers Ferry Armory in what was then the state of Virginia, but his order dated February 28 directed not only the alteration of four model 1841 Harpers Ferry rifles to Morse's design at Springfield Armory but also the production of two hundred cartridges for each of the four carbines and four M1841s.[16] Erskine S. Allin, master armorer in temporary command at Springfield, wrote on March 5 that he had received the order to alter four Harpers Ferry rifles but that he had none on hand and requested that they be forwarded to him from Harpers Ferry Armory.[17]

Craig informed both Morse and Whitney of the new order on March 1, 1860.[18] Morse wrote to Allin on March 2, 3, and 4. Though the letters are not preserved, it is clear from Allin's responses that Morse was anxious to get the work moving along.[19] Allin responded to Morse on March 5, confirming that he had written Craig earlier that day, requesting that he send four Harpers Ferry rifles to Springfield so they could be altered. Allin told

Morse that he would immediately begin making preparations to facilitate the work, and as soon as the four firearms arrived he would "proceed to the work with as much dispatch as is possible."[20] In another letter, on March 6, Allin told Morse that the Springfield Armory did not have a sample firearm from which to base his work and asked Morse to send one to Springfield. Allin did not specifically state what type of firearm he wanted, M1841 alteration or the new carbine.[21] Morse was planning to visit Springfield in early April, but Allin was not ready for him yet. Allin wrote to him on April 2, reassuring him that "we are doing all that we can to advance your work," and suggested that Morse delay his visit by a week.[22] Allin reassured Morse that he would not allow the work to be retarded by Morse's absence, and that he was planning to order the cartridge cases that Morse had asked for.[23]

Morse raised an old issue with Floyd in early April. He was anxious to get the fifty-four altered M1822 muzzle loaders from Springfield into service. Anticipating that they would stand up well under field conditions, he had written to Floyd on December 24, 1859, asking that all of the guns and cartridge cases be forwarded to the Washington Arsenal so that the cases could be charged with powder, and guns prepared for issue. Floyd agreed with Morse and gave the order to Craig. Morse complained to Floyd on April 5, 1860, that though the guns had been shipped to Washington, they had not been issued because they were not quite "ready for service."[24] Morse said that Craig had persistently refused to give the order to Major G. D. Ramsay, commander of the Washington Arsenal, to carry out Floyd's instructions. Craig responded to Morse's claims on April 7. He stated to Floyd that he had fully carried out all of his orders, including rifling the guns. He actually prepared only fifty-three altered muskets, because one had been sent to Lowell, Massachusetts, to be used in preparing twelve thousand cartridges, which had not yet been manufactured. Craig wrote that Morse was now asking for a new alteration to the guns so that they could accept an improvement to his cartridge. Because the budget was already exhausted and new tools and labor would cost even more money, Craig asked for Floyd's direction prior to his beginning any work. Craig closed by writing, "This reference for your further instructions in the case I consider the more essential that you may be fully advised of the imperfect state of Mr. Morse's invention and the difficulty of adapting it to military use."[25] Floyd responded on April 9 by ordering Craig to direct Ramsay to prepare the arms for service under Morse's supervision at the Washington Arsenal, and Craig immediately issued the order.[26]

In mid-1860 the U.S. Congress became aware that the secretary of war had purchased the patent rights from Morse. Senator Jefferson Davis was particularly unhappy and inserted a clause into an appropriations act for the fiscal year ending June 30, 1861. It stated that no arms or military supplies of a patented invention, nor the right to use any patented invention, could be purchased without being authorized by law and an appropriation explicitly set forth.[27]

In June 1860, Morse was living in Washington, D.C., with Marianne and their three children as well as his stepson, Burdett, who was recently accepted as a cadet at West Point.[28] George was forty-eight years old and described his occupation to the census taker as a cotton planter worth $80,000 in real estate and $80,000 in personal property, approximately $3,000,000 in today's currency. He owned one plantation with slaves and two or three additional tracts of land in Louisiana, all of which netted $4,000 to $5,000 annually.[29] He wrote in 1872 that in 1861 his total worth was $120,000, which he had

accumulated from his planting interests as well as his jobs as surveyor, engineer, and commissioner of swamp land in Louisiana.[30]

Craig wrote to Morse on June 19, telling him that the Ordnance Department wished to purchase one Morse carbine and one hundred cartridges. Though no reason was stated, there was some sense of urgency, and Craig requested that Morse respond without delay and supply the firearm to the Washington Arsenal by June 25 or to the New York Arsenal no later than June 27.[31]

Work on Morse's firearms progressed slowly that summer, but Floyd was determined to begin the work at Harpers Ferry on both the alteration of Harpers Ferry rifles and the new carbine. He wrote to Craig on July 5: "With a view of altering rifles to Morse's plan at Harpers Ferry Armory, you are requested to have sent to that armory from Springfield one of the last models of altered rifles with its appendages, together with all of the tools which have been made at Springfield for the purpose of making such alterations, and also the drawings, or copies of them, by which the work has been done, *as well as copies of the drawings of the new carbines, now in course of manufacture at the Springfield Armory*" (emphasis added).[32] Craig wrote to the new superintendent at Springfield, Isaac H. Wright, on July 7 to inform him of Floyd's directions.

Probably due to McCulloch's influence, the state of Texas expressed interest in Morse's firearms in the summer of 1860. In a letter to Texas Senator Louis T. Wigfall on July 13, Floyd said that the normal firearm quota for the state of Texas for the year 1860 was 169 muskets, but that an equivalent number of Harpers Ferry rifles could be substituted if he wished. Floyd even offered to supply Texas with 341 altered rifles—170 for the present year and 171 for the next year's quota in advance. Floyd presented three options: (1) The governor of Texas could take 341 unaltered Harpers Ferry rifles, or (2) he could take 237 as the full quota altered to Morse's design, or (3) he could take 341 and pay four dollars per gun to have them altered to Morse's design at Harpers Ferry Armory. Texas Governor Sam Houston, who felt that Morse's breechloader was the best gun for Indian warfare, had begun negotiating with the War Department for guns and supplies to outfit several new companies of Texas Rangers who protected the frontier.[33] Houston wrote to Floyd on July 28, expressing his preference for using Whitney rifles altered to Morse's plan and asking how many would be supplied by the government as a fourth option, to which Craig responded, 268. Texas Secretary of State E. W. Cave told Floyd on October 23 that Texas had a large number of Mississippi rifles on hand and asked Floyd about possible terms for having those altered to Morse's plan.[34] The order was never completed, but the interest was clearly there.

Floyd instructed Craig on July 13 to order Alfred M. Barbour, superintendent of Harpers Ferry Armory, to begin as soon as possible after receiving the altered rifle and tools from Springfield to convert the .54 caliber Harpers Ferry model 1841 rifle to Morse's breechloader design. He also instructed Barbour to "go on with the alterations according to the Springfield model to the extent for which the right has been obtained. To ensure correctness in doing the work, Mr. Morse will be permitted to see it, as it progresses, and to make any suggestions he may deem proper to aid the master armorer in carrying out the alterations in the best and most effective manner."[35] A. M. Ball, master armorer at Harpers Ferry, received two boxes from the Springfield Armory on July 19. One contained a Harpers Ferry .54 caliber rifle altered to Morse's breech-loading design, and the

other held tools and gauges needed to modify identical rifles.[36] Ball told Craig that there were currently 99 Harpers Ferry rifles being stored at Harpers Ferry Armory. They had originally been sent there to be converted using Storm's plan, but when the budget ran out they were placed in storage. Ball did not have a clear idea of how many Harpers Ferry rifles Craig wanted to alter to Morse's plan and requested that Craig inform him of such so that he could make adequate preparations.[37] Ball did not mention the carbine work going on there. Craig acknowledged Ball's query and responded to Barbour on July 25 that the government had originally purchased from Morse the right to alter 2,000 arms and that Ball should limit his plans to that number. Craig consented to the use of the rifles previously designated for the Storm conversion and asked how many more rifles Barbour needed to do the work. He also ordered that the work be conducted in such a manner that, if it were to be interrupted before 2,000 were completed, as small a number as possible would be left in a partial state of alteration.[38] On August 9, Craig sent Barbour 300 rifles from the New York Arsenal to be altered using Morse's plan and asked Barbour how much it would cost to alter that number.[39] Barbour was absent, and Ball estimated that the necessary facilities, tools, and other start-up costs would be $1,500 to $2,000. An additional $4 per gun would be needed to cover the costs of materials and labor.[40] In his detailed itemization of costs, Ball clearly showed that this version of Morse's gun had a lever for ejecting the cartridge, most consistent with his unaltered Patent 20503, even though it was the Munck hybrid they were planning to use.

Ball wrote about the alterations to Harpers Ferry rifles on September 18, 1860, calling Morse's gun "unexceptionable," meaning that it was faultless, and "ne plus ultra," the highest degree of quality. He wrote, "The tools are advancing steadily, and I feel that we will be able to turn out a very superior lot of arms during the winter."[41]

By September 1, 1860, Augustus Brown of Waterbury, Connecticut, had manufactured 6,000 cartridge cases of the original 12,000 ordered the previous December and expected to have another batch of 6,000 completed within a week.[42] He experienced much trouble in making the tools used to manufacture the cartridge cases, causing delays in completing the order. The final price was thirty dollars per 1,000 cartridges. Brown hoped that the next order would be large enough to warrant the construction of better machinery for making the cartridge cases. Morse wrote to Craig on September 25, asking for direction concerning the shipment and inspection of the cartridges.

Craig responded immediately the same day in a frustrated tone. He referred Morse to the instructions of the secretary of war from December 28, 1859, in which the secretary instructed Morse to inform the Ordnance Department of the cost of the 12,000 cartridge cases *before* he had them made, but he had failed to do so.[43] Morse wanted to discuss making more cartridge cases, but Craig refused to do so until Morse complied with the original order. Because the cases had already been made, it was impossible for Morse to inform the government of the cost beforehand, so he shipped them to the Washington Arsenal, where they were made up into cartridges and inspected. Apparently this satisfied Craig.[44]

As the original 12,000 cartridge cases were undergoing inspection at the Washington Arsenal, Morse wrote to Floyd on November 10, telling him he had reason to believe that some of the altered rifles would be complete by the time the cartridge cases were fully ready for use. Morse suggested putting out advertisements for proposals for the

manufacture of 600,000 cartridge cases for his rifles then being altered. He did not mention making any cases for the carbines.[45] Craig disagreed with Morse's proposal in a letter written the same day. He felt that 12,000 cartridges would be satisfactory for the number of Morse's firearms proposed to enter service in the near future. He recommended that future production of Morse's cartridge cases be done at the Frankford Arsenal, which was already making metallic cartridges for Burnside and Maynard carbines and would likely produce Morse's cartridges at less cost and to a higher standard of quality.[46] Floyd agreed with Craig until he received Morse's explanation on November 16 that the 12,000 cartridges being inspected were made for the altered .69 caliber muskets at the Washington Arsenal and would not fit the .54 caliber rifles being altered at Harpers Ferry, nor would they fit the carbines being manufactured there. Morse repeated his request that the new cartridge cases be made as soon as possible. Floyd changed his mind and agreed with Morse, approving his request on November 23.[47] According to the Springfield Armory web site, the "Frankford Arsenal made some Morse ammunition in 1860 in calibers 0.54, 0.58, and perhaps 0.69."[48] The Frankford-made cartridges were described as the typical Morse type, tinned, and with flanged heads and rubber bases.

Floyd requested that a set of drawings for Morse's rifle alteration and its associated tools be supplied to the War Department on November 14, but he did not mention the carbine. Morse continued to have frequent communications with Wright in November 1860, when they discussed the production of cartridges and a new type of cartridge wadding which had alternating layers of cloth and rubber.[49]

William Maynadier, captain of ordnance, wrote to Captain Josiah Gorgas, commanding officer at the Frankford Arsenal, on December 7 explaining what had happened. He said that the 12,000 cartridges made for the M1816/1822s would not fit the carbines and rifles "which have been ordered to be made and altered" at Harpers Ferry Armory.[50] Maynadier ordered Gorgas to make preparations to manufacture 600,000 cartridge cases and to communicate with Morse in Washington. Maynadier also notified Morse that Floyd, who had only a few weeks remaining in office, had approved his proposal and had ordered Gorgas to fabricate 600,000 cases with Morse's technical input.[51] Gorgas wrote to Morse on December 17, asking for two cartridges to be used as models and any suggestions Morse might have for making alterations to the cartridges. Gorgas said that, as soon as he received the cartridge models and an altered rifle, he could begin making 600,000 cartridges.[52] Morse sent Gorgas the cartridges five days later and suggested that he contact the Springfield Armory to obtain the gauges used in inspecting the cartridges so that he would not have to wait on a gun before commencing work.[53] Harpers Ferry Armory shipped one rifle altered to Morse's design to Frankford Arsenal on December 29.[54] In late December 1860 the Ordnance Department approached Augustus Brown, asking his price for manufacturing 100,000 cartridges for Morse's firearms, but on January 9, 1861, Craig cancelled the inquiry.[55]

As the manufacture of the new carbines progressed at Springfield Armory into the autumn of 1860, Morse remained intimately involved with the process. He testified in 1877, "I went backward and forward to and from the arsenal, remaining there a great deal of the time, until these five arms were completed." Morse also stated that Erskine Allin and others at Springfield were very satisfied with the five carbines and spoke very highly of them.

Meanwhile, Morse continued to tinker with a modification of the trigger mechanism. He sent a letter to Wright on October 15, 1860, describing a new design for a hair trigger for the carbine's side lock. Wright responded on October 18 that Morse had neglected to include a drawing with the letter, and that his written description was not practical, given the current design of the lock.[56] Wright wrote back on October 19 to discuss the hair trigger and to include a drawing of a better design, but also wrote that "the work on three carbines is now progressing with good speed."[57] Wright must have agreed to another modification which they called the "guard lock" because he told Morse that those changes could be made within twenty days if Morse agreed to them and if he made no further alterations. The side lock on one of the carbines was already completed, and Wright was waiting on Morse's decision about the attachment of the hair-trigger modification, but the work on this one carbine was too far advanced to make a conversion to the guard lock without the loss of a great deal of time.[58] Wright wrote to Morse again on October 26, asking him to make a decision soon about the application of the hair-trigger changes made to the guard lock on two carbines and to the side lock on the third.[59] This guard lock, probably the same as that seen on Morse's brass-frame carbine, was so termed because it was placed internally within the frame and thus "guarded" by it, as opposed to the externally mounted side lock previously used on all of Morse's designs. The discussions concerning a guard lock versus side lock continued for several months. Wright wrote to Morse on December 28 that a draftsman was hard at work on the drawings of both the side-lock and the guard-lock carbines. Wright anticipated that he could provide Morse with the drawings of both carbines within a week to two. On January 8 and February 4, 1861, Wright sent drawings of the carbines to Morse and referred to them as "your guard lock carbine" and "your side lock carbine."[60] Eventually, Morse chose to manufacture the brass-frame carbine with the guard lock.

A Springfield armorer, S. Adams, wrote on September 17, 1860, that he had personally watched the progress of the carbine being manufactured, making it very clear that this was not a Muzzy-built carbine. The foreman of the machine shop at Springfield, Samuel W. Porter, referred to the recently built carbines as "your cavalry carbines" in a letter to Morse.[61] Wright wrote to Morse on October 31, "We will now proceed to finish up the remaining two carbines as fast as possible."[62] Allin wrote on November 8, 1860, that he had "just completed" five carbines for the government and had thoroughly test-fired them.[63] Allin wrote, "I have now to inform you that they are a very compact, well-proportioned and beautiful arm, better adapted for the cavalry service than any breech-loading arm that I have seen, as they can be more readily loaded without interfering with the trappings of the horse furniture, etc., and for accuracy, rapidity of firing, penetration, and safety are not equaled by any breech-loading carbine that has yet been made for the public service."[64] When presented with a Morse firearm at his December 1875 deposition, the sixty-three-year-old Allin confirmed that it was "intended for a carbine" and that it, along with four others, were made at the Springfield Amory in 1860.[65] Supporting Allin's statements is the fact that a set of tools for making Morse's carbine was captured at Harpers Ferry in April 1861 and was sent from Richmond to the Tennessee State Armory in Nashville, where Morse was superintendent, in the summer of 1861.[66] V. D. Stockbridge also supported the statement when he described Morse's cartridge extraction mechanism in a postwar deposition and stated that all three of his guns, the 1856 Muzzy model, the 1858

alteration models, and the "carbines of 1860" had modifications of the same design.[67] It is highly likely that "Morse's new gun," described by Allin, Adams, and Porter, was actually the first prototype carbine that was eventually made with a brass frame in Greenville, South Carolina, between 1862 and 1865.

More evidence comes from Superintendent Wright, who wrote to Morse on November 6, 1860, "You are correct in supposing that we are making three carbines; one upon the order of the Hon. Secy. of War, and two upon your own. I now understand that you want an extra barrel of .54 calibre, to accompany the guard lock carbines, which are to be of .44 and .50 cal. respectively. This shall be attended to."[68] Morse still had one of the .44 caliber carbines in his possession in 1863. H. Newton Reid, a worker at the State Works, described it in a letter in September 1863: "Col. Morse has one of his carbines that was made at the Springfield Armory as a modle [*sic*] that he will sell for five hundred dollars. The caliber is the same as the Colts Navy Pistol [actually Colt's *Army* pistol]. It is a very finely finished Gun (or you might say an extravagantly finished Gun). It has a nice case and everything complete."[69] Morse retained the same gun as late as 1877.[70]

Wright wrote to Morse again on November 14, 1860, that he was pressing the work on the carbines. His men were working on them night and day, and he expected them to be completed on November 20. Wright also said that he was working up a cost estimate and would send that to Morse, showing that Morse was planning to pay for some of the work.[71] Two of the five carbines were Morse's personal property. Wright delivered the other three, two .44 caliber and one .50 caliber, in a five-case shipment to Craig in Washington on November 22, 1860.[72] The first case contained the two carbines with guard locks, one .44 and one .50 caliber. Also in that box were appendages and an extra .54 caliber barrel to be adapted to the .50 caliber carbine. The second contained ammunition for the .50 caliber carbine. In the third case were one hundred cartridges for the .44 caliber carbine. The fourth case held a carbine the secretary of war had ordered, along with all appendages and a hundred cartridges. By process of elimination, it can be determined that the fourth case held a .44 caliber carbine with a side lock. Finally, the fifth box contained two hundred .54 caliber cartridges. Wright also provided Morse with Allin's estimate of the cost of manufacturing the carbines and suggested that, if he left off the patch box, he could save $1.07 per gun.

Morse recalled after the war that all five carbines were sent down to the Washington Arsenal.[73] When Floyd resigned in late December 1860, Morse remembered that McCulloch got control of all five of the carbines. He gave one or two to Morse, and Floyd had one or two in Richmond. McCulloch carried one of the carbines until the day he was killed at the Battle of Pea Ridge on March 7, 1862. One wartime comrade wrote of McCulloch, "I often saw him riding through our camp alone in citizen's garb [he never had a uniform], with a small breech-loading rifle swung across his shoulders, going in the direction of the enemy."[74]

We do not know definitively when Morse began to work on the idea of his brass-frame carbine, of which at least 1,035 were ultimately produced. Based on the above evidence, it was probably 1858 or 1859. The carbine was a weapon completely different from both his Muzzy-built carbine and the converted muzzle loaders. It shared a Munck-type breechblock design with most of the altered muzzle loaders, but it was a new-manufacture

weapon, not an alteration of an old one. The carbine had extraction nippers like the Muzzy-built carbine, but lifting its breech cover did not automatically cock the hammer as with the Muzzy. Unique to the carbine were a brass receiver, a short, twenty-inch barrel, and a hammer which Morse called a "guard lock." Basically, the carbine was the combination of Munck's hybrid variation of Morse's breech-loading mechanism and a brass-frame receiver with a carbine-length barrel. Morse never attempted to register it with the U.S. Patent Office.

Work continued on the alteration of the Harpers Ferry rifles as well. Floyd wrote on December 27, 1860: "The Ordnance Bureau will direct that the alteration of the rifles at Harper's Ferry Armory to Morse's plan, ordered last July, be carried out with all possible dispatch, and 1,000 of them be finished as soon as possible, and 100 cartridges for each of the altered rifles be procured from the manufactory at Waterbury, Connecticut, and sent to Harper's Ferry, to be put up and forwarded with the rifles to the United States Arsenal at San Antonio, Texas."[75] Floyd, a southerner, was criticized for shipping arms to Southern states and resigned only two days later, after which time the flow of guns to Southern states stopped. The order for one thousand rifles shows up in the Harpers Ferry ledger for December 26 with a note dated January 31, 1861, that says the arms had not yet been prepared and that in any case no more than three hundred of them were to be altered. Floyd did not mention the carbines in his December 27 order, which was never carried out. We know that no Morse muzzle-loader alterations took place at Harpers Ferry prior to October 30, 1860, and, though some alterations were started there, likely none were completed in late 1860 or early 1861.[76]

In late 1860, a Columbia, South Carolina, physician, James Morrow, knew Morse and had been aware of his inventions for some time. Morrow was impressed with the breech-loader and felt that it was an ideal weapon to be distributed to citizens who could use the smoothbore version for both general civilian purposes and for military defense if necessary. Morrow had talked to Morse, who said that he would attend to the preparation of the gun if Floyd would allow Morrow to borrow one to take back to South Carolina for demonstration purposes. On November 2, 1860, Morrow made the request to Floyd, who replied on November 12 that, though he would gladly comply, he had no guns at his disposal to loan to Morrow. This was likely Morse's first connection with anyone from South Carolina, but it was not to be his last.[77]

CHAPTER 7

Morse in Late 1860 and Early 1861

During the last few months of 1860, Morse solicited and received endorsements from several key men at the Springfield Armory who had played a part in manufacturing his firearms and ammunition. He added a few letters from other dignitaries and published them in a pamphlet, "Morse's Breech-Loading Fire-arms," in late December 1860.[1] The text of the pamphlet makes it clear that its intended audience was the U.S. government, and that Morse was once again lobbying for the production of his guns. The exact reason for the timing of his action is unknown, but most likely he realized he would have very little future contact with the skilled workmen at Springfield because the manufacturing process of his inventions was being moved to Harpers Ferry. It is also likely that Morse, knowing that an imminent war could offer rapid implementation of his inventions, was positioning himself to be ready. It appears that Morse was gathering testimonials from knowledgeable men, many of whom were southerners, to be used as leverage with the most interested government, whether that might be the United States, individual states, or a new nation to the south. All of the responses from the Springfield men were dated between September 18 and November 8, 1860.

In the opening paragraphs of the pamphlet, Morse wrote:

The inventor, Mr. G. W. Morse, submits the following letters relative to his fire-arms, under the conviction that the officers of the United States Armories, who have had them under experimental trial for the last three years, who have also seen and examined all of the different kinds of new inventions in that line, and who have passed the best part of their lives in the construction of arms, are certainly the best qualified judges in reference to the mechanical and scientific combinations requisite for the construction of a weapon for war purposes, and that the other letters from persons well-skilled in the practical use of arms, are entitled to much consideration.

The altered arms received the favorable award of the Board of Ordnance Officers convened at West Point, composed of Col. Ripley, Maj. Ramsey, and Capt. Maynadier, in the fall of 1858; since which time these arms have been much improved, *but the new carbines, made upon the same plan, to which some of these letters refer, have never been submitted to any regularly organized board of officers.* (emphasis added)[2]

Here Morse makes it crystal clear that his "new carbine" had never been tested by a board of ordnance officers. The Searles-built gun had been thoroughly tested in the 1857 trials, and Morse probably used a Morse-built carbine in the 1858 trials. Thus the untested "new carbine" to which he referred could only have been the one which ultimately became the brass-frame carbine made in South Carolina during the war.

Morse opened the pamphlet with a letter written by Joseph Lane on December 19, 1859. Lane had been a major general in the Mexican-American War, the first governor of Oregon Territory, and one of Oregon's first two senators. Supporting both slavery and secession, he was nominated as John C. Breckinridge's vice-presidential running mate for the southern wing of the Democratic Party in 1860. Lane wrote: "I have examined your breech-loading rifle with much care and interest and also the Government musket and rifle altered by you to breech-loaders, and have no hesitation in saying that I regard those weapons as vastly superior to any now in use, and to any ever yet invented. Troops armed with these would be equivalent as one to five armed with ordinary muzzle loaders."[3]

Morse also included letters from several Springfield Armory men, Isaac H. Wright, superintendent; E. S. Allin, master armorer; another armorer named S. Adams, who was later master armorer at the Richmond Armory; W. G. Chamberlain, foreman of the Filing Shop; and Samuel W. Porter, foreman of the Machine Shop. Morse also included previously discussed letters written in August 1858 by Benjamin McCulloch, A. M. Ball, and Senator L. W. Powell.

Wright, Allin, Adams, Chamberlain, and Porter all agreed that Morse's carbine was the best breechloader for cavalry purposes they had ever seen. Wright, who had considerable opportunity to observe and evaluate Morse's firearms, wrote the longest, most comprehensive endorsement on October 31, 1860:

> Having been requested by G. W. Morse, Esq., of La., to give my opinion as to the value of his invention for breech-loading arms, I beg leave to say, that from my observation of the facility of manufacturing arms with the Morse improvement; of the simplicity and strength of the work; of the cheapness as compared with the result obtained; of the accuracy of the arm; of the handiness and rapidity of loading; of the ease with which they can be kept in order even by common soldiers; of the length of range at which they are effective, and I may add also, of their surprising cleanliness after many discharges, I am satisfied that they are greatly superior to any other fire-arms that have yet been made, and altogether the best adapted for military service.
>
> For the same reasons I consider Mr. Morse's plan for altering muzzle-loading muskets to breech-loading, the best that has been offered, and one that will render a rifled musket of the present style far more efficient at a very moderate expense.[4]

Morse concluded the pamphlet by extolling the virtues of his carbine for both the navy because of its waterproof qualities and the cavalry because of its ease of loading while on

horseback. He closed by writing of the firearm, "A careful examination of it will not fail to convince all sensible and unprejudiced minds that any people who first get accustomed to its use, will be infinitely stronger than double their number armed in the ordinary way."[5]

Wright had not changed his mind about Morse's gun a decade later. He gave a very similar opinion in 1870 in support of Morse's application for an extension on his patents. Wright wrote that the prejudice against breechloaders could not be overcome in the prewar years because there was no special demand for the gun. But the value of the breechloader began to be appreciated after the war broke out, and by then Morse had revolutionized the art by proving a practical application of the automatic breech-loading mechanism.[6] Though Wright's assessment was accurate, Morse was, unfortunately, no longer personally involved in the development of firearms in the United States after the war broke out, and for that reason, he failed to receive credit for his accomplishments.

In the long run, Morse and the development of his inventions would have fared better if he had remained loyal to the Union. Though the U.S. military was slowly moving toward the use of breech-loading firearms and did employ some during the Civil War, that conflict was fought primarily with muzzle loaders. It was immediately after the war that the U.S. Army began to make the switch to breechloaders as its primary infantry weapon, and, had Morse been a part of that establishment, he and his inventions would have been closer to the forefront of their research and development. Secretary of War John B. Floyd issued a prophetic endorsement of breechloaders in his last annual report, though he did not mention Morse by name: "Very frequent and numerous experiments have been made under my direction of breech-loading arms, and inventions for this purpose are wonderfully numerous. Many have been rejected, but some plans for breech-loading have been approved, after very numerous experiments, and are now conceded by all who are familiar with them, and capable of judging, to be by far the most efficient arms ever put into the hands of intelligent men. Immediate steps ought to be taken to arm all our light troops with the most approved of these arms."[7]

Floyd wrote in the same report that "the highest efficiency of a body of men with firearms can only be secured by putting in their hands the best breech loading arm." He also predicted accurately that the breechloader would supersede the muzzle loader and that the preferred arm for cavalry would no longer be the saber, but the revolver and breechloader.

Floyd resigned about December 29, 1860. Joseph Holt replaced him as secretary of war but served only until March 4, 1861, when he was succeeded by Simon Cameron. On January 7, 1861, Daniel A. Butterfield, colonel of the Twelfth Regiment, New York Militia, asked Holt if he could purchase two breech-loading carbines made at Springfield Armory for the purpose of conducting experiments in the militia service. Craig responded to the request on January 11 to say that Butterfield's application could only be filled through a request from the governor of New York. Further, Craig said the only breech-loading carbines made at Springfield were of Morse's design, and, to his knowledge, only pattern models had been completed, implying that no Morse carbines had been made at Harpers Ferry as of early January 1861.[8]

Morse continued to be closely involved with Wright in research and development of his inventions at Springfield into early 1861. Since the autumn of 1860, Morse and

Wright had been ordering different types of rubber cloth for cartridge wadding from N. Hayward Company, Manufacturer of India Rubber Goods, in Boston. They had been frustrated that Hayward's responses were slow and the quality of the product unacceptable. Wright wrote a stinging letter to D. C. Hayward on November 14, 1860, saying that the most recent shipment "does not prove satisfactory" because it was too tender and it broke after several discharges.[9] Morse's goal was to develop a rubber substance that was strong enough to prevent gas from escaping from the rear of the cartridge when fired, yet pliable enough to withstand many firings and still fit snugly in the base of the cartridge after multiple reloadings. Morse's specifications called for a piece of material that was not more than one-fifth of an inch thick. One version was to be made of one-half rubber and the other half from alternating layers of stout rubber-coated cloth and rubber. Another experimental wad of the same thickness was to be from one-half hard leather and one-half rubber.[10] Hayward filled Wright's order for the above and had a sample ready by early January 1861. Wright wrote to Morse on January 1 to say that the Hayward Rubber Company had sent a sample of new rubber wadding for the base of the cartridge and that it worked better than anything he had previously used. Hayward had tried without success to create a combination rubber and leather wad, and Wright felt Morse would be inclined to adopt the newer plain-rubber type.[11] Wright wrote to Morse again two days later that Hayward had supplied yet another type of rubber wad, which Allin deemed inferior, and Wright was anxious for Morse to choose one so that he could order cartridges.[12]

Colonel Craig wrote to A. M. Barbour, superintendent at Harpers Ferry, on January 19, 1861, requesting to know how many muskets had been taken in hand at Harpers Ferry for alteration to Morse's plan. Craig wrote that an unnamed state, which happened to be Texas, had requested 258 of the firearms, and he told Barbour to limit the alterations to that number for the present time unless he had already begun work on more than that.[13] Barbour's reply is not preserved, but the answer is probably that none had been made. In fact, throughout all of the motions and petitions of Morse's prolonged lawsuits after the war, numerous depositions were given by men who had worked at the Springfield Armory and none by workers at Harpers Ferry. Additionally, there were several requests made during the trials to produce as evidence Morse's firearms that had been made at Springfield and ammunition from the Frankford Arsenal, but none for arms or ammunition from Harpers Ferry Armory.[14]

South Carolina seceded from the Union on December 20, 1860, followed by Mississippi on January 9, 1861, Florida on January 10, Alabama on January 11, Georgia on January 19, Louisiana on January 26, and Texas on February 1. The Confederacy was established on February 8. Virginia, in which Harpers Ferry Armory was located, did not secede until April 17. All of the machinery used to manufacture Morse's inventions had been moved from Springfield to Harpers Ferry Armory by the time the Confederacy was established.

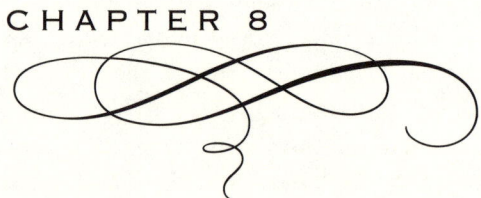

CHAPTER 8

Harpers Ferry, Nashville, and Atlanta

The U.S. government had contracted with Morse to manufacture one thousand of his new carbines and to alter two thousand muzzle loaders using his design, and by early 1861 the production line was gearing up at Harpers Ferry Armory in Virginia. The outbreak of the Civil War stopped the production of all Morse designs by the federal government, but the move to Harpers Ferry significantly affected Morse for the rest of his life.

We know that Morse was actively working with both Springfield and Harpers Ferry armories in early January 1861, but we do not know why he chose to side with the South by March. We can, however, draw inferences from his actions, events, and other circumstances. First, though Morse was born in New Hampshire, Louisiana had been his adopted home for the past twenty years. He accumulated substantial personal wealth there, and his only means of financial support other than his inventions was his property in that state. Morse testified after the war that in early 1861 he was opposed to the dissolution of the Union and was in favor of settling all difficulties between the North and South but that he feared the Louisiana state government might confiscate his property if he remained loyal to the Union. Because the U.S. government was unable to protect him "in the possession of his property, and in the imminence of threatened confiscation of his property, . . . he proceeded to undertake the care and protection of his said property, and also of his wife, children and two widowed sisters [Priscilla and Rebecca] in person, leaving his patent interests, as he supposed, in abeyance for a short time only."[1]

This postwar argument is weakened by the fact that, first, Morse was in the service of the Confederacy no later than March 1861 and that, in the same deposition, he stated, quite disingenuously, that he never assisted the Confederacy in any way. Second, Morse's wife, children, and in-laws were all natives of Louisiana. His brother, Peabody, and his two sisters and their families also lived there. It is true that Morse's abolitionist father and brother, Isaac, still lived in Massachusetts, but George was much closer to the Louisiana

side of the family. Third, and most important, the demand for Morse's firearms was greater in the South, especially Texas, which had an active order at Harpers Ferry for his altered arms at the time. Several Texans, including Governor Edward Clark and Ben McCulloch, were strong supporters of Morse's firearms designs, as were the governors of Alabama and Mississippi. In addition, Morse's chief supporter, John Floyd, former secretary of war, highly favored the development of Morse's design and had recently returned to his native Virginia. Morse's primary antagonists were the conventional military planners in the North. The U.S. Ordnance Department and, specifically, Colonel H. K. Craig, were not yet convinced of the military usefulness of breechloaders, and the ones they did favor—at least the carbines—were not designed by Morse. Morse testified on several occasions after the war that most of the ordnance officers in the U.S. Army were opposed to the concept of breech-loading shoulder arms, and that the implementation of his designs, both the Harpers Ferry rifle alterations and the manufacture of his new carbine, was moving very slowly. With what appeared to be much stronger support in the South than in the North, Morse gambled that his best chance of success was with the Confederacy.

In addition to the overwhelming support for his firearms being in the South, several other factors came into play but are more difficult to assess. First was the issue of contracts with the U.S. government. Morse testified after the war that, in late 1860 or early 1861, he was considering an extremely lucrative deal which, if concluded, could have caused him to remain with the Union. He wrote, "The United States Government paid me five ($5) dollars each, for the right to alter two thousand arms to my plans, and were in treaty with me for the right to alter one hundred thousand (100,000) more, when the war broke out, which of course put an end to the transaction, altho they offered me every inducement to remain, and go on with the work."[2] This is the only evidence we have that there were discussions involving 100,000 Morse alterations, and we know that Morse's postwar recollections were not always accurate. We do not know how serious these negotiations were, if they existed at all, or if Morse cancelled the talks out of his loyalty to the South, or if the U.S. government cancelled them.

Second, he personally possessed the intellectual properties necessary to carry on the production of his carbine, namely the drawings that Wright had sent him in early 1861, at least one example of his new carbine, and a wealth of experience in manufacturing his inventions. This element was highly portable and would not have swayed him either way. Third, the specialized machinery for making his firearms remained in federal hands at Harpers Ferry Armory in early April 1861. Access to the machinery was certainly preferable to Morse's future plans, but it was not absolutely critical. It could be replaced, but not without significant time and effort as seen in Greenville, South Carolina, in 1862 and 1863. He preferred not to abandon the machinery, and he worked hard to retain it. For Morse, the machinery issue was a gamble. Though it remained under the control of the U.S. government in early April 1861, Morse guessed correctly that it would eventually fall under the control of either Virginia or the new Confederacy, and by March he was working with Jefferson Davis toward that end.

Fourth, we do not know where his personal political loyalty lay. He might have been truly devoted to the Southern cause in early 1861 when he wrote to Jefferson Davis in March that he hoped to have the satisfaction of knowing that his exertions and honest endeavors to benefit the Confederacy would be crowned with success.[3] On the other hand,

he might have been faithful to the Union, as he stated in postwar depositions. His post-war testimony paints him as anything but a Southern patriot, vehemently denying that he ever assisted the Confederacy and stating that the only reason he went South was to protect his personal property in Louisiana. He might have been a quiet man of principle, but the abundance of evidence suggests that he was essentially an apolitical opportunist, not a political ideologue.

Why, in the end, did Morse side with the South? Though his family and land might have been factors, the stronger influences were both the likelihood that the machinery at Harpers Ferry would end up under Southern control and supporters like McCulloch, Floyd, and other political leaders in Southern states. The cumulative effect of these in-fluences provided him the best chance of manufacturing and selling his firearms in large numbers. Morse was forced to choose between the North and the South in early 1861, and it appears that he did so based primarily on the odds of personal success rather than any sense of loyalty to his adopted state or to the South.

Morse cast his lot with the Confederacy within weeks after the new nation was estab-lished, and its president, Jefferson Davis, sent him to Washington, D.C., in early March 1861 to investigate the status of arms-related machinery, which was still under the control of the U.S. government at Harpers Ferry. Morse was unaware that Davis had also sent Ra-phael Semmes on a duplicate mission. Morse was in Washington when he reported back to Davis on March 6:

> In pursuance of my understanding with you respecting the machinery for arms, I immediately, on my arrival here, went to work to find out the facts relative to the business, and had prepared a letter to you as the result of my investigations, which I took on Sunday evening to Captain Semmes for delivery, as I learned he was going direct to Montgomery. Much to my surprise, he informed me that he had been sent here fully authorized to transact the same business, and instead of going to Alabama he was going to the East to see Mr. Ames. Finding myself thus completely ignored in the transaction, I, of course, withhold as useless the communication, as no doubt Captain Semmes has kept you well informed upon the subject. I regret that I had no knowledge of his appointment, as that would have prevented my placing myself in an unpleasant position with all of the parties with whom I had been for some time in intercourse in reference to the propositions which I had the honor to submit to the military committee of the Congress. I hope that Captain Semmes may succeed in the enterprise, *for then I shall have the satisfaction of knowing that my exertions and honest endeavors to benefit the Confederacy will have been crowned with success.* I still believe, however, that as I had taken the initiative in this business I could have been of some service in its execution. (emphasis added)[4]

Morse must have felt like he was starting all over again, attempting to convince a new government that his inventions were worth manufacturing, but, as was his nature, he persisted. Employing his political skills, he wrote to Leroy Pope Walker, Confederate sec-retary of war, in March 1861:

> I enclose herewith some letters in reference to my breech-loading arms, to which I desire to call your attention. Presuming that you will require some arms of this class, I

respectfully request your examination of some models of mine, which I left in charge of the Honorable J. P. Anderson. If success does not attend the efforts now being made for the establishment of a Southern armory, these arms can be manufactured in Europe or anywhere else, as readily, at about the same cost as those of the ordinary kind. I shall be happy to make a trial of them before you at any time you may indicate. P.S. It seems to me that the adoption of a single plan of breechloader, with one caliber, for both the army and navy, would be advantageous.[5]

Ben McCulloch was appointed Texas state commissioner on March 9, 1861, and was authorized to purchase a thousand Colt revolvers and a thousand Morse rifles for Texas to use fighting Indians on the frontier. McCulloch had recommended those two firearms as the best for such irregular warfare.[6] He immediately went to Richmond, where he secured a contract for the revolvers. He had hoped to procure Morse rifles from Harpers Ferry Armory but was unable to do so because they had not yet been manufactured.[7] McCulloch sought the approval of the new Texas governor, Edward Clark, to give Morse a contract to manufacture a thousand carbines. McCulloch wrote that Morse was "very anxious to introduce his arms into the frontier service," but that he required a contract for at least two thousand copies.[8] Morse suggested that McCulloch try to persuade another state or perhaps the Confederate government to submit a similar contract, bringing the total up to two thousand. McCulloch wrote that the cost per gun would be thirty-five to forty dollars, numbers consistent with Morse's 1858 contract for the manufacture of a hundred carbines. McCulloch also told the governor that an upfront fee of ten thousand dollars would be necessary if the guns were to be made in Europe, but in spite of that, he favored the proposal and felt that the arms could be delivered within six months. He wrote, "We want this arm in Texas to be used against the Indians and I hope the State will do all in her power to obtain them at once."[9] He felt that this was the best means for Texas to obtain a good cavalry carbine and that the state should act soon because of the possibility that a federal blockade might prevent such endeavors.

Hoping to fill the order by using a private manufacturer, Morse submitted a proposal to a Philadelphia gun dealer, Joseph C. Grubb & Company, on April 6.[10] Morse must have changed his mind about the minimum number he could afford to produce, and in keeping with McCulloch's wishes, Morse told Grubb, a strong Union man, that he wished to contract for the production of one thousand of his breech-loading carbines for the state of Texas. Both Morse and McCulloch preferred to have them manufactured in Europe, and Morse asked Grubb if he could assist them in accomplishing the task. Morse specifically suggested Belgium and asked how long it would take before the guns could be delivered. He also inquired about the cost of manufacturing them there and how much the advance payment would be. Offering as references two Confederate commissioners, Payne and Harrison, Morse wrote, "I enclose you a circular in reference to this gun, and I doubt not that if I can get its manufacture under way, that any number of them will be sold in the South, where some models of it have been in use for some time past by Major [Benjamin] McCullough [*sic*], and shown all over the country."[11]

Grubb responded on April 8, telling Morse that he had seen models of Morse's firearms and that the high opinion given to them by military men was well founded. Grubb did indeed have experience working with Belgian gunmakers and would have been happy

to oblige Morse in usual times but given the "unsettled or rather revolutionary condition of our country we must decline in taking any steps in that direction."[12] The next day Grubb forwarded copies of the correspondence to Simon Cameron, and the relationship between Morse and Grubb ended there. Reluctantly, McCulloch agreed to substitute Colt and Sharps carbines for Morse's, and that was the end of his efforts to produce the carbine overseas. The exchange between Morse, Grubb, and McCulloch strongly supports the theory that Morse had a fully developed carbine, not a muzzle-loader alteration, ready for manufacture in early 1861.

Morse's brother, Isaac, wrote in 1870 that both he and George were in Washington, D.C., on April 12 when Fort Sumter was fired upon and that he was trying to aid George in his efforts with the officers of the U.S. government. Isaac was a respected attorney who supported the Union. It is unclear exactly how he was supporting George. Maybe he was trying to negotiate the proposed contracts for 100,000 firearms, or he might have viewed this as a simple business deal in trying to help George acquire the Harpers Ferry machinery. Isaac wrote that George was unsuccessful in the endeavor. Seeing that war was probable and because all of his property and family were in Louisiana, George went home. As they parted ways, Isaac told his brother, "Under no circumstances do you allow yourself to raise your hand against the flag of our country."[13] In a most solemn manner George assured Isaac that he would not, a promise that he had already broken a month earlier and would continue to break over the next four years. George left Washington between April 6 and 12, 1861, and "went South."[14] Within a few months he was the superintendent of the Tennessee State Armory, manufacturing war materiel for the Southern cause.

The Virginia Convention adopted its Ordinance of Secession on April 17, 1861. U.S. Army First Lieutenant Roger Jones was in command of Harpers Ferry Armory and a detachment of about fifty mounted riflemen. Anticipating an attack by an overwhelming force of Virginia state troops, he made thorough plans to destroy both the armory and the arsenal's 4,287 stand of small arms and could not imagine anything other than its complete destruction.[15] Shortly after 10:00 p.m. on April 18, Jones's position became untenable and, before withdrawing his small garrison into Pennsylvania, he set fire to two arsenal buildings and the armory.[16] Virginia militia captured Harpers Ferry that night. The arsenal, along with its thousands of small arms, including the model 1841 Harpers Ferry rifles designated for alteration to Morse's design, were destroyed in the fire.[17] Harpers Ferry residents, many of whom made their living at the facility, were able to put out the fire at the armory, saving most of its contents. By capturing Harpers Ferry, the state of Virginia now possessed "one of the largest and best sets of arms-making machinery in the world."[18] Most of the equipment salvaged there was used in the manufacture of rifles and muskets, including the machinery and tools used to make Morse's inventions.[19]

A decision then had to be made, either to leave the machinery in place or to move it. Men knowledgeable about its operation opposed moving the machinery because some of it was immense in size and much of it could not be moved without total loss or, at the least, rendering it unfit for future use. In fact, there was so much machinery that it was estimated at the time that it would take six weeks to move it all.[20] The master armorer stated that, if the equipment remained in place, he would require only a three-week cleanup period, after which Harpers Ferry Armory would be capable of producing fifty shoulder

arms a day, but if it were moved, it would require one or two years before it could be put back into operation.[21]

The Virginia legislature passed an ordinance on April 25 authorizing Governor John Letcher to remove as much of the machinery as might be useful for the manufacturing and repairing of muskets from Harpers Ferry to the armory at Richmond. They also authorized him to move the remainder of the equipment to Lynchburg, where it could be put to use as quickly as possible. The legislature requested that Robert E. Lee be placed in charge of security for the machinery at Harpers Ferry and as it was being transported.[22] On April 27, Letcher ordered Lee to direct Colonel Thomas J. Jackson to proceed to Harpers Ferry to assume command of that post and to organize volunteer troops in the area. Letcher ordered the machinery to be moved up the Shenandoah Valley deeper into Virginia in spite of the large cost. Lee also ordered Jackson to expedite the transfer of the machinery designated for the Richmond Armory, to complete the manufacture of any guns or rifles, if possible, and to protect all other materials at the armory.[23]

Morse, having personal and intimate knowledge of some of the machinery at Harpers Ferry Armory, was certainly anxious to save the machines that could be used to manufacture his inventions. Clearly conveying his Southern sentiments, he was in Richmond when he wrote to Lee on April 29: "Permit me to call your attention to the fact that there are some parts of the machinery at Harpers Ferry—for instance, that portion adapted to stocking and rifling the guns and profiling machines—which it would be exceedingly difficult to have replaced in case of their loss. I would respectfully suggest the removal of such machines beyond the reach of danger at the earliest possible moment. I can add that these machines are of a lighter kind and hence the easiest moved, and also the easiest destroyed with a sledge hammer in the hands of a strong man. I am ready to serve you, if you desire it, in selecting with the aid of the master armorer and in removing it."[24]

On May 1, Lee ordered Jackson to move all the machinery away from Harpers Ferry as fast as possible. Only two weeks after capturing Harpers Ferry, he burned the remaining armory buildings, abandoned the site, and moved the equipment away. In all, 430 machines and 57,000 hand tools were removed from Harpers Ferry to be distributed among various Confederate armories, arsenals, and factories.[25] The U.S. government estimated the total loss to be $1,207,668.[26] The majority of equipment, including Morse's, was designated for Richmond, and other pieces were taken to Strasburg for temporary storage.[27] In the transfer process, some of it was lost and other pieces were damaged or broken. The losses were relatively minor, however, and did not affect the overall completeness and value of the property.[28]

The Advisory Council of Virginia recommended on May 21 that the colonel of ordnance be instructed to have the musket-making machinery from Harpers Ferry set up in the Virginia State Armory, and to use the utmost dispatch in getting it into operation.[29] Because there was so much equipment, the State Armory was not able to use it all immediately. On May 22 the Advisory Council suggested that the governor loan some of the Harpers Ferry machinery to the State of North Carolina, specifically that the former United States arsenal at Fayetteville, North Carolina, should receive as much of the machinery for manufacturing rifles as could be spared without inconvenience to the State of Virginia. Consequently, before June 19 the governor sent about one-third of the Harpers Ferry machinery to Fayetteville, where the armory was prepared to put it to immediate

use.[30] A similar arrangement on a smaller scale was soon made with the Tennessee State Armory at Nashville. In making these distributions, the Committee on the Subject of the Stores, Machinery, and Property Captured at Harpers Ferry expressed a sentiment echoed by the governor of Tennessee, Isham G. Harris, in April 1862 when he allowed the same equipment to be loaned to South Carolina. The committee wrote: "The public necessity, and immediate demand for arms, justified any course likely to effect the desired object."[31] By late May 1861, Morse was living in Richmond, and his wife and family soon followed.[32]

On April 24 the State of Virginia and the Confederate government entered into an agreement which stipulated that, after Virginia became a member of the Confederacy, it would turn over to the Confederate States all the public property it had acquired from the United States on the same terms previously used by other states. A point of confusion arose about the exact terms the other states had used, but President Davis addressed that question in a letter to the committee on June 24 in which he stated that several states, including Georgia, South Carolina, Texas, Arkansas, and others, had transferred all forts, navy yards, arsenals, and other public sites, as well as the arms therein, to the Confederate government.

To clarify its recommendation, the Committee on the Subject of the Stores, Machinery, and Property Captured at Harpers Ferry passed a three-part resolution in late June, and the state convention passed it on June 29. Titled "Harpers Ferry machinery, stores, &c," the first part read, "That the Governor of the Commonwealth of Virginia turn over and transfer to the Government of the Confederate States, for use during the war, all the machinery and stores captured by the Virginia forces at Harpers Ferry, now in the possession of the State, reserving the right of property in the same."[33] Second, the committee recommended that the permanent location of the machinery be within the Commonwealth of Virginia, that the Confederate government pay for the transportation expenses from Harpers Ferry to Richmond, and that the Confederate government pay for the loss of any of the equipment during the war. Finally, the committee authorized Letcher to allow the Confederate government to use the armory buildings in Richmond for operating the Harpers Ferry machinery.[34] The issue of ownership of the Harpers Ferry machinery that was soon to be transferred from Richmond to Nashville would remain unresolved even after South Carolina came into the picture a year later and would come up repeatedly over the next three years.

Morse testified after the war that he turned down an offer to supervise the manufacture of his firearms at the Fayetteville Armory, but there is no independent confirmation of that. By mid-July 1861 he was superintendent of the fledgling Tennessee State Armory at Nashville, which was created exclusively for the repair of small arms.[35] How and why Morse ended up in Nashville is a mystery. He wrote to L. P. Walker, Confederate secretary of war, on July 18 requesting that Walker allow certain machinery captured at Harpers Ferry to be loaned to the Tennessee State Amory. The content of the letter indicates that Governor Harris planned to manufacture guns, probably Morse's guns, at the Tennessee State Armory. Morse wrote:

> In accordance with your suggestion I submit the following as the list of machines which I hope to obtain for the purposes set forth in a letter from the Governor of

the State of Tennessee to His Excellency President Davis: One trip-hammer, with such special tools for welding gun barrels as are at hand; 2 small planers; 1 screw machine; 1 cone machine; 2 small lathes; 1 propelling machine; 2 drilling machines, with 3 or 4 spindles each; 8 milling machines; 1 rifling machine; 1 nut-boring machine; 1 smooth-boring machine; 1 barrel-turning lathe; 1 punching press; 1 horizontal milling machine for ramrods, &c.; 1 old breech screw-cutting machine; 1 old index machine.

It is the loan of these tools only which is asked for, the value of which may be fairly estimated at from $8,000 to $10,000. There are several good reasons why the request should be granted, and one of them is, that under the representations of General Polk that it would be done, the State of Tennessee has purchased buildings and grounds for an armory. Another is that at Nashville workmen from Louisville and Saint Louis are easily obtained to duplicate them and make more of the same kind. Still another reason is that the State has purchased large supplies of war material, and the Confederate Government has not only already availed itself of a part of this in the form of percussion-caps, but will want large supplies of powder from her mills.[36]

Morse spent three weeks in Richmond in late July and early August inquiring about the equipment, and he returned to Nashville by mid-August. Though Walker's response to Morse's letter and personal visit are not preserved, we know that he agreed to send at least some of the machinery from Richmond to Nashville. Within two months, the equipment, valued at $2,309.97, began to arrive.[37]

William S. Downer, superintendent of the Richmond Armory, wrote to then-colonel Josiah Gorgas, chief of ordnance in Richmond, on November 26, 1862, in response to a query from Gorgas about the equipment sent to Morse in Nashville in the summer of 1861. Downer said that the machinery was turned over to Morse as an agent of the State of Tennessee prior to Downer being appointed military store keeper at the armory in Richmond. Downer was unable to find any receipts for the articles, but he was able to create a list of the machines that were sent to Tennessee from the order book of J. H. Burton, who had been in charge of the armory in 1861. The machines were: one rifling machine for barrels (new pattern); one fine-boring machine for barrels (old pattern); four milling machines (old pattern); one milling machine (new pattern); one vertical drilling machine (eight spindles, old pattern); one hand machine for threading breech screws; one profiling machine, vertical (new pattern); one compound planing machine (new pattern); one vertical drilling machine; and one set of tools for manufacture of Morse's breech-loading rifle.[38]

Most of this equipment eventually made it to the State Military Works in Greenville, South Carolina, by the summer of 1862. Particularly critical to Morse's operation was the profiling machine, a cutting tool guided by the contour of a model used to mill irregular patterns. He had written in April that the profiling machine was among those pieces of machinery that would be exceedingly difficult to replace if lost.[39] When the machinery from Nashville was inventoried and transferred to Greenville in 1862, only parts of the profiling machine were shown on the inventory, one possible reason that it took more than a year for Morse to produce his first carbine in Greenville.

Morse was back in Nashville when he wrote to General Polk in Memphis on August 12, 1861, expressing both his disappointment at only modest success and hope for better

productivity in the future: "I obtained but few machines at Richmond, owing to the de-cided stand taken against me by Mr. [J. H.] Burton, the Superintendent of the Armory, and his master armorer, Mr. Adams. I am convinced that they could have shared some machines which they positively refused to loan, without much detriment to the Armory there, although they did not think so. They may be right and I am not disposed to blame anyone; for all parties there seemed disposed to help us as much as possible, but the result is that I lost nearly three weeks time, without receiving any adequate return."[40]

By mid-August 1861 the Tennessee State Armory occupied four floors of a brick build-ing that measured eighty by forty feet, and a blacksmith shop, nearly as large, held six forges and was almost ready for use. Morse planned to repair and rebore old firearms and to make new ones. He expected to have two trip hammers, a drop hammer, a lathe to turn the outside of gun barrels, and another machine to rough-bore the inside of barrels working soon. He also expected the machinery from Richmond to arrive soon, but he still lacked an index machine for cutting gearing, an iron planer, and some small lathes. Richmond could not spare any bayonet machinery, but Morse planned to manufacture a .50 caliber rifle with a sword bayonet attachment. In what is possibly the first reference to his inside-lock design produced in South Carolina in 1864, Morse wrote, "I shall push with all my power the means to make gun barrels, and parts of locks, but have hardly any doubt that my lock must be adopted, as it is so much more easily made."[41]

Morse was positive about his chances of success in Nashville, writing, "We have so good an establishment started that I have great hopes of its success. . . . I believe that we had better push all of our energies upon this one establishment for the manufacture of small arms. We have the power, the room, and can collect all of the tools, and men in the surrounding country, and besides, have more tools, or machines of the proper kind, or soon shall have, than can be found anywhere else about this country."[42]

Morse also needed skilled workers. He told Polk in mid-August that he urgently needed six barrel welders and some gunsmiths or good machinists. Morse advertised in the *Nashville Patriot* on September 5 and 7 for blacksmiths, gunsmiths, iron finishers, and gun stockers to work at the State Armory.[43] He also sent out one of his men as a purchas-ing agent to acquire tools that month. On September 11, Morse requested that the agent try to locate engine lathes from six to eight feet, planers from four to eight feet, and an index machine or one similar for cutting gear. Morse suggested that, if the man were un-able to locate these objects, he contract for at least two lathes and one planer in Florence, Alabama. Morse also requested that the man go to Chattanooga, Tennessee, and inquire about two large boxes marked for G. W. Morse that were being shipped from Richmond to Nashville. It is clear that Morse was in the process of preparing for production at Nash-ville using a combination of existing equipment, the Harpers Ferry machinery, and newly purchased items.[44]

As superintendent of the Tennessee State Armory, Morse performed a job in Nash-ville that was virtually identical to that of David Lopez in Greenville, South Carolina, nine months hence. In an ironic turn of events, the less experienced Lopez would end up being Morse's superior, creating friction between the two men. Throughout the remain-der of 1861, Morse was occupied with managing the day-to-day operations at Nashville, such as signing a receipt for the purchase of steel, objecting that a man from Georgia hired away one of his employees, complaining that the coal he purchased was full of dirt,

and worrying that he needed more lead.[45] Morse was at Nashville for only about seven months, from mid-July 1861 to mid-February 1862. While it is possible that he fabricated some parts for his brass-frame carbine in Nashville, there is no evidence to support that, nor that he completed any carbines there. In February 1862 he was forced to evacuate Nashville, taking with him as much of the machinery as could be salvaged, and hoping to set up shop elsewhere.

Nashville was a significant shipping port on the Cumberland River and bore symbolic importance as the state capital. Federal troops captured Fort Henry on the Tennessee River on February 6, 1862, and Fort Donelson on the Cumberland River on February 15, threatening Nashville. The capital was evacuated and, in late February, became the first Confederate state capital to fall to Union troops. Both armories in the city, the Tennessee State Armory under Morse's direction, and the Nashville Arsenal, which had been transferred to the hands of the Confederacy on September 16, 1861, were evacuated.[46] Captain Moses H. Wright, commander of the Nashville Arsenal, wrote that he evacuated Nashville on February 19, and arrived in Atlanta on March 5 to set up the Atlanta Arsenal.[47] It is likely that Morse and the Tennessee State Armory evacuated about the same time as Wright. Prior to the evacuation of Nashville, Morse was able to pack up some of the machinery and stock from the State Armory and ship it to Atlanta. Some of it was incomplete, much was damaged, and some parts were lost while it was being removed from Nashville. Nevertheless, Nashville's loss turned out to be a significant benefit to the South Carolina State Military Works.

The historical record of Morse's time in Atlanta from about mid-February 1862 until he moved to Greenville later that year is relatively silent. Though physically displaced, Morse continued to sign receipts as superintendent of the Tennessee State Armory as late as May 6, 1862, and his workmen continued to be productive. On March 31, April 30, and May 6 his workers cleaned and repaired a total of 573 muskets, rifles, shotguns, and Hall and Joslyn carbines and delivered them to the Ordnance Department in Atlanta. There is no mention of any Morse carbines on these invoices.[48] Morse was still in Atlanta on August 1 when he wrote to S. R. Mallory, secretary of the Confederate Navy, to tell him he felt that the Red River should be closed to gunboat traffic before the next annual rise in the river. Morse expressed concern that a large amount of cotton and slaves would be lost in a Union invasion of western Louisiana and eastern Texas. He explained that he had extensive experience in clearing the Red River and offered his services to the secretary in closing off the river to Federal naval traffic.[49] He did not mention that some of the cotton at risk was on his plantations there. Mallory did not accept Morse's offer, leaving him free to accompany the machinery to the State Works in Greenville, South Carolina.

CHAPTER 9

South Carolina State Military Works and David Lopez Jr.

David Lopez Jr. built Institute Hall in 1854. It could seat three thousand. South Carolina's Secession Convention signed the Ordinance of Secession here on December 20, 1860. The Circular Church is to the left of Institute Hall. Both structures were destroyed by fire on December 11, 1861. From collection of Bob Zeller.

William Henry Gist was the pro-secessionist governor of South Carolina from December 10, 1858, to December 14, 1860. At the end of his term, he felt that the state was well supplied with arms. Gist estimated that the state had over one hundred pieces of ordnance and 32,000 small arms, which included access to 22,000 rifles and pistols at the U.S. Arsenal in Charleston. In addition, there were many small arms in the hands of the militia, and some of those were modern types.[1] Historian Charles E. Cauthen wrote that the state had only about 17,000 pounds of gunpowder, 1,000 shot and shell, about 150 artillery pieces of various calibers and ages, and about 6,500 various rifles and muskets at that time.[2] Several events soon improved the situation. South Carolina seized the U.S. Arsenal in Charleston on December 29, 1860, taking possession of the arms stored there, and the state purchased 650 modern Enfield rifles in March 1861, bringing the total to more than 28,000 small arms. In addition, the new governor, Francis W. Pickens, purchased 300,000 pounds of powder from Hazard's Mills in Connecticut in December 1860 and January 1861. The state's needs were temporarily satisfied.[3]

The signing of the Ordinance of Secession inside Institute Hall on December 20, 1860. The size of the structure indicates that Lopez was a leading contractor in one of America's major cities in the 1850s. Courtesy of Museum and Library of Confederate History, Greenville, S.C.

After South Carolina seceded from the Union on December 20, 1860, the state's troop strength grew rapidly throughout early 1861, placing a new demand on existing stores of arms and ammunition. Up to November 1861 Pickens shared generously—perhaps too generously—with other states. To Virginia he sent 11,260 small arms as well as a considerable number of swords and pistols.[4] To Florida he provided 6,000 arms and 5,000 pounds of powder; to North Carolina, eleven heavy cannon and 30,000 pounds of powder; and to Lynchburg, 1,000 arms and 100,000 cartridges. By the autumn of 1861 there was, once again, a scarcity of arms and ammunition in South Carolina. By the end of the year, the state had about 27,000 men under arms and another 6,000–7,000 ready to go into service, and Pickens's appeals for assistance to the secretary of war became frantic.[5]

Edward Manigault, who had been chief of ordnance for South Carolina until he resigned on October 1, 1861, held the rank of major in command of the Sixth Battalion, South Carolina Volunteers, when he made a report to the Ordnance Board on November 21. Clearly stating the reasons that South Carolina needed its own arms-making capabilities, Manigault argued that the state was sovereign, and as such had a right to retain the ability to defend itself by force of arms. He said it was the duty of the state to "retain the means of defence, and not give up to any other power whatever all her military material, and in so doing shift from her own shoulders the duty and labor of her defence."[6] He specifically recommended that the state always have on hand one hundred pieces of modern, heavy artillery, eight batteries of field artillery, ten thousand small arms and a proportionate number of carbines, swords and pistols for the cavalry. Manigault concluded: "The occurrences of this summer and autumn demonstrate clearly that the state should not rely entirely upon the arms of the Confederate Government for her protection against a foreign enemy."[7] Gist agreed with Manigault and summarized the need for South Carolina to have its own armory when he wrote in August 1862: "With an armory to keep the State at all times supplied with good arms, and with the materials for making gunpowder at our command, we may feel confidence in our means of defence."[8]

Facing a serious state of affairs only six weeks after Union forces established a strong base on South Carolina soil at Port Royal in early November 1861, the General Assembly responded to Pickens's plea to increase the production of cannon, small arms, and gunpowder with an insipid proposal. On December 21, the Committee on Commerce and Manufacture concurred with the governor's desire to encourage more production of those items and recommended that the government should "extend a liberal patronage to all home establishments, whether of private individuals or incorporated companies, for the manufacture of cannon, small arms, gunpowder and woolen fabrics, having a proper regard to price and quality."[9] Soon, however, South Carolina would create her own armaments factory in the Upcountry village of Greenville.

By the end of 1861, many South Carolinians had lost confidence in Pickens's ability to defend the state, which had suffered recent setbacks and was under serious threat. Port Royal had fallen, much of Charleston had burned in early December, and there was a dire need for new troops, but few men were volunteering. The Secession Convention, which had been created by the state legislature in early December 1860, primarily to consider the issue of secession, was still a functioning entity and soon took drastic corrective action. Meeting on December 26, 1861, and again in an emergency session on January

7, 1862, the convention created a five-member Executive Council, which virtually supplanted the governor as the state's executive authority.[10] In addition to Pickens, who was generally antagonistic to the entire concept, the members of the Executive Council were Lieutenant Governor W. W. Harlee and three others elected by the convention—former U.S. senator James Chesnut Jr., Attorney General Isaac W. Hayne, and former governor William Henry Gist. The Executive Council, and especially Gist, would soon set out to fill the needs that Manigault recommended, resulting in the creation of the South Carolina State Military Works, or, simply, State Works.

The Executive Council was much more than a cabinet of advisers to the governor. It was a radical concept which bore characteristics of both the legislative and executive branches, and it possessed extraordinary powers to conduct the war. The governor was merely one member whose presence was not necessary to conduct business; the lieutenant governor and two other members constituted a quorum. The Executive Council was given "not only the ordinary executive power, heretofore exercised by the governor but almost unlimited war powers including full control of the military organization of the state, the power to declare martial law, arrest and detain disloyal persons, appropriate private property with compensation, appoint such agents as necessary, and draw money from the treasury for public purposes."[11]

"A striking feature of the work of the Executive Council was the energetic and intelligent effort made to obtain war supplies under blockade conditions."[12] The council sent an agent abroad in March 1862 to purchase medical supplies, rifles, and ammunition. Materiel was brought in through the blockade at Charleston and Wilmington and was purchased from privately imported stocks. In addition, the council made efforts to improve the sources of supply by using labor and raw materials available within the state. The state acquired a lead mine in Spartanburg District. Banks, churches, colleges, and individuals donated thirty-three tons of lead. Churches in Sumter, Cheraw, Georgetown, and other towns donated their bells to be recast into cannon.[13] In mid-1862 the Executive Council established a saltpeter plantation on the grounds of the state hospital in Columbia and placed it under the Department of the Military. By combining saltpeter, sulfur, and willow, the facility was expected to produce 1,333 pounds of gunpowder within a year. Gist wrote in August 1862 that the plantation promised to supply gunpowder in quantities sufficient to meet the current needs of the state and could be increased as needed.[14] As of the end of 1863, the plantation had produced nothing.[15]

By late 1861, conditions were gloomy. Richmond, Nashville, New Orleans, and Charleston were all threatened. The prevailing opinion, soon proven correct, was that Federal forces would use their new base at Port Royal to attack Charleston in the spring or summer of 1862. Concerned that Charleston might fall, along with its industrial capabilities, the Executive Council sought to protect the state's ability to manufacture shot and shell, and deemed it advisable to move the machinery for such manufacture to a safer place. Gist summarized his thoughts: "If the city [Charleston] had been captured, there was no other place in the state where shell or shot could be immediately cast, and if the machinery in the city had been lost, it could not have been supplied without great difficulty and at an enormous expense. . . . [T]he prospect was that the Confederate Government could not supply the troops with arms and ammunition. Under these circumstances, and with the danger of invasion extending even into the interior, it would have been culpable

negligence but to prepare for it, by endeavoring to supply ourselves with the means of defense."[16]

At a meeting of the Executive Council on February 19, 1862, Pickens made several proposals relative to arms production that presaged the creation of the State Works. Many of his suggestions were never adopted, but they demonstrate that he and the council were seriously considering the ordnance and arms needs of the state. One of the governor's ideas was that the Executive Council would appoint two competent men to take control of two powder mills in the upper part of the state and that the council assume direction over all powder already produced as well as all aspects of ongoing production. Further, the governor proposed that two men be appointed to take over control of existing iron works in Spartanburg and York Districts and that they use all resources and labor to cast fifty cannon as soon as possible. Another of the governor's resolutions was "That all gunsmiths and artisans in brass and iron be collected and employed in such foundries and workshops as may be designated, for making and repairing all small arms that can be made, and to execute the above resolutions, the Chief of the War Department [James Chesnut Jr.], in consultation with the Adjutant General, is authorised to employ and use such agents as he may think proper."[17]

At about this time, a prominent Charleston carpenter and builder, David Lopez Jr. (see color plate 4), who had contracted to do military work for the state up to that time, was given larger, statewide responsibilities. The tenth of twelve children, Lopez was born in Charleston on January 16, 1809, into a family whose ancestors were Sephardic Jews. By the 1850s, he was a pillar of Charleston's Jewish community.[18] Gist described Lopez as "a gentleman of great ability and very conversant with manufacturing in all its branches."[19] Gist chose the fifty-three-year-old Lopez as general superintendent of the State Works for those reasons and because he was a respected building contractor, overseeing the construction of residences, a large synagogue, and several commercial structures in Charleston, where he enjoyed an excellent reputation. He was a "master builder of high caliber in the American South," and his obituary called him "a remarkable man with a fine mind, great originality of ideas and wonderful broadness of grasp."[20]

David Lopez Sr., born in 1750, was founding president of Charleston's Hebrew Orphan Society and president of the city's Kahal Kadosh Beth Elohim (KKBE) Synagogue. He died in 1812 when his namesake was only three years old. David Lopez Jr., was raised by his mother, Priscilla Moses Lopez (1775–1866), older siblings, and slaves.[21] They raised him in the Jewish faith, worshipping at KKBE Synagogue.

Lopez married a Christian, Catherine Dobyns Hinton (June 30, 1814–June 17, 1843), of Edgefield District, in 1832.[22] David and Catherine had six children, two of whom died in infancy. The four surviving children were raised in the Jewish faith and traditions. The first child, John Hinton Lopez (1833–1884), was born in Georgia.[23] Within a year the family returned to Charleston, and David started a carpenter and joiner's business. Clearly, he was successful; by the end of the year he was advertising to hire five or six good carpenters.[24] Another son, also named David, was born on August 30, 1834; Moses E. Lopez was born on July 10, 1836; and Priscilla, in 1839.[25]

Lopez purchased his first five slaves in 1837; one was a carpenter named Kit. Later that year and into 1838 Lopez built his earliest known building, 153–155 Queen Street. A fire started on King Street on April 28, 1838, and swept through Charleston, destroying 150

David Lopez Jr. gained a reputation for building large structures in Charleston, one of which is Kahal Kadosh Beth Elohim, also known as KKBE, in 1839–40. Founded as a Sephardic Orthodox congregation in 1749, KKBE is the second oldest synagogue building in the United States and the oldest in continuous use. Courtesy of KKBE Museum.

William Henry Gist (1807–1874) was governor of South Carolina from December 1858 to December 1860 and, as chief of the Department of Construction and Manufacture, oversaw the creation of the State Military Works in 1862. He is shown here in an 1845 oil painting by William Harrison Scarborough. Courtesy of Greenville County Museum of Art, purchased with funds from 1988 Museum Antiques Show, sponsored by Elliott, Davis & Company, CPAs, and donated by the Museum Association.

acres in the heart of the commercial district. Reconstruction of the city over the following years launched Lopez's career. About that time, he built apartments on Wentworth Street and a house for himself on South Street that included hot and cold running water on all three floors. His experience with the use of piped water probably influenced some of the construction techniques he later incorporated at the State Works. In 1839 Lopez won the contract to build a new synagogue on Hasell Street to replace the one KKBE lost in the fire. It was his first large project, and he completed it in 1840.[26]

Catherine died on June 17, 1843, and David married thirty-two-year-old Rebecca Moise in 1846.[27] David and Rebecca also had six children, two of whom died in infancy. In the 1840s Lopez was active in the local Democratic Party, the Apprentice Library, and the Work House. He was on the committee to oversee the funeral procession and burial of John C. Calhoun, and he was chairman of Charleston's Board of Fire Masters in 1846. This activity, along with his personal loss of property in the fires of 1838, 1851, and 1861 likely led to fire-control methods that he later incorporated at the State Works.[28]

Throughout the decade of the 1850s, Lopez's career soared. In 1852 he built a residence at 192 Queen Street called the Lopez House, and later that year he built the four-story Browning & Leman Department Store at 225–227 King Street. In 1853 and 1854 he worked on the Farmers' & Exchange Bank at 141 East Bay. Also in 1854 he assisted other Charlestonians in building a convention center called Institute Hall, where the Ordinance of Secession was signed on December 20, 1860.

Rebecca Lopez died in 1858, leaving Lopez a widower to care for their four young children. He continued to work hard after Rebecca's death, collaborating on construction at Zion Presbyterian Church, St. John's Lutheran Church, the Charleston County Court House, and others.[29]

By 1860, the fifty-one-year-old Lopez was a professional success with a large family to support. Living with him were two children by his first marriage: twenty-five-year-old David, who was an invalid, and twenty-year-old Priscilla, who would marry Daniel S. Hart that December. Also in the home were the four surviving children by his second marriage: twelve-year-old Aaron (1848–1899), eleven-year-old Eugene, eight-year-old Julian (1852–?), and six-year-old Edward (1853–1914). Lopez also owned fourteen slaves in 1860, seven of whom were working for him at the State Works in the summer of 1863. Jim was a fireman, and the others were carpenters—Kit, Barber, George, Marcus, Jacob, and John.

Soon after South Carolina seceded from the Union, Lopez's talents as a carpenter, building contractor, and manager were immediately recognized, and the state government contracted for his services. When Major Robert Anderson evacuated Fort Moultrie on December 26, 1860, he burned the gun carriages left behind. Postwar mayor of Charleston, William A. Courtenay, wrote of the respect that Lopez enjoyed and how his skills were utilized to remount Moultrie's cannon: "Governor Pickens sent in haste for Mr. David Lopez, the leading carpenter contractor of the city. New heavy gun carriages had to be constructed forthwith. No such thing had ever been built in Charleston. Mr. Lopez replied 'If I can see one, or the remnants I can build them.'"[30] By early January 1861 Lopez provided a cost estimate for seasoned oak necessary to remount Moultrie's Columbiads, and he was given the authority to acquire the timber as expeditiously as possible.[31]

Lopez's work continued for at least several months. General P. G. T. Beauregard, who assumed command of Charleston Harbor on March 3, 1861, issued orders stating that

John Hinton Lopez and his wife, Maria, bought this house in 1862. In June 1866 they sold it to Thomas B. Thruston for two thousand dollars, and it remained in that family until it was demolished in the 1960s. It was located 0.4 mile from the State Works at 506 Augusta Street, near its intersection with Dunbar Street and next door to the former Claussen's Bakery. Courtesy of Gordon Thruston.

"Mr. Gwyn will order of Mr. Lopez the pintle-blocks, traverse circles, and platforms required for the new batteries on Sullivan's Island."[32] Lopez's sons, John and Moses, as well as his son-in-law, Daniel Hart, assisted him in this and similar work at Mount Pleasant and Morris Island.[33] Moses was trained as a civil engineer and was working in St. Louis when South Carolina seceded. He returned home and joined a militia company called the Palmetto Guard, which was attached to the Seventeenth Regiment, South Carolina Militia. He went on duty with his unit on December 27, 1860, and served with it for over a year. On January 11, 1862, Moses was relieved from duty with the Palmetto Guard, and he joined his father shortly afterwards.[34] John was also a member of the Palmetto Guard, which was assigned to the Iron Battery at Morris Island in early 1861, participated in the bombardment of Fort Sumter in April, and occupied the fort after that engagement. It is not clear exactly how much of the first year of the war John spent with the Palmetto Guard and how much he was on detached service assisting his father.[35] There are no records of Daniel Hart serving in the Confederate military. All three men eventually followed David to Greenville when he became general superintendent of the State Works in 1862.

It was during 1861 that Lopez established himself and solidified his reputation with individuals at the highest level of state government. Pickens recalled nearly two years later that Lopez was the "energetic man" who did the good work at Cummings Point, and he was rewarded for it with the position at the State Works.[36]

CHAPTER 10

1862

The Executive Council began to take more serious action when on February 22, 1862, it considered a proposal that James Chesnut Jr., chief of the Military Department, be allowed to use any means he deemed necessary to employ smiths and laborers in speeding up the repair of guns and arms in the state. Pickens recommended on February 24 that Lopez be officially appointed to superintend the task.[1] At the governor's suggestion, the council approved the proposal on February 25 and authorized Chesnut to instruct Lopez to purchase all the machinery and stock that was available and suitable for such purposes. In an August 28, 1862, report, Gist referred to the figure of $476.65, most likely the sum that Lopez spent on the material he moved by rail to Columbia to get it away from the imminent dangers threatening Charleston. It was very valuable machinery but ultimately insufficient for the purposes of the State Works.

In his first action in the new position, Lopez proposed to the governor and Executive Council on February 24 that they join Governor Thomas O. Moore of Louisiana in sending $100,000 to the West Indies to purchase arms being stored there. Pickens wrote to Moore the same day, recommending that they cooperate in the plan. Unfortunately, Lopez had misunderstood Moore "in some respect relative to the amount of arms expected in the West Indies."[2] Moore had already dispatched three steamers to the West Indies in hopes that they would return with loads of arms and ammunition. He proposed to Pickens that they cooperate on similar plans in the future in the event that the current endeavor failed, but it appeared that Lopez's first attempt at procuring arms was not a success.

It would soon become evident that South Carolina's initial efforts to become self-sufficient in the arms industry were inadequate, but it was the best the state could do without better facilities. The governor and Executive Council confirmed Lopez as "Superintendent of all State Works for making cannon and so forth" on March 1, 1862.[3] The facility in Greenville did not yet exist, and Lopez was essentially in charge of a nascent department using temporary work-space. He recommended to the Executive Council on March 12 that Greenville would be a suitable place for the establishment of the State Works but also proposed that he work out of the temporary shops on the State House grounds until the necessary structures could be built in Greenville.[4] The same day Lopez asked for

instructions on how to handle several railroad cars loaded with machinery from Charleston and being held at the depot in Columbia. The council directed him to unload the cars and place the machinery in the workshops on the State House grounds.[5] Eventually, after the State Works was set up in Greenville, Lopez moved the machinery there.

At about that time Lopez expanded the setup in Columbia using workshops which had been built for the construction of the new State House as a temporary facility for the State Works where he intended to repair and alter state-owned small arms. The Columbia workshops were able to repair and put into working condition a considerable number of arms by converting flintlock muskets to percussion, altering bayonets to fit properly, and making new stocks when necessary. Lopez also put his machinists and blacksmiths to work in Columbia performing tasks that were preparatory for the workshops soon to be built in Greenville. By the time they moved the entire setup to Greenville, the Columbia operation had repaired and altered to percussion 1,620 muskets, 213 rifles, and 2 carbines, and had repaired 24 double-barreled shotguns. In addition, it had manufactured 874 bayonets. The workmen in Columbia also built one battery wagon and six caissons complete with limbers, poles, and spare wheels. They also manufactured 4.5 tons of spikes for the gunboat *Palmetto State* as well as 1,000 pikes and shafts.

Pickens proposed on February 19 that the Executive Council adopt a plan for either manufacturing or purchasing the most effective pike and authorize Chesnut to order 1,000 of them as soon as the plan was approved. The pike was an ancient defensive weapon, essentially a long shaft of wood tipped by an iron or steel blade. They could be manufactured quickly, and Pickens envisioned that they might actually be used by troops in the field. With the advent of gunpowder, the use of pikes had significantly declined over the preceding centuries but was not completely obsolete. The Confederacy was facing shortages of firearms for troops already in the field, and leaders such as Thomas J. "Stonewall" Jackson advocated the use of the pike, and the governor of Georgia ordered them by the thousands. Recognizing that they were not equal to firearms, Gist, nevertheless, did consider them to be a better defensive weapon than no arms at all. On February 25 the governor contracted with David Lopez's son, John, for 1,000 pikes and the council confirmed the agreement on March 1. This is the first mention of John in the council's minutes.

Also on March 1 the council authorized the use of state-owned ash wood being stored on the State House grounds for pike shafts. Sometime in late February or early March 1862, the temporary workshops on the State House grounds began to produce pikes and shafts, the first arms manufactured by the State Works. Chesnut instructed John Lopez to contract for the shafts only and not for the pikes, but somehow the governor and council directed him to contract for 2,500 shafts *and* an unspecified number of pikes. After 1,000 pikes and shafts were made, the council cancelled the contract for both pikes and shafts on March 10. In late 1864 at least some of the pikes and shafts were still on the grounds of the State Works in Greenville, and the state auditor called them "unusual weapons" that were "useless to the State."[6] Also at that time 1,203 pikes were being stored at the State Arsenal in Columbia, but it is unclear if these were the same as those manufactured in 1861.[7]

On March 14 the Executive Council instructed Lopez to select a location for setting up the machinery he had previously been authorized to purchase. The council also gave him the authority to select the location for erecting other works for the manufacture of

arms, ordnance, and other weapons of war. In addition, he was authorized to purchase these locations for the state subject to the approval of the governor and Executive Council before any contract was finalized. Finally Lopez was required to submit estimates of all probable costs to the governor and Executive Council.[8]

At Gist's suggestion, the Executive Council created several administrative departments, including Justice and Police, Treasury and Finance, and Military. Under the Department of the Military, the council had begun the work of manufacturing arms and ordnance. Recognizing that the state faced an enormous task in attempting to manufacture its own heavy ordnance and small arms, the governor and council created the Department of Construction and Manufacture on March 24, 1862, and named Gist as its chief.[9] It was this new department that further developed the State Military Works, which was already functioning at a temporary location on the State House grounds but would soon end up in a permanent facility in Greenville. Gist wrote of his new responsibilities, "The most important duty devolving on me was the establishment of a foundry and workshops for casting cannon, making gun carriages and the manufacture of small arms."[10] It was very clear from the beginning that the purpose of the State Works was to cast cannon and to repair and manufacture small arms. In fact, Gist ambitiously anticipated the State Works would be ready to turn out shot and shell by October 1862 and cannon soon thereafter. [11] He authorized Lopez to put the foundry and workshops into operation, and he gave Lopez the authority to employ workmen and others necessary for getting the State Works into full swing. The state of South Carolina guaranteed payment of all purchases and wages for such purposes.[12]

Sometime in early or mid-March, Gist sent a special agent, whose name is unfortunately not recorded, to Richmond. With the assistance of authorities there, the agent was able to procure the temporary services of a Mr. Campbell, who worked at Tredegar Iron Works and was an intelligent man with experience in the manufacture of arms and ammunition.[13] Campbell agreed to visit South Carolina for the purpose of assessing the quality of its iron ore.

Several iron mines and blasting furnaces were scattered across the Upcountry of South Carolina at the time, and Gist sought to determine if the iron was suitable for casting cannon. Gist collected relevant information about several iron works in the state, and he, Campbell, and Lopez visited the two most promising iron mines. One was the Nesbitt Iron Manufacturing Company, located near Limestone Springs in Spartanburg District. It had been in existence for about fifty years and was the largest in the state. They also inspected King's Mountain Iron Works in York District. Campbell's opinion was that the pig iron at King's Mountain appeared to be well adapted to casting cannon, but he recommended that a sample be sent to Richmond, where it could be cast into a cannon and test-fired. Gist arranged for ten tons of King's Mountain iron to be sent to Tredegar, and he received the assurance of then-colonel Josiah Gorgas that he would cast and test-fire the cannon.

After a delay due to General George B. McLellan's appearance on the Virginia Peninsula below Richmond that spring, Gorgas cast a twenty-four-pounder using South Carolina iron on April 21 and subjected it to a "severe and unusual test," the results of which were "highly satisfactory and established beyond doubt that the iron was entirely suitable for casting ordnance."[14] Gist allowed Gorgas to keep the twenty-four-pounder.[15] Tredegar

also cast a twelve-pounder gun using South Carolina iron on April 21, but its fate is unknown.

Campbell informed Gist that there were only a few places in the Confederacy where suitable iron was being mined and that it was necessary to protect the supply. Gist anticipated casting cannon at the State Works when he wrote: "Although a limited quantity of this iron is now made [in South Carolina], yet if necessary it can be largely increased, the ore being abundant and the facilities for making it at our command."[16] Lopez purchased significant quantities of iron from King's Mountain Iron Works for a variety of purposes over the following year. In early 1863 alone he bought 1,000 pounds of iron bar and 29,000 pounds of pig iron. Ultimately, the repair and manufacture of small arms was a success at the State Works, and, though the facility did produce some shot and shell, it never cast any cannon.

Lopez and Gist had much work to do identifying and developing a permanent site, the location for which had not yet been determined in mid-March. Additional machinery had to be acquired, workmen hired, a location established, and buildings constructed. It would require the skills of Gist, Lopez, and others to make all the necessary decisions. Their idea was to begin operations with the resources immediately available and expand if necessary as the needs of the state and capabilities of the facility dictated. Gist wrote, "The Council having decided to establish a foundry and armory on a small scale, authorized me to search out a suitable location and procure a site."[17]

In selecting the best location for the State Works, Gist and Lopez first had to consider multiple challenges, including the safety of the location, type of power to use, modes of available transportation, and access to raw materials and labor. Lopez recommended Greenville about the time he was authorized to select a location on March 14, and, certainly, Gist also deliberated these issues in the weeks prior to the creation of his new department on March 24.

Gist and Lopez made an exploratory trip to Greenville in late March. After discussing the issues with Lopez, Gist selected Greenville Court House as the best place for the location of the State Works, and the governor and Executive Council approved his recommendation on March 27. Though the official name of the facility remained the "State of South Carolina, State Military Works," it was commonly called the State Armory early in its existence. Lopez also referred to it as the Greenville and Columbia Railroad Camp and Shops, and others called it the Greenville Armory. Its most popular name was the "State Works," or simply, the "Works."

Gist enumerated several reasons for making Greenville his choice. First, it was desirable to locate the State Works as close as possible to the iron mines. It was equally important to locate the facility on a railroad line so as to facilitate the import of raw materials and the export of finished products. Specifically, the transportation of heavy gun carriages, shot and shell, and large cannon such as a ten-inch Columbiad, which Gist foresaw casting, was a problem. Transporting heavy cannon was extremely expensive and time-consuming when done overland. Transportation by rail required less time and was cheaper, but some of the bridges on existing railroads could not support the weight of cannon like the ten-inch, eight-ton Columbiad.

By the start of the war, South Carolina had a fairly extensive rail system. In 1862 Greenville was the terminus of the Greenville and Columbia Railroad, which ran for 164

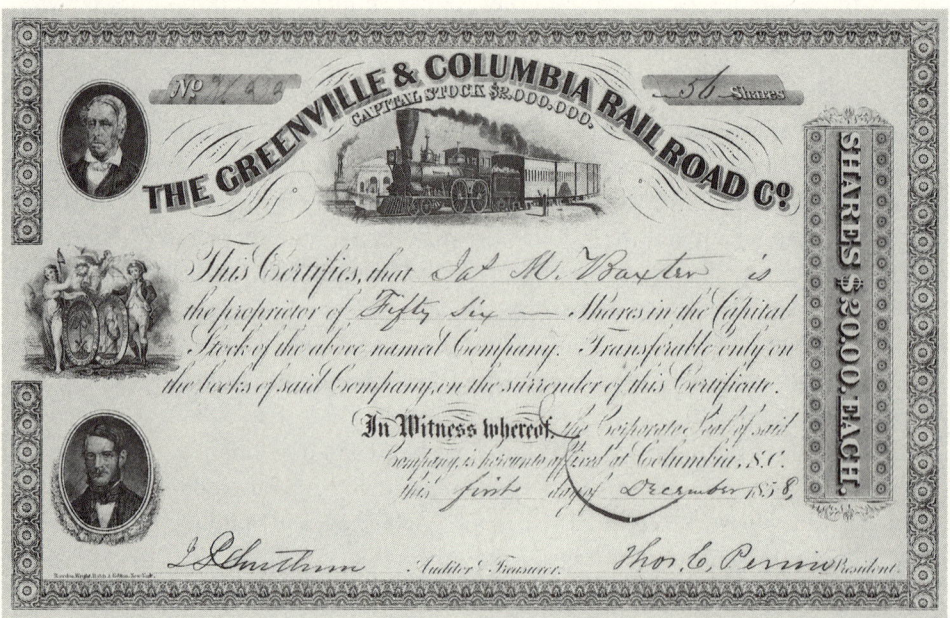

Stock certificate of the Greenville and Columbia Railroad Company. Courtesy of Museum and Library of Confederate History, Greenville, S.C.

miles between those two places via Belton, Greenwood, Newberry, and Alston. After several years of planning, construction of the Greenville and Columbia Railroad began in Columbia in 1849, and, largely due to the efforts of Vardry McBee, it was completed at Greenville in December 1853.[18] The line terminated in Greenville near the intersection of Augusta and Pendleton roads, where the railroad built passenger and freight depots. The passenger depot was established in a residence on Augusta Road between Vardry and Beattie (now Field) streets. The railroad track ran behind the house, and the front lawn and carriage driveway that opened onto Augusta Road were preserved so that passengers could come up the drive and into the house without unnecessary exposure to the train itself. Railroad business offices occupied the second floor of the house. A brick freight depot measuring 125 by 50 feet was built on the opposite side of the tracks between Beattie (Field) and Pendleton streets. The *Southern Patriot* reported, "This arrangement will be the most pleasant and convenient of any depot in the South."[19]

Several spurs of the Greenville and Columbia Railroad were built between the two towns. One, known as the Anderson Branch, ran from Belton to Anderson, and from there the Blue Ridge Extension, part of the incomplete Blue Ridge Railroad, continued to Walhalla. Another spur, the Abbeville Branch, ran from the Greenville and Columbia Railroad at Cokesbury to Abbeville. A separate line, called the Laurens Railroad, ran from the Greenville and Columbia Railroad at Newberry to Laurens. Greenville was not directly connected to Spartanburg by rail during the war. The Spartanburg and Union Railroad ran from Spartanburg through Union to Alston and connected with the Greenville and Columbia Railroad there. All of these lines shared a connection with the rest of the state at Alston via the Greenville and Columbia Railroad.

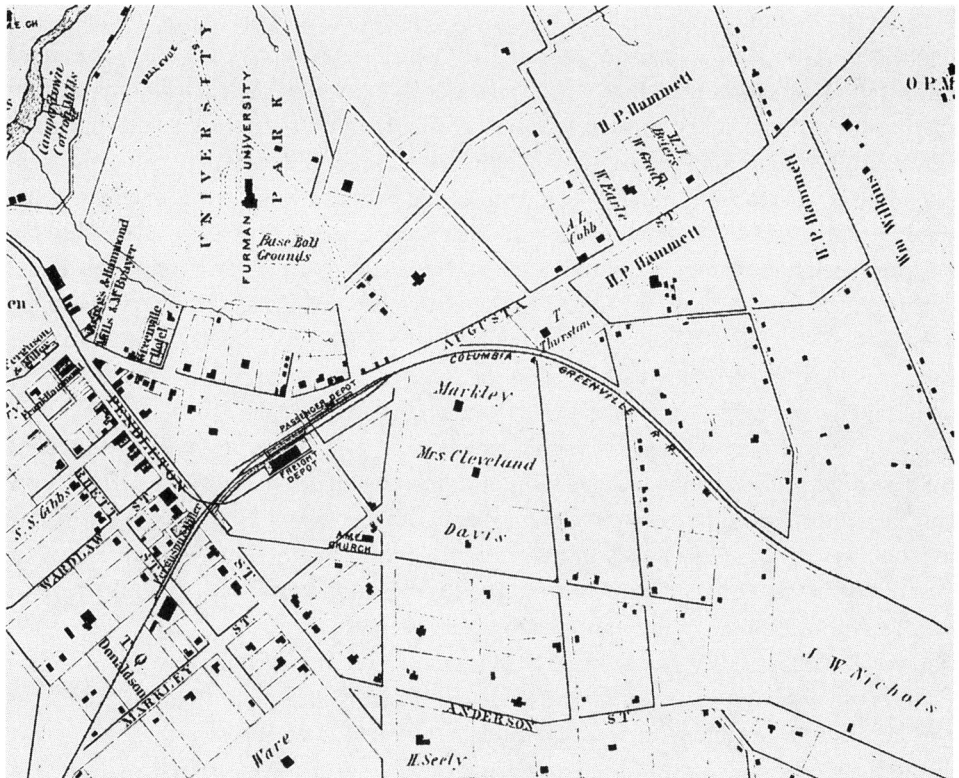

This is Gray's New Map of Greenville, dated 1887. It shows the Greenville and Columbia Railroad entering Greenville from the right side. The passenger depot, a converted residence, was located between the track and Augusta Street to the left of Markley's property between Vardry and Beattie (Field) streets. A freight depot was built on the opposite side of the track between Beattie (Field) and Pendleton streets to the left of Mrs. Cleveland's property. The two depots were built at the terminus of the railroad when it was completed in 1853. The State Works was located on the property labeled "J. W. Nichols." The name of the railroad was changed to Columbia and Greenville in 1882. Map courtesy of South Carolina Room, Greenville County Library System.

Northward from Columbia ran the Charlotte and South Carolina Railroad with a separate line, called the King's Mountain Railroad, running from Chester to Yorkville. Also from Columbia running southeastward to Charleston was the South Carolina Railroad, which had two branches, one from Branchville to Hamburg and across the Savannah River to Augusta, Georgia, and the other from Kingville to Camden. Between Kingville and Camden, the Wilmington and Manchester Railroad connected to the South Carolina Railroad west of Manchester and ran northeastward through Florence to Wilmington, North Carolina. At Florence a separate line called the Cheraw and Darlington Railroad ran northward to Cheraw, and the Northeastern Railroad ran south from Florence to Charleston. The Charleston and Savannah Railroad, which was so critical to the defense of the coastline, connected those two cities. All rail lines in the state at the time shared a common five-foot gauge, known as "5-0."

A logical location for the State Works would appear to be one of the iron-manufacturing facilities Gist and Lopez had visited, both of which were on rail lines. Gist's argument against the King's Mountain Iron Works was that its finished products would have to be carried down the King's Mountain Railroad to Chester and transferred to the Charlotte and South Carolina Railroad to get to Columbia. His reason for not choosing the Nesbitt Mine in Spartanburg District was similar; the material would have to travel down the Spartanburg and Union Railroad and be transferred to the Greenville and Columbia Railroad at Alston. Another argument against the Spartanburg District mine was that a bridge on the Spartanburg and Union Railroad was deemed unsafe for heavy loads like a ten-inch Columbiad. In choosing Greenville, Gist ensured that smaller, lighter loads of iron could be delivered to Greenville, and the finished, heavier products could be placed on a train in Greenville and sent directly to Columbia on a single rail line.

Gist's second reason for choosing Greenville was that nearby growths of abundant timber could be used in the construction of numerous buildings and shops. Third, Gist considered the type of power for the State Works. Water power was abundantly available in South Carolina, and the Reedy River was only a mile away from the Greenville site. But Gist ultimately chose to go with coal-powered steam engines, a decision that would create serious supply problems and financial strains in 1863 and 1864.

Fourth, Lopez argued in a letter to the Department of the Military that the location's remoteness and obscurity would be less likely to attract the enemy's attention. Additionally, he felt that the employees would remain better focused on their work by not having a nearby city with vices such as gambling and other "vulgar recreational activities."[20] At the time, the population of the town of Greenville was only about eighteen hundred and the district nearly twenty-two thousand.[21] It had active Baptist, Episcopal, Methodist, and Presbyterian churches, as well as Furman University, the Greenville Baptist Female College, and the Southern Baptist Theological Seminary, most of which had built new buildings in the 1850s.[22]

Gist's fifth and most important reason for choosing Greenville was that Vardry McBee donated twenty acres to the state for the purpose of building the State Works. McBee was about eighty-seven years old and was the largest land owner in Greenville District at the time. A philanthropist, he had previously donated land to the four major Greenville churches as well as for male and female academies.[23] Additionally, McBee played a major role in bringing the railroad to Greenville, and it ran through his land about half a mile from its terminus closer to town.

McBee's donation was likely the single most significant factor in Gist's decision. In a report to the Executive Council on March 28, Gist stated that he and Lopez had recently visited Greenville.[24] It must have been on this trip that McBee offered the land. Gist introduced the following resolution to the Executive Council on March 28: "Whereas, Mr. Vardry McBee has generously and patriotically made a donation to the State of twenty acres of land near the village of Greenville, for the purpose of erecting a foundry and workshops, Resolved, that the governor and council, on behalf of the State of South Carolina, tender to Mr. Vardry McBee their profound thanks for his valuable gift."[25] Gist later wrote that he made "certain recommendations," surely that the council accept McBee's offer, and the council agreed.[26]

The *Carolina Spartan* reported on April 10 that McBee had "made a donation of twenty acres of valuable land, about half a mile from the town near the railroad, on Brushy Creek, to the State, for the purpose of establishing at this place a gun factory and machine shop."[27] Probably quoting Lopez, the *Greenville Enterprise* wrote, "The manager states that some time will elapse before small arms can be manufactured—only repaired—while preparations will be made at once for turning out all kinds of ordnance."[28] The council approved the preparation of necessary papers and execution of a title on the property in mid-May.[29] McBee's deed, which was signed on August 11, 1862, and recorded on September 2, states that he gave a twenty-acre tract of land near the village of Greenville to the state of South Carolina "for and in consideration of the interest I feel in the establishment of Southern Independence."[30] The deed is clear that McBee donated the parcel to the state without restrictions, and the understanding of the state's leaders was the same. In late 1864, the State Works was put up for sale, and a legal challenge to the ownership of the land arose. The state auditor, James Tupper, addressed the issue in a letter to Governor M. L. Bonham on November 7, 1864, writing, "[I]t cannot be reasonably supposed that the state would have expended the large sum which these works cost upon lands that were subject to revert in the event of a change (from unforeseen circumstances) of the use originally contemplated by the state and the donor. . . . An *absolute* title to the state was an indispensable condition to the location of these works upon the land in question."[31]

Thirty-four-year-old Charleston architect, civil engineer, and surveyor John A. Michel surveyed the tract of land on August 8, 1862, at McBee's request.[32] Lopez, who lived next door to Michel in Charleston in 1860, probably recommended the surveyor to McBee.[33] Lopez later employed Michel as a draftsman at the State Works; available payrolls show that he worked there from mid-1863 through at least late February 1865.[34] The twenty-acre tract was located adjacent to the Greenville and Columbia Railroad about half a mile from its terminus. The property is bounded today by the same railroad bed, Nelson Street and Anderson Road. During the entire existence of the State Works, from the summer of 1862 until the end of the war, passengers on every train traveling to and from Greenville got a close-up view of the facility and, after the war, were able to witness its demise. Michel's detailed description of the quadrilateral-shaped property is recorded in McBee's deed. The property line ran parallel to the Greenville and Columbia Railroad for nearly 1,400 feet. Turning away from the railroad track to the northwest, the line ran north for 547 feet to the center of Anderson Road. From there the line paralleled Anderson Road for nearly 1,100 feet to the corner of Mrs. Jane Austin's lot. Finally, the line between Austin and McBee's property ran about 850 feet east to a stake near the railroad. Both Anderson Road and the railroad track are still there, allowing the outline of the property to be easily recognizable today.

In early 1862, McBee's land seemed to be ideal for Lopez's purposes, but its deficiencies soon became apparent as the war progressed. In the final analysis, it was probably McBee's donation of the twenty-acre tract that swayed Lopez and Gist to choose Greenville for the location of the State Works. McBee's property offered two major advantages—the land was free, and it was located immediately adjacent to a railroad, facilitating the importation of coal, iron ore, and other raw materials and the exportation of finished products.

Placing the facility on McBee's tract had disadvantages as well, foremost of which was

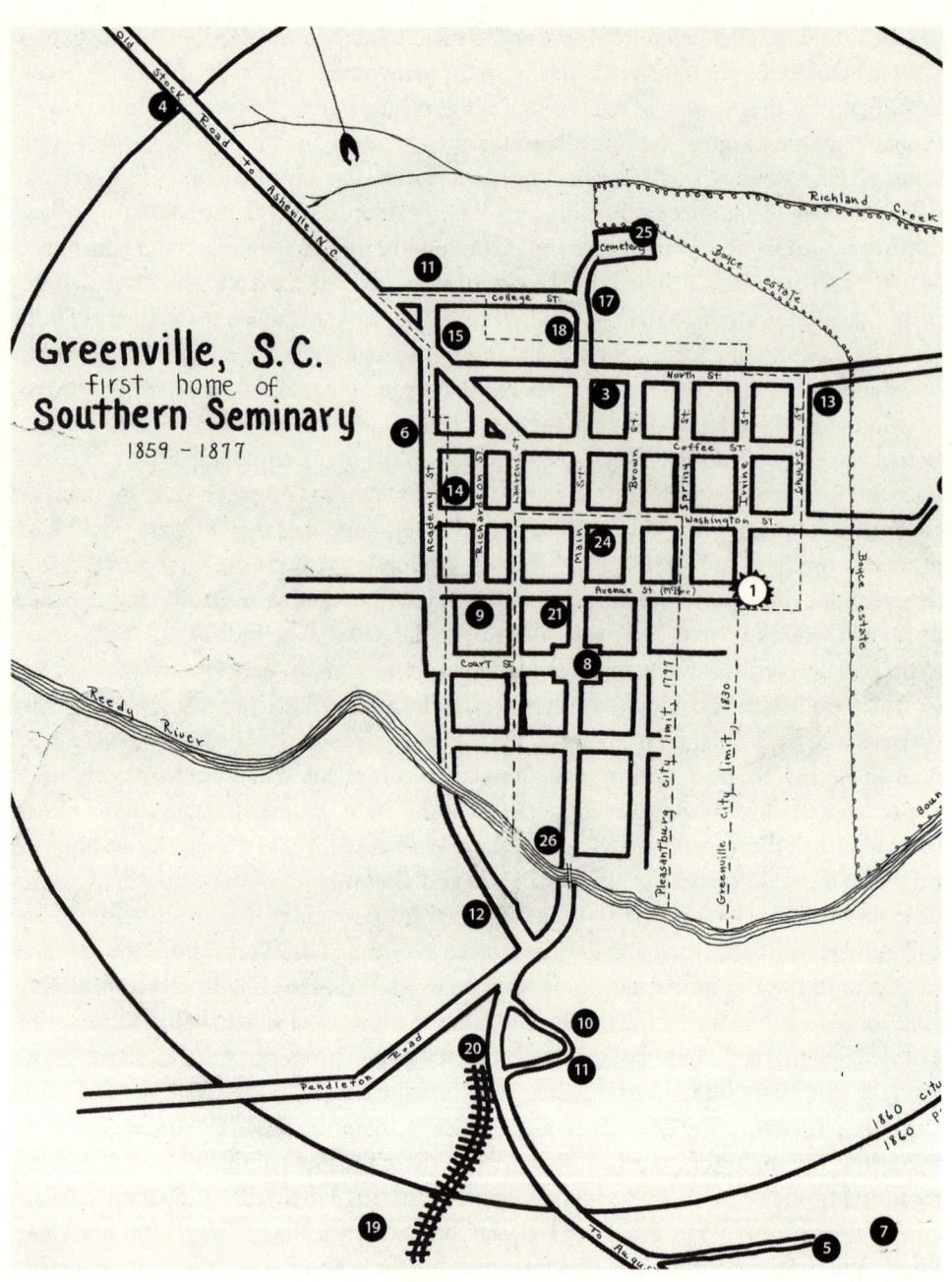

Greenville, S.C.
first home of
Southern Seminary
1859 – 1877

(FACING) Greenville, S.C. First Home of Southern Seminary. Courtesy of Special Collections and Archives, Furman University

1. Classroom building of the
   Southern Seminary
2. Boyce Lawn, home of James P. Boyce
3. Home of John A. Broadus
4. First home of Basil Manly Jr.
5. Second home of Basil Manly Jr.
6. Third home of Basil Manly Jr.
7. Home of Dr. William Williams
8. Record Building, designed by Robert Mills
9. The Baptist Church
10. Richard Furman Hall
11. President James C. Furman's home
12. Colonel G. F. Townes's home
13. Christ Episcopal Church
14. Presbyterian Church
15. Buncombe Street Methodist Church
16. Greenville Female College
17. Home of C. J. Elford
18. Julius Smith's home
19. South Carolina State Works
20. Terminus of the Greenville
    Columbia Railroad
21. Beattie's Store
22. Confederate Fort
23. Confederate paper mill
24. Goodlett House
25. Springwood Cemetery
26. Gower, Cox, and Markley

that the location forced Lopez and his successor, J. Ralph Smith, to deal with major transportation costs. Because steam power was the only option at this location, not only did Lopez have to purchase steam engines, but he had to buy and transport coal to operate them. By choosing McBee's land and coal-powered steam engines, Gist and Lopez committed the State Works to dependency on a rail system that became increasingly expensive and less reliable as the war progressed. The transportation of iron ore was another problem. Similar to coal, iron ore had to be transported by rail from Upcountry mines to Greenville over several different rail lines, an endeavor that became so expensive by late 1863 as to result in a negative financial impact on the State Works.

Because McBee's land was in a relatively isolated geographical location at the end of the rail line, many additional items required by State Works had to be imported over long distances by rail, driving up Lopez's operating expenses. For example, some of the necessities that came in on the Greenville and Columbia Railroad in March 1863 were iron, lumber, spokes, a 270-pound machine, miscellaneous packages, and several passengers doing business with the State Works. Lopez's cost for rail freight that month alone was $190.04.[35] Another example comes from raw material transported by rail to the State Works in August 1864. The list included fifty tons of coal, twelve carloads of wood, and one carload of sand for casting. Other items were nails, rice, fodder, lime, and salt.[36] In early 1862, however, neither Lopez nor Gist could anticipate that by late 1863 a failing railway system would drive up the operating costs so far as to make the facility unprofitable. The legislature's Committee on the Military reviewed the desirability of either selling the State Works or moving them to Columbia in late 1864, and in its report dated December 24 the committee called the location of the State Works "unfortunate," yet not so bad as to recommend its removal.[37]

Though Lopez was highly respected for his abundant professional talents, nearly his entire life had been spent in the civilian sector, and he had some difficulty adjusting to working for the government. The first hint of discord came shortly after he was appointed general superintendent of the State Works. Something happened in late March 1862 that

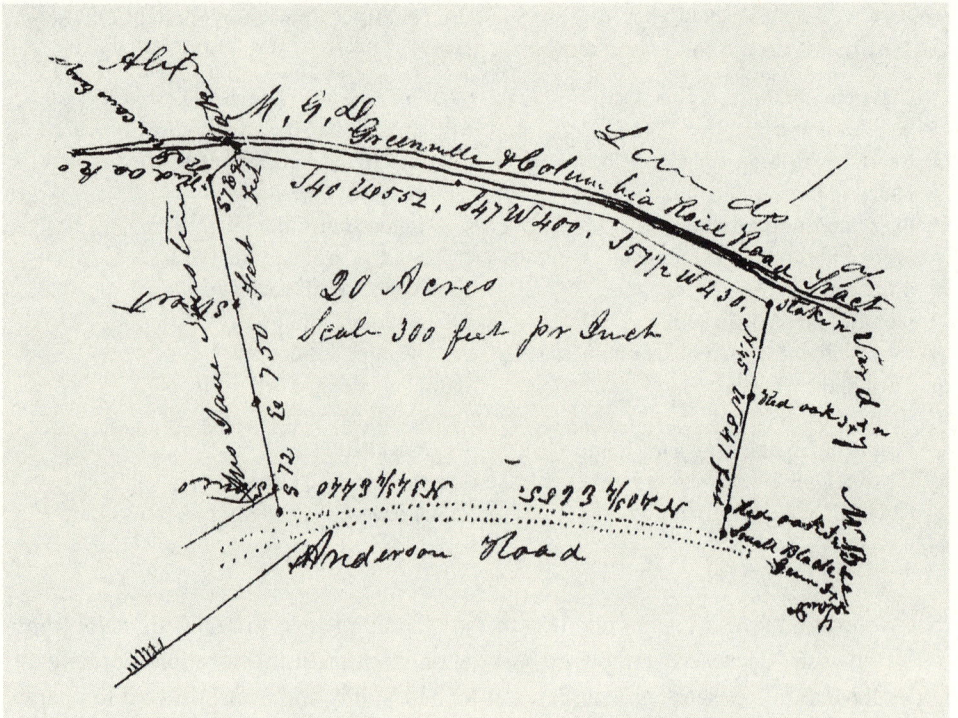

The twenty-acre tract of land which Vardry McBee donated to South Carolina for the State Military Works as it appears in the 1862 deed. It is bounded by the Greenville and Columbia Railroad to the southeast, *top of the image*; Anderson Road to the northwest, *bottom*; Mrs. Jane Austin's land to the east, *left*; and Vardry McBee's land to the west, *right*.

made Lopez think the governor and Executive Council were about to terminate him. Gist had been corresponding with Lopez in relation to Lopez's salary, and Lopez gave Gist the impression that he felt his services were no longer necessary. Possibly sensitive about the failed arms deal with Louisiana, Lopez took offense that the council placed certain limitations on his autonomy when, in fact, it was requiring of him no more than other department heads. The council had no intention of terminating Lopez and conveyed its high opinion of him in a resolution on March 27: "That the Chief of the Department of Construction and Manufacture [Gist] be requested to confer personally with Mr. Lopez, and inform him that he is entirely mistaken in supposing that the Governor and Council have desired to employ any other person in his stead, and that they are at a loss to understand why he should have thought so, that they have a high appreciation of his ability, energy, and patriotism, and that they only have proposed that he should be dealt with, as they deal with all other heads of departments, requiring estimates of contracts, and arranging a fixed rate of compensation for services rendered."[38]

By the end of April, Lopez's annual salary was fixed at $3,000.[39] His son, Moses, was hired as a clerk in the office and was paid $1,200 annually. Most if not all of the Lopez family moved to Greenville during the war. David purchased property on the corner of Anderson Road and Vardry Street, about half a mile from the State Works. Moses's wife

moved with him to Greenville and joined the Greenville Ladies Association in Aid of the Volunteers of the Confederate Army.[40] David's son, John, bought a house at 506 Augusta Road in 1862, also about a half-mile mile from the State Works. In June 1866 he and his wife, Maria, sold it to Thomas B. Thruston, and it remained in that family until the house was demolished in the 1960s.[41] David's sister, Louise, wife of Abraham Moise, bought property on Pendleton Street.[42]

On April 2, Lopez provided his estimate of operating expenses, $7,101.49, and Gist presented the figure to the Executive Council, which approved it. Neither Lopez nor Gist offered a time frame for the estimate, but it would have been grossly inaccurate even if it were only a month's expenses. By the end of August, only four and a half months hence, Lopez's expenses had already amounted to $82,000. Both Lopez and Gist soon learned that the actual cost of operating the State Works was much higher than either man expected.

CHAPTER 11

Morse Comes to South Carolina

At the direction of Governor Pickens and the Executive Council, Lopez traveled to Atlanta in April 1862 to personally negotiate a deal for the machinery from the Tennessee State Armory and, if successful, to supervise its transfer to Greenville. Lopez telegraphed Tennessee's Governor Isham G. Harris from Atlanta on April 23: "I am here as an agent of the State of South Carolina, with authority to treat with the State of Tennessee for the machinery and stock saved from the Nashville Armory [Tennessee State Amory]; will you authorize your agent to turn it over to me, subject to future settlement with South Carolina?"[1] By April 26, Lopez had not received an answer and wrote to Harris, who was then at Corinth, Mississippi.[2] Lopez reiterated his credentials and request and wrote that, since Nashville had been abandoned and some of the machinery saved, he hoped to obtain it for South Carolina on terms that would be equitable between sister states engaged in a common cause provided Harris did not plan to put it to immediate use.[3] Lopez told Harris that South Carolina was about to start an armory in the Upcountry and was deficient in many machines, a situation that would be ameliorated by the use of the Nashville machinery. Lopez urged Harris not to permit such valuable tools to be idle while there was a place waiting to use them. Harris responded in a telegram addressed to Morse, who was still in Atlanta: "Turn over the stock and machinery to agent of South Carolina, subject to future settlement, taking inventory."[4] A year later, Harris wrote that he made the decision "because I thought it improper and impolitic for the machinery to be idle in times like the present."[5] Morse conducted the required inventory, which Lopez reproduced for Gist. Morse did not, however, assign a monetary value to the inventoried items, an omission that created problems in later years. Also left out of Lopez's communications with both Harris and Gist was the fact that Morse retained personal ownership of a considerable number of the tools used in making his new breech-loading carbine. More than two years later, in late 1864, Morse would offer to sell his tools to South Carolina. He ended up renting them to the state for $300 for their use in making the first thousand carbines, and in 1866 he submitted a

$150 claim for their use at the State Works and another $150 for damage they sustained there.[6]

The transaction in Atlanta in April 1862 was the first interaction between Lopez, the builder, and Morse, the inventor, and their paths merged for the next sixteen months. The acquisition of the Nashville machinery eventually allowed the State Works at Greenville to produce Morse's new carbine, a weapon that President Davis considered to be the best breechloader of the time and one that became the most significant article manufactured at the State Works.[7]

Through his communications with Harris, Lopez was satisfied "that the State of Tennessee would place at the disposal of this State [South Carolina] all the machinery and stock saved."[8] Lopez reported to Gist on May 1, "As you will perceive (from the above telegrams) Governor Harris adopted my suggestion for turning everything over to the State of South Carolina, subject to future settlement."[9] Lopez's report to Gist also included a complete inventory of the articles obtained from the State of Tennessee. (See appendix 1.) He also informed Gist that Morse had obtained the most valuable pieces of the Nashville machinery from the Confederate government. Lopez recognized that some of this machinery was not Tennessee's to sell, and he proposed a transfer instead of a sale. Had a formal sale taken place, much of the useful machinery, that is, what belonged to the Confederate government, would have been left out of the transaction. By agreeing to Lopez's plan, Harris was able to turn over all of the Nashville equipment to South Carolina without distinction as to whether it belonged to Tennessee, Virginia, or the Confederacy. Though the arrangement was beneficial to South Carolina because the State Works could immediately put the machinery to use, the arrangement would create confusion about its ownership over the next three years.

Morse, along with his machinery and some of his skilled workers, came with the deal. While in Atlanta, Lopez was able to secure the services of several workmen who had previously been employed by Morse in Nashville. Lopez sent them to Columbia so that they could immediately begin work there altering and repairing small arms while the State Works in Greenville was being constructed. Morse soon became intimately associated with the State Works. His annual salary at the Tennessee State Armory had been thirty-five hundred dollars, but he accepted three thousand per year as the first, and only, superintendent of the Small Arms Department in Greenville. His salary was never increased. Morse wrote in 1866 that Lopez had offered him thirty-five hundred, but that he "refused to accept more than he, as General Superintendent, was allowed."[10] Morse was still in Atlanta in May 1862 but soon moved to Greenville, certainly no later than August.[11] His understanding of the agreement between the two governors was that he would function semi-independently, but friction soon developed with Lopez. Recalling his original arrangement with Lopez, Morse implied that the two men never got along when he wrote in July 1863: "When I made my contract with Mr. Lopez as agent of this State, for the transfer of the Tennessee State Armory machines, tools, stock, etc. etc. it was agreed that I should be superintendent of the Small Arms Dept., having my master armorer [George T. Brooks] with me together with many of my men, all of whom were to be under me, as they had been at Nashville. I had my draughtsman and patternmaker, and and [sic] expected to control my own business. When I found that I could not have such controls under Mr. Lopez, I demurred, and after a time, he has resigned."[12]

The Executive Council unanimously passed a resolution on May 1, 1862, approving Lopez's action "in procuring such valuable and indispensable machinery, and upon such favorable terms."[13] Further, the council promised "that the State of South Carolina will account to the State of Tennessee in future settlement for everything received" in the transaction.[14]

Gist called Lopez's efforts to procure the machinery for use at the State Works indefatigable, and he estimated the value of the machinery at twenty-three thousand dollars, but its market value at the time was inflated by 50 percent. Gist recalled correctly in August 1862 that the terms of the agreement in April were that Governor Harris allowed South Carolina to take and use the machinery, and Gist issued a receipt for it, subject to settlement at a fair price after the war. It was a very good deal for South Carolina. If Harris had insisted on a standard sale, South Carolina would have been forced to pay an inflated price and would have been deprived of cash which could be used for other purposes. With the addition of the Nashville machinery to that already collected from Charleston and Columbia, the State Works was much closer to carrying out its mission.[15]

Shortly thereafter, in May 1862, Lopez began "erecting a shop at these works [Greenville] to receive and put in operation the machinery."[16] Lopez reported to Gist on August 15 that he had begun to clear land in Greenville and to erect buildings on March 20 of that year, probably meaning that he began to clear land on March 20 and to erect buildings in the following weeks and months.[17] He was appointed general superintendent of the State Works on March 1 and recommended to the Executive Council on March 12 that Greenville would be a suitable location. We know that the machinery that Lopez purchased in Charleston in late February was moved first to Columbia and later to Greenville before there were any buildings at the State Works site. We know that Vardry McBee agreed to donate the land in March. The Department of Construction and Manufacture was not created until March 24, with Gist as its chief. His report indicates that he painstakingly evaluated sites before finally choosing the Greenville location, but he might have made this decision before March 24. Lopez agreed with Gist when he wrote that there was some delay in choosing the site due to discussions and examinations, but he did not say when or how long this delay took place.[18] Though Lopez provided input into the location, the actual decision was not his to make. It was Gist who made the final decision on Greenville after considering Lopez's recommendation and the Executive Council that gave final approval of the Greenville site on March 27 or 28. Lopez wrote that he "*was directed* [emphasis added] to locate the Works on a tract of land donated to the state by Vardry McBee, Esq., adjoining Greenville, and directly on the line of the Greenville and Columbia Railroad."[19] The most likely scenario is that Lopez decided that Greenville was a good location as early as March 12, and he might have begun some basic site work there before the creation of the Department of Construction and Manufacture on March 24 and final approval of the location a few days later. After the site was ready, buildings went up over the next few months.

Soon after deciding on the Greenville site and clearing the land, Gist and Lopez faced additional obstacles and delays in getting started. In part, this was due to a shortage of available labor during the spring planting season, which was unusually wet that year.[20] Farmers were delayed in preparing their fields, and the lack of labor slowed down the availability of raw materials with which to build the workshops. The problem was

alleviated as soon as the spring crop was planted, and laborers began to seek other employment.

Lopez soon began to construct the buildings necessary for a foundry and workshops, and the work progressed rapidly. As of mid-June, however, the skilled workmen were still in Columbia where their work repairing arms was considered to be of enough importance to excuse them from performing required street maintenance.[21] Throughout the spring and summer of 1862, Lopez purchased necessary items to fulfill his mission. He bought more than three tons of castings, as well as oak lumber and charcoal. He also had some of his men working on a pattern maker.[22] In late July the workmen familiar with the Nashville machinery were still in Columbia, but a great deal of construction went on at the Greenville site that summer.

The acquisition of machinery and stock from the Tennessee State Armory forced Lopez to rethink his goals. In mid-1862, he had begun work on the shops necessary to carry out the original intent of the State Works, that is, the repair of small arms and the manufacture of shot, shell, cannon, and other war materiel. In order to fulfill his mission, he was in the process of purchasing machinery to augment that acquired from the Tennessee State Armory. The Executive Council authorized him to purchase machinery for $29,379.39 on May 12.[23] Lopez asked for and received an additional $10,165 on May 21, and in mid-July he requested even more funds.[24] When the machinery and skilled labor from Nashville came under his supervision by early May, he decided to put them to use as soon as possible. In doing so, he was forced to slow down completing the shops which were already being built. Though some of the Nashville machinery was missing altogether and had to be replaced, and other incomplete or broken pieces required repair, Lopez constructed a workshop specifically for the Nashville machinery that summer. It measured one hundred by fifty feet and was powered by a fifteen- to twenty-horsepower steam engine.[25] The new workshop was complete and in operation by August 15.

Though some products had been completed by August 15, Lopez was still in the process of making and otherwise obtaining new machines necessary to commence manufacturing arms. He wrote at that time: "[I] hope soon to have [the workshops and buildings] all finished, and give my undivided attention to manufacture."[26] Recognizing that setting up the Nashville machinery had delayed his original plans, Lopez decided to prevent that from happening again by establishing Morse's small arms shop as a separate administrative department within the State Works. His intention was "to put [the small arms factory] upon such a basis as will enable the State to manufacture arms to a limited extent at first, but so arranged that the facilities can be increased at any future time."[27] His future accounting ledgers, as well as those of his successor, J. Ralph Smith, show two separate accounting systems, one for Morse's small arms department and another combining all other departments at the State Works.

As Lopez pursued his work that summer, a legal issue arose with the Nashville equipment in July. Colonel Josiah Gorgas, chief of the Confederate Bureau of Ordnance, wrote to Governor Pickens on July 15 inquiring about the machinery and expressing an opinion that some of it might belong to the State of Virginia and thus be held in trust by the Confederate government. The Executive Council authorized Pickens to respond to Gorgas, explain to him the conditions of the agreement with Governor Harris, and tell Gorgas that the council was unwilling to take any steps in the matter until it could confer with Harris.[28]

The Executive Council passed a resolution on July 31 requiring the State Works to limit its output to its original mission, that being the casting and equipping of cannon, the casting of shell, and the repair of small arms.[29] This temporary abeyance in the use of Morse's machinery and the manufacture of small arms was probably because the council did not want to go to the expense of putting the Nashville equipment to use until the legal issues were resolved. It also gave the council time to consider if it wanted to add the manufacture of small arms to the State Works' mission.

As of early August the only output from the State Works came from the temporary shops in Columbia, but everyone involved was anxious to bring the Greenville facility into production mode. The Executive Council continued to support the State Works when it approved Lopez's request for $6,000 to meet his August payroll, $5,500 for the September payroll, and another request for $11,573 in early September.[30]

On August 11 the council, acting on a communication it had received from Gorgas about the tools and machinery from Tennessee, instructed Attorney General Hayne to prepare the required bonds.[31] On about August 21, the council required Lopez to give bond in a penalty of $10,000, setting aside the issue of ownership of the Nashville machinery for the time being.[32]

Gist and the Executive Council continued to wrestle with the direction and focus of the State Works. The original intent was for the State Works to cast cannon, shot and shell, to *repair* small arms, and to manufacture other items necessary for war. With the acquisition of the Nashville machinery, the State Works acquired the ability to actually *manufacture* small arms. Therein lay the question; that is, was the manufacture of small arms necessary in Greenville? Gist asked Lopez to make a presentation to the council on August 4, and the council's response was to ask Gist to make a formal report answering specific questions, which were clearly directed toward addressing the issue of whether manufacturing small arms in Greenville was desirable. The council required Gist to answer the following questions:

1. What was the cost of construction of everything up to this time?
2. How much of this amount is invested in machinery and preparation for manufacture of small arms, and which cannot be used for casting cannon and equipage or for repair of small arms?
3. How much would be necessary to complete the works for manufacture of small arms according to the scheme of Mr. Lopez?
4. How soon could a turn of small arms be made?
5. What is the value of the machinery and stock in possession which has cost us nothing; and how much of this is exclusively useful in the manufacture of small arms?

Gist asked Lopez to issue a report addressing the council's questions, which he did on August 15, and Gist read Lopez's report to the council on or about August 27.[33] Incorporating Lopez's report, Gist issued his own on August 29. These two documents give us the most insight into the early development of the State Works.[34] The council was satisfied that the manufacture of small arms could be performed effectively at Greenville and on September 19 rescinded its July 31 order prohibiting it.[35]

By September 1862, the basic design of the State Works was set and would remain in place for the remainder of the war. In addition to a front office, there were seven depart-

ments or shops, each with its own foreman. The small arms shop worked exclusively on Morse's breech-loading carbine under his direct supervision. An entirely separate gun-repair shop, sometimes called the musket-repair shop, performed work on those weapons. The other departments were a carpentry shop, machine shop, blacksmith shop, foundry, and a separate department of laborers and watchmen.

Now that Morse's "small arms factory" was officially established, Lopez refocused his efforts on the original purpose of the State Works.[36] The gun-repair shop remained very busy repairing a variety of weapons in the late summer and early fall of 1862. Between August 5 and September 30, the shop worked on many muskets, including model 1822s, English muskets, and others, primarily converting them to percussion. The shop also worked on a variety of rifles, including Yager and model 1841 "Mississippi" rifles made at both Harpers Ferry and the Palmetto Armory in South Carolina. They also continued to repair other articles such as bayonets and cavalry pistols. At this time the State Works was issued a mounted, two-pounder brass cannon. It is not clear if the purpose was to repair the cannon or to use it for some other function, perhaps as a signal gun or for local defense.[37]

In fact, Lopez had accomplished quite a bit at the Greenville site since the creation of the State Works in late March. At least half of the twenty-acre tract had been cleared, and Lopez designated a section measuring 600 by 700 feet, or about ten acres, for the initial phase of construction. He wrote that this section was enclosed, implying that he had erected a fence around it. Supporting this is an undated abstract showing that the State Works purchased 252 fence posts, which would allow for about one post every 10 feet.

Lopez erected no fewer than fifteen buildings by mid-August 1862, and many of those were immediately operational or would be very soon.[38] First to be constructed were buildings to house laborers. After building the shop for the Nashville machinery, Lopez erected a carpenter's shop measuring 100 by 40 feet. It contained machinery suitable for making field and siege gun carriages, caissons, battery wagons, forges, sabots, tents, and all other articles necessary for military field use. A twenty-five-horsepower steam engine provided power for the carpenter's shop, which was fully operational by mid-August, and it was used primarily at that time to complete the construction of other buildings on the site. Lopez wrote on August 15 that the carpentry shop would soon be used to manufacture the war materiel for which it was designed.

Lopez built a second machine shop on the same dimensions as the first, 100 by 50 feet. By mid-August, a thirty-horsepower engine was in place, and heavy machinery was being set up in the building. Lopez anticipated that this machine shop would be in full operation within three or four weeks. A blacksmith shop, measuring 100 feet by 50 feet, was soon completed and occupied. It contained eighteen forges and enough room for twenty more. It also held three nearly-completed trip hammers, two for welding gun barrels and a third for heavy smith work. Adjacent to the blacksmith shop was a brick foundry. It measured 115 feet by 55 feet and had a 28-foot ceiling. The foundry was finished except for its roof, the completion of which was delayed while waiting on lumber to arrive. Also on the site were a 20- by 40-foot office, a storehouse measuring 60 feet by 25 feet, and a receiving and storage building for iron measuring 35 by 16 feet. Stables and a tool house were also constructed.

Additional buildings were built to support Lopez's workmen, which included white civilians and soldiers, many slaves, and a few free black men. A 40- by 18-foot smoke-house was erected, as was a 60- by 16-foot living house for "negroes." A hospital and three kitchens complemented the structures at the site. Lopez concluded his discussion of the structures by writing, "&c," implying that there might have been even more structures on site that he did not include in his report.[39]

Lopez had acquired experience with fire-control measures and indoor plumbing during his decades of contracting experience in Charleston. Perhaps drawing on this knowledge, he erected a complex water-distribution system at the State Works. Lopez's men dug a well that was capable of supplying the entire complex, which was nearly com-pleted by mid-August 1862. Water was pumped from the well through a 600-foot-long set of wooden pipes into an underground brick reservoir. The reservoir held thirty-five hundred gallons and was located at the highest point on the property, on an elevation about 40 feet above the well. The exact locations of the well and reservoir are not known, but the railroad line runs along the low point of the site, the most likely spot for the well, and about 600 feet away is the high point along Anderson Road, making it the logical site of the reservoir. From the reservoir, water was delivered under pressure to the various workshops for both general use and fire suppression. Lopez arranged to borrow an excel-lent fire engine from Charleston's Vigilant Fire Engine Company and was anticipating its arrival in mid-August 1862.[40] In June 1863 he purchased a fifty-two-pound fire-engine hose from M. H. Nathan, chief of the Charleston Fire Department, probably to be used with the fire engine.[41]

Gist reported on August 25 that the State Works had accumulated a considerable quantity of raw materials. A large amount of coal had been transported from Chatta-nooga, Tennessee, when that city was in danger of falling to Union forces in February, and the State Works had used very little of it by late August. Also on hand was a con-siderable quantity of steel, iron, copper, brass, files, and other materials for the manu-facture of arms and munitions. A large amount of lead had been amassed there as well, some of which had been donated as gifts and some purchased. Most of it, however, was subsequently turned over to the Confederate government, but Gist did not seem very concerned about the loss. He wrote that a sufficient supply of lead was readily available in the Upcountry in the form of privately owned water pipes. His recommendation to the council in August 1862 was to leave the lead in the hands and homes of its owners until an emergency required its removal.

As of August 15, 1862, the entire production of the State Works had come from its temporary facility on the State House grounds in Columbia. Nothing had been manu-factured in Greenville up to that time. Gist estimated that the products of the temporary workshops in Columbia had saved the state a total of $22,893 as of mid-August. Repair-ing, cleaning, and altering 1,620 muskets saved $12,960. The operation had also repaired and cleaned 213 rifles, 2 carbines, and 24 double-barrel guns and had made 874 bayonets, saving the state $817. The workmen in Columbia built six caissons, complete with limbers, poles, and spare wheels at a savings of $3,600, and one battery wagon at a savings of $750. The Columbia works had also made one thousand pikes and shafts, saving $3,000, and 4.5 tons of spikes for the gunboat *Palmetto State,* saving the state $1,766. Lopez wrote at the same time that a large amount of miscellaneous firearms, which he described as being

irreparable, had been turned over to the State Works from the State Arsenal, and that he was making preparations to put them in good working order for field use. Even with this level of production, the State Works showed a deficit of $66,633.12 during its first five months of operation.

In his report to the Executive Council dated August 28, 1862, Gist recorded that, as chief of the Department of Construction and Manufacture, he had received from the Treasury Department $95,212.02 as of August 15.[42] Reflecting the great majority of the work of the Department of Construction and Manufacture done by Lopez and the State Works, $82,539.92, or 87 percent, was accounted for by Lopez as expenses for the State Works between its inception in March 1862 and August 15. An additional $982 was spent by the department on iron to make spikes for the Confederate ironclad ram, *Palmetto State,* and was to be refunded to the state. Flag Officer Duncan N. Ingraham, of the Confederate Navy, was supervising the construction of the *Palmetto State* by Cameron and Company in Charleston. On March 12, 1862, the Executive Council authorized that $300,000 be placed under the control of Ingraham, Captain J. R. Hamilton, and George A. Trenholm for the purpose of building an ironclad, marine ram for the defense of Charleston Harbor. Her keel was laid down in January 1862, and she was commissioned on October 11. Ingraham had requested the spikes, and had paid Lopez $1,766.06 for them by August 15, 1862.[43] Lopez had also sold some iron to J. M. Eason for $518.11 to be used on the gunboat *Chicora.* The remaining $11,600 was spent on miscellaneous items not related to the State Works.[44]

Lopez balanced his books on August 15, 1862, and itemized his expenses for the months between March and August: machinery and tools—$18,946.01, building materials—$5,617.44, stock—$29,350.92, provisions—$6,001.40, salaries and payrolls—$19,634.06, incidental expenses—$1,764.94, and balance on hand—$3,509.32.

Gist explained the difficult economic situation of operating the State Works during the Federal blockade of Southern ports. Prices of raw materials and tools were enormous. He wrote, "The expense of carrying on the 'State Works' has been necessarily very considerable, owing to the high prices of skilled labor, material and provisions."[45] The cost of iron, steel, and other materials had quadrupled since the beginning of the blockade. Cannon powder had gone up from 25 cents per pound to $2.25 per pound, and rifles had increased from $15 to $75. Some items, such as files, could be obtained only by running the blockade, and he was compelled to pay the asking price for them.

In late August 1862 Gist reviewed his plan for the State Works for the next six months. He continued to envision the facility primarily as a manufacturer of cannon, shot, and shell. In spite of the economic challenges in operating the State Works, Gist anticipated casting shot and shell no later than October 1862 and cannon soon thereafter.[46] He also discussed the manufacture of small arms but anticipated they would not be produced before February or March 1863, "unless the exigencies of the service should require them," in which case they could be produced much sooner.[47] To Gist it appeared to be a matter of correctly utilizing the skilled manpower already employed. He wrote: "It must be remembered that the same hands to a considerable extent necessary to carry on a foundry and work-shops, can at the same time make the machinery and tools necessary to manufacture small arms; and in that way the work can be done much cheaper than by employing hands exclusively to make machinery and fit up tools for small arms."[48]

Gist's goal was not to establish a large armory such as the ones at Fayetteville or Tredegar in Richmond, but rather to develop one on a scale commensurate with the current means and needs of the state and to increase the size and output if necessary in the future. He summarized his thoughts:

> If the war should continue for some time, and more especially if the blockade of Southern ports should become so effective as to prevent further importation of arms, it will be absolutely necessary for every armory that can be put into operation to be engaged in the manufacture of small arms; and even if the war should end in a short time, we will to a considerable extent be compelled to keep up our military organization and keep arms in the hands of our people to be at all times prepared to repel the aggression of bad neighbors. With an armory to keep the State at all times supplied with good arms, and with the materials for making gunpowder at our command we may feel confidence in our means of defence.[49]

Gist's hopes and predictions soon began to materialize, but only in part. The State Works was functioning by the autumn of 1862, but its output was limited to the repair of small arms, not the production of Morse's carbine or cannon. In early October the council ordered one thousand percussion caps to be sent from the arsenal in Charleston to Lopez in Greenville for the purpose of trying small arms.[50] In mid-October the council approved Lopez's request for forty-five hundred dollars.[51]

In late October, Governor Pickens, who was nearing the end of his two-year term, visited the State Works. The reason for the visit is not known. Possibly it was because the estimated date for casting shot, shell, and cannon had come and gone with none of those items having been produced. More likely, the reason for Pickens's visit was to prepare for his end-of-term summary reports. Pickens made a report to the Executive Council on October 24, recommending that Major W. G. Eason, ordnance officer of the state, personally inspect or delegate an expert to inspect the Greenville facility. Specifically, Pickens wanted a knowledgeable person to conduct a general examination of the State Works and to report to the legislature on the number of arms repaired and manufactured, the machinery on site, and the fitness of the foundry's intended purpose. Interestingly, Pickens referred to the State Works as a facility for the repair and manufacture of small arms, but he did not mention the casting of cannon, shot, or shell.[52] Lopez confirmed to Pickens on December 12 that the original plan to cast shot, shell, and cannon had not changed. He wrote, "we are now making all the castings required for the different departments of the establishment, and are ready to cast shot and shell and will soon be ready to cast cannon."[53] Pickens wrote to Beauregard the next day: "It [State Works] is just being prepared for casting cannon on a large scale, and at any time that you may make requisition, state authorities would, I have no doubt, be glad to aid you in anything you may want. It is from this that you are able to arm the eight regiments ordered to you."[54] Pickens was referring to the Second, Third, Fifth, Sixth, Seventh, Eighth, Ninth, and Eleventh Regiments of Reserves, which enlisted for ninety days service on November 18.

By early November, Lopez was preparing to manufacture wagon wheels. He sought and received permission from the Executive Council on November 3 to seize a spoke machine belonging to James Gibbs of Abbeville. Lopez was authorized to pay Gibbs seven hundred dollars for the machine, and if Gibbs refused to accept the offer, to provide him

compensation in the usual manner offered when private property was appropriated for public use.[55]

Pickens delivered a speech to the South Carolina Senate near the end of his two-year term in November. He said that, in previous communications to the legislature, he had called attention to the importance of establishing a small arms factory, and recalled that there had been one in Greenville during the War of 1812. He said that he had also drawn attention to the iron in Spartanburg District as being eminently suitable, because of its great adhesive qualities, for large cannon. He praised the Executive Council for creating such an establishment at Greenville, including a foundry where cannon could be cast. He said the location was well selected, and Gist's practical judgment—aided by Lopez's energy, mechanical talent, and knowledge—forced the establishment into rapid maturity.[56]

Though the Executive Council was effective in its work in the first months of 1862, public sentiment turned against it as early as that spring. Public outcry against perceived excesses and unpopular decisions of the council in its management of the war effort spilled over to include a general disaffection of the Secession Convention itself. The convention met from September 9 to 17, 1862, and determined that both the Secession Convention and the Executive Council had performed their responsibilities admirably. Bending to public opinion, however, the convention voted to dissolve itself as of December 17, 1862, and that the legislature would determine whether to retain the Executive Council. The legislature abolished the Executive Council on December 18 and "declared invalid all its acts, proceedings, resolutions, and orders, except contracts."[57] Also on December 17, the legislature elected a new governor, Confederate States Representative Milledge Luke Bonham. The governor's office, now free of the Executive Council, once again bore a more conventional executive role.

By late 1862 the South Carolina state government had not specifically authorized any payment to Morse for the use of his patents in the production of his carbine, even though Lopez had made Morse's shop functional by mid-August. Some work was going on there, but it was probably the repair and manufacture of necessary machines and tools, and possibly the fabrication of carbine parts and a prototype or two. Morse's actions reflect a degree of frustration as he began to look for other manufacturers of his carbine. He was naturally aggressive in pursuing his goals, as demonstrated by his dealings with Nathan Muzzy, the U.S. War Department, President Jefferson Davis, and the Tennessee State Armory. In late 1862 he approached both an Atlanta arms maker, H. Marshall & Company, and the Confederate government in an effort to interest them in making the carbine. Simultaneously, the South Carolina state government began to take meaningful action on the carbine. (See color plate 5.)

In late 1862 and early 1863, early versions of Morse's brass-frame carbine began to appear, one in Atlanta, Georgia, and two or three more in Greenville. The *Intelligencer,* of Atlanta, reported on December 13, 1862:

MORSE'S PATENT IMPROVED BREECH-LOADING CARBINE

One of these fine carbines, made in this city, at the manufactury for arms of H. Marshall & Co., was exhibited to us a day or two ago by Mr. Marshall, the head of that enterprising firm. We were struck with the simplicity of its construction, and its power of execution. The Government, we learn, is already impressed with its value,

and in all probability we shall soon see manufactured here a large quantity of this valuable arm for the defense of our homes and firesides, with which to drive the enemy from Southern soil. These arms, we are advised, can be made here with comparative rapidity. If so, we see no good reason why the large work shop of the enterprising firm referred to above should not be engaged by the Government to turn them out by the thousands. That our readers may form some idea of the simplicity and the value of this carbine, we present them with a description of it, and the cartridge with which it is loaded:

Its full length from the butt-plate to muzzle is thirty-four inches, its weight seven pounds, its length of barrel twenty inches; it shoots a conical ball of fifty-four-one-hundredths of an inch in diameter eight hundred yards with perfect accuracy; while if held at an elevation of forty-five degrees, it will throw the ball *one mile and a quarter without being strained.*[58]

The butt is of wood, but the remainder of the stock, (with the exception of a rib passing up the under side of the barrel to receive the wiping rod) is of brass.

Immediately in rear of the breech is a chamber hollowed out of the stock to receive the cartridge, (hereinafter described,) which is so constructed as to find its correct position. This chamber is opened by lifting a cover on the top of the stock; this in opening draws back a slide, which withdraws from the barrel and drops into the hand the cartridge which has been discharged, leaving the chamber clear for the insertion of the new cartridge, which is forced into the place in the breech by the advance of the slide caused by closing the cover; the piece is then ready for firing, as the cartridge contains the cap within itself.

The cartridge, which can be repeatedly re-loaded, consists of a cylinder of brass or copper containing in the centre of its diameter and about three-eighths of an inch from the end, a cone to receive the cap. The cap (which is the ordinary chapeau cap of the army) is inserted in a small hole punched in the centre of a wad of India rubber, which is then placed in the lower end of the cylinder, so as to bring the cap down on the cone; the cylinder is then filled with powder and the ball, having been inserted in the opposite end, is driven home by a blow from a mallet. This completes the manufacture, which is so simple and rapid that a man can easily make up 100 in an hour, while they are so perfectly water-proof that they may lie under water almost any length of time without detriment.

The piece is cocked in the usual way, and fired by a pin which is driven forward and strikes the cap inserted in the end of the cartridge.[59]

The *Intelligencer* stated that the prototype was manufactured at H. Marshall and Company in Atlanta in December 1862 and that there was a desire to put the carbine into mass production there. Hammond Peletiah Marshall was a forty-eight-year-old Maine-born dentist who eventually settled in Atlanta, where he operated a sword factory known as H. Marshall and Company, also called the Atlanta Sabre Manufactory.[60] The company received an advance of $10,000 from the Confederate Ordnance Office in May 1862 to make 4,000 cavalry and artillery sabers, and it actually produced 5,832 in that year alone. In January 1863 the company contracted to make 3,000 sabers at $18.50 each, and it also

produced belt plates, horseshoes, pikes, axes, and other miscellaneous brass products. In all available records from 1862, there is no documentation of H. Marshall and Company producing any firearms whatsoever.[61]

While it is certainly possible that H. Marshall and Company put together a single Morse carbine in late 1862, it is highly unlikely that it manufactured more than one. Morse did live in Atlanta for a few months in early 1862, but he was employed at the State Works sometime between May and August. His skilled workers had been set up in Columbia by late spring or early summer, and they, along with all the machinery from Nashville, were in Greenville by August. A comparison of the names of thirty-one men employed at H. Marshall and Company in August 1862 with the names of those men who worked in Morse's small arms shop at the State Works in early 1863 shows no commonality. Also in late 1862 the South Carolina state government was just beginning the process of authorizing the construction of the first thousand carbines. The most likely explanation for the construction of a single carbine in Atlanta in late 1862 is that Morse was exploring all of his options, including trying to drum up business with Marshall in Atlanta. Though some researchers have speculated that several hundred Morse carbines were manufactured in Atlanta, no evidence exists that any other than the single prototype was produced there. A Morse carbine with no serial number survives today at the Greensboro Historical Museum and is generally felt to be the Marshall-built prototype.

In late 1862 the South Carolina state government began to take action on Morse's carbine, and it was at that time that the State Works produced its first two examples. Prior to the dissolution of the Executive Council and the change in governors in mid-December, Pickens and the council instructed Lopez to manufacture a Morse breech-loading carbine. Lopez wrote on December 22 that, in response to this order, "I manufactured a breech loading carbine of Morse's patent; took it to Columbia."[62] Because Lopez referred to this particular model as a pattern, it was most likely a prototype with no serial number, but we cannot say that for certain.[63] Probably referring to this same carbine, Morse wrote to the legislature a year later, December 1, 1863, requesting consideration of reimbursement for the use of his patents: "The peculiar model upon which these arms are being manufactured, was made here [at Greenville] at my own cost, the authorities of the state here, being unwilling to incur even this preliminary expense in the matter."[64]

Lopez took the carbine to Columbia and, at Pickens's request, appeared before the Military Committee of the Executive Council sometime in late November or early December 1862 and left the gun with the governor. On December 9, "Mr. Hutson, [chairman of] the Committee on the Military, made a report on a resolution as to Morse's patent breech loading carbine, which was severally ordered for consideration tomorrow and was ordered to be printed."[65] Sometime between December 9 and 18 the resolution was referred to the Military Committee, either by Pickens or the Executive Council, directing the committee "to inquire and report upon the expediency of having manufactured one thousand or more Morse's Patent Breechloading Carbines, according to the pattern manufactured at the State Works and left in the Governor's office."[66] The members of the committee examined the weapon and were very pleased with its performance. They found that the carbine was simple and easy to load, not liable to get out of order or foul, and could be made at the State Works for thirteen to fifteen dollars each. Before he left

office, Pickens urged Lopez to begin manufacturing the carbines, but Lopez was unwilling to do so without the legislature's approval.

On December 18 the House of Representatives approved the following resolution and sent it to the Senate: "Resolved that the Governor be authorized to have one thousand of Morse's patent breech loading Carbine made at the State Works for the use of the State; and that Fifteen thousand dollars be appropriated therefor if so much can be necessary."[67] Lopez wrote to Bonham on December 22 that a resolution was offered in the House ordering one thousand or more carbines, "but the Legislature adjourning, no action has been taken."[68]

Lopez asked the same day if he could submit an example to Bonham for his personal inspection. He wrote, "It is a very cheap and efficient arm and can be easily made. Shall I proceed in the matter or await the action of the Legislature?"[69] About a week later, on January 3, 1863, Bonham's private secretary responded, "The Governor desires that you should proceed with the manufacture of this arm and concurs with his predecessor in his estimate of its value."[70] Bonham's executive authorization appears to be the order that Lopez was looking for to start making components of the carbines. Even with Bonham's endorsement and the Senate's final approval on January 31, it would be another eight months before the State Works turned out its first batch of production carbines.[71]

Meanwhile, Morse also approached the Confederate Ordnance Department in late 1862 or early 1863, hoping to stimulate its interest in manufacturing his carbine. Morse submitted one of his carbines about that time for evaluation by William S. Downer, who had worked at Harpers Ferry before the war and was currently military store keeper at the Richmond Armory. We do not know exactly which carbine Morse shipped to Downer. It might have been a prototype without a serial number, or it could have been stamped with a very early serial number, or it is possible that it was the carbine made by H. Marshall and Company. It is probable that Morse made his initial inquiry to Downer before Bonham gave Lopez approval to begin making the carbine on January 3, 1863. Downer, who would have a relationship of another sort with the State Works later in 1863, rejected the carbine because of two concerns. First, he was worried about the old problem of the breech cover rising up when fired. This preproduction model was a Type I in which the latch was weak, and the breech cover could even fall open if the weapon were raised at more than a 45-degree angle. Second, Downer was worried about issues involving the manufacturing of the carbine's unique ammunition. One might wonder if it was Downer's criticism that prompted Morse to explore modifications to the latch and breech block, which ultimately led to the Type II and Type III carbines. Downer reported to Colonel Josiah Gorgas, Confederate chief of ordnance, in February 1863:

> I beg to say in regard to Morse's Breechloader that for a carbine aside from the fault of requiring metallic cartridge cases, and the fault of the cartridge, it is a very clean and serviceable gun. The specimen sent to me is much less complicated that what I have seen formerly, but I do not think that the change has benefitted the arm. I think there seems to be a fault in the gun in the arrangement of the plunger or breech piece in connection with the piece acting as a stop, and the part for raising the stop and opening the Breech. I think there is sufficient play between the stop and the

Breech piece to bring the force of the explosion on to the pins which hold the arms connecting the Breech piece and Stop (This nomenclature is my own). This fault may be seen by snapping the gun, which throws the finger piece of the breech arrangement slightly up.

The Cartridge case I think is the weakest point about the gun. The arrange[ment] for the cap cannot be made (of the material of which the case is made) stiff enough to resist the blows of the hammer yielding to it, so that very often there is not sufficient resistance to explode the cap after the first discharge.

There will also be great difficulty in getting the material for making the wad in the rear or base of the cartridge.

I think the price is fair if the arm is approved.

Please do not understand me as being opposed to the gun on any but the grounds named above. I have often fired the old one and found it, except the faults I mentioned, a very clean, accurate, and pleasant weapon to shoot. But at the same time I think that without a modification of the cartridge, it cannot prove a reliable arm in the service.[72]

Lopez certainly gave the impression that he manufactured only a single example of the carbine when he wrote on December 22, 1862, that he manufactured *a* breech-loading carbine of Morse's patent. He wrote to Bonham again on January 20, 1863, that he had left the carbine in Columbia when he made the presentation to the legislature in December, but, by January 20, it was no longer in Bonham's office. For that reason, Lopez shipped *another* Morse carbine to Bonham along with a Burnside carbine.[73]

When Lopez made his presentation of the Morse carbine to the Military Committee of the Executive Council in late 1862, one of the members, Captain Perrin, had acquired a Burnside carbine at the surrender of Harpers Ferry and loaned it to Lopez for his examination. Lopez had never seen one before, and Perrin wanted him to analyze it and compare the two weapons because he thought that Burnside's carbine was superior to Morse's. Lopez analyzed the two and sent both back to Bonham, along with his written opinions, on January 20.

Lopez left no doubt that he preferred Morse's carbine to Burnside's. He wrote to Bonham that Burnside's would cost 75 percent more than Morse's to manufacture, and cartridges for Burnside's carbine could not be made in any quantities because of the scarcity of materials required. Further, Morse's carbine required only two movements to unload the spent cartridge and reload a new one, and no priming was needed, whereas Burnside's carbine required four movements plus priming, or capping, to accomplish the same task. Lopez discussed the advantages of Morse's center-fire design. Because there was no nipple or cone in Morse's design, the opening in the cone could not become obstructed or corroded. The cap, or primer, which was built into Morse's cartridge, "explodes in the midst of the charge, it can never snap [misfire], the explosion of the cap ensures the discharge of the gun."[74] In comparison, Burnside's carbine had the traditional cap on a cone and was subject to the possibility of misfire. He wrote, "Morse's gun will as certainly discharge under water as it will in a clear atmosphere. Burnside's gun had a chamber around the cone seat which could catch the water, and the hammer did not protect the cone from

corrosion, but freely let the water into the chamber."[75] Lopez referred to Morse's gun having the ability to function under rigorous, often wet, field conditions when he suggested that the weapon could theoretically be fired under water. Lopez closed his letter to the governor by writing, "I am progressing in preparation for the manufacture of the arms ordered but the want of funds to purchase materials of which I wrote your Excellency on [January] 9th delays me very much."[76]

CHAPTER 12

Labor

Securing adequate labor, both in quantity and quality, was a perpetual issue at the State Works. Lopez needed skilled craftsmen such as gunsmiths, blacksmiths, machinists, brick masons, and carpenters. He also needed unskilled laborers to tend the fires, to work as night and day watchmen, and to perform general labor around the armory. He utilized both free men and slaves; many of his carpenters were the latter.

Available information about labor at the State Works in 1862 and early 1863 is scarce. About two-thirds of Lopez's white workforce was civilian and one-third current or former servicemen. Of the 35 known servicemen he used, some were on detached duty, others had been honorably discharged, and even a few were absent without leave. Bonham had a working relationship with Beauregard, who in early January 1863 had written to Bonham, "If however a suitable man can be selected for service in the State Works, either in the Cavalry or Infantry of this Department, and I am furnished with his name, I shall be most happy to have him detailed at once."[1]

Available payrolls offer insight into labor at the State Works.[2] Both Lopez and his successor, J. Ralph Smith, kept separate payrolls for white and black workers. The earliest extant payroll for black workers, both free and enslaved, is titled "Pay Roll of Negro Mechanics and Laborers at the State Works for the month ending March 15, 1863." The earliest known payroll for white workers is "Pay Roll of Mechanics and Laborers at the State Works for the month ending July 15, 1863."[3] The names of 107 white men, 8 free black men, and 116 slaves appear on these rolls, which are available through February 1865. The names of 7 additional white employees come from other sources. There were likely more workers than that, but records for 1862 and early 1863 do not exist.

Free black men held jobs such as carpenter, gun stocker, pattern maker, fireman, and laborer. White men held all jobs, from foreman of skilled laborers to helpers and night watchmen. Many slaves, especially those belonging to David Lopez, worked as carpenters. Some were blacksmiths, brickmen, firemen, and wagoners, but most were laborers and helpers. Between March 1862 and October 1, 1863, the state expended $252,546.03 on the State Works, and its single largest expense, $102,089.72, was for labor and services.

Labor was a statewide problem throughout the war. The Executive Council recognized the problem when it approved a resolution in mid-March 1862. Acknowledging

that some employees of factories and foundries were essential to the public service, the council agreed that, if a man were drafted, he could be withheld from Confederate service if the head of the manufacturing establishment certified that he was essential to its business.[4] Lopez and Smith utilized this option several times.

Prior to October 9, 1862, General John C. Pemberton, commanding the Department of South Carolina and Georgia, frequently detailed gunsmiths to work at the State Works. Pemberton's successor, Beauregard, initially did not feel authorized to do the same. Gist proposed and the Executive Council approved a resolution on October 9 requesting the chief of the Military Department to write to the secretary of war to clarify the question of whom the state must approach in such cases.[5] The issue must have been resolved to Beauregard's satisfaction since several soldiers were later detailed to the State Works. Smith wrote to Brigadier General Thomas Jordan, Beauregard's chief of staff, on December 5, 1863: "Being very much in need of workmen for the manufacture of shell and shot for [the] Confederate Government, I would respectfully ask that Private Edward E. Tavel Co. B, 27th Regiment, S.C.V. (now on duty at James Island) be detailed for service at these works."[6] Citing the basis of his appeal as General Orders #273, Smith repeated his request for Tavel on January 4, 1864, and added an application for another man, W. H. Halsall, stating that the services of both men were badly needed for the casting of shot and shell. Both men were soon employed at the State Works.

The Executive Council also addressed the issue of using slave labor for public works, meeting even more opposition and only modest success. The slave population of the state was a major source of labor for the construction of earthworks in the coastal areas, particularly around Charleston. Bonham was in favor of this practice, but many planters were adamantly opposed to it. Legislation addressing the issue was ineffective, and leading military authorities bypassed civil channels and impressed the enslaved population under their own authority. Near the end of his term, in November 1862, Pickens, who supported the practice, endorsed a plan to provide a permanent force of about four thousand slaves, but it was never adopted.

In spite of ongoing efforts, after the abolition of the Executive Council, the supply of slave labor virtually dried up.[7] Another law, this one dated December 18, 1862, divided the state into four divisions and allowed the governor to requisition as many slaves as were needed for up to 30 days service. Owners were paid eleven dollars per month per slave but were allowed to buy their way out for one dollar per day per slave. The plan was suspended pending the Confederate government agreeing to pay for any slave who escaped to, or was captured by, Federal forces. Bonham then reverted back to the plan as it existed under the Executive Council, and a call was placed on certain districts for the needed labor force. On February 6, 1863, an amendment to the December 18 law allowed its implementation. Bonham faced many difficulties, including finding a state agent to run the system, strong opposition from the slave owners, and a persistent demand from military authorities for more labor. Slave owners complained that their slaves were not treated well, that they were not returned home on time, that the frequent calls interrupted normal agricultural practices, and that some districts were called upon more frequently than others. Though the slave labor force at the State Works diminished as the war progressed, a large number of slaves worked there until the end of the war.

In late 1862 Lopez advertised for skilled labor, both free and slave, in newspapers across the state. In Columbia's *Daily Southern Guardian* of December 31, 1862, Lopez advertised: "Six to eight stockers or carpenters capable of Stocking Guns can find steady employment at the State Works in Greenville, where good wages will be paid. The same Exemption from Conscription will apply as in other Government Works."[8] He advertised for gun stockers in the *Daily South Carolinian* on January 17, 1863, and again on February 17: "To brass workers, an experienced hand, accustomed to light work, can find good employment at the 'State Works' Greenville by application." Appearing in the *Mercury* of March 21, 1863, was the following advertisement:

To Machinists, Finishers and Gunsmiths
    Good employment and wages will be given at the state works, at Greenville, for machinists and gunsmiths. The same exemption from Conscription and Military Duty apply to men engaged in "these works" as in other Government Works. Provisions and living is much cheaper than in other sections of the state.

Another advertisement appeared in the *Mercury* on March 3, 1863:

To Owners of Negro Mechanics
    Wanted at the state work in the interior of this State, in a healthy and safe location, carpenters, blacksmiths, machinists, and wheelwrights. They will be taken care of and good wages paid.

Lopez's successor, J. Ralph Smith, also had labor problems, and he placed an advertisement in the *Tri-weekly Carolinian* on October 17, 1863: "Wanted: at these works good machinists and gunsmiths to whom liberal wages will be paid. Good references required."[9]

A special session of the legislature in April 1863 resulted in little more than strengthening the fines associated with the slave labor system. In the face of an imminent attack on Charleston during the first half of 1863, Beauregard requested 3,000 slaves but received only about 330. The Federal attack on Morris Island on July 10, 1863, precipitated a new crisis around Charleston Harbor, to which Bonham responded by appealing to slave owners to voluntarily send a work force, and he authorized the mayor of Charleston to impress free blacks. The result was that 2,850 slaves and free blacks were available for work by the end of July.[10]

This temporary success lasted only about a month; by the end of August, Beauregard's labor force had once again dwindled. He implemented a plan, approved by Bonham, to impress slave labor, and by the end of September his workforce was back up to 2,850. Another special session of the legislature met on September 21, 1863, and considered Bonham's request that fines be abolished, that the governor be allowed to impress slaves into service when he deemed necessary, that the length of service be increased to two months, and that free blacks be impressed into service. Rebuffing the governor, the legislature's only action was to increase the fine to two hundred dollars. Another new law on December 17, 1863, abolished the system of fines and required sheriffs to arrest and deliver those slaves who failed to report. Public opposition remained strong, generally supporting sheriffs' unwillingness to strictly enforce the plan. The problem of an inadequate labor force persisted throughout 1864.[11]

The state was finally prodded into action by the Confederate government. Circulars issued by the Bureau of Conscription on September 23 and December 12, 1864, called for conscription officers to proceed under the Conscription Act of February 17, 1864, to impress for twelve months' service South Carolina's quota of 20,000 slaves needed for military labor. On December 23, 1864, the legislature passed yet another new law that finally allowed the governor to impress one-tenth of all male slaves between eighteen and fifty for twelve months' service.[12]

Black Labor

In March 1863, 4 free black men and 69 slaves worked for Lopez at the State Works. By July there were only 3 free blacks and 50 slaves. Based on available records from March 1863 to January 1865, the number of free black men employed at the State Works remained constant between 2 and 4. In contrast, the number of slaves was highest in March 1863, but steadily fell off by that summer. Between the summer of 1863 and the summer of 1864, the number of slaves fell from 50 to 35, but it rose to 49 by January 1865. At least 8 free black men were employed at the State Works during it existence. Two, Henry Bulkley and Joseph Huggins, were carpenters from Charleston who likely had an association with Lopez before and after the war.

TABLE 1. Free Black Employees

Henry Bulkley	gun stocker	March 15, 1863, to July 15, 1863
William Black	pattern maker	March 15, 1863
Zion Collins	fireman	March 15, 1863, to 1865
Joseph Huggins	carpenter	March 15, 1863, to January 15, 1865
Baylas or Baylus Holley	fireman	January 15, 1864, to August 15, 1864
Zack Holly	laborer	July 15, 1864, to January 15, 1865
Reuben Wilson	laborer	January 15, 1865
John Arthur	laborer	January 1865

SOURCE: Comptroller General State Auditor, Records of the State Works, Payroll (Negro), "Pay Rolls of Negro Mechanics and Laborers," South Carolina Department of Archives and History, S126182. Available payrolls are: March 15, 1863; July 15, 1863; August 15, 1863; November 15, 1863; January 15, 1864; February 15, 1864; March 15, 1864; July 15, 1864; August 15, 1864; October 31, 1864; December 31, 1864; January 31, 1865.

No fewer than 116 slaves were employed at the State Works between 1862 and 1865. Typically, between 35 and 70 slaves worked there in any single month. Of the 69 slaves employed at the State Works in March 1863, 21 were carpenters, 6 of whom were owned by David Lopez and another by his son, John. Thirty-six of the slaves were laborers or helpers, 4 were blacksmiths, 3 were brickmen, 3 were firemen, and 3 were wagoners.

By July 1863 the number of slaves had fallen to 50, of whom 16 were carpenters, 27 laborers or helpers, 3 blacksmiths, 2 firemen, and 2 wagoners. For most of the year 1864 the number of slaves at the State Works in any single month was in the mid-30s; the decrease was attributable to the fact that there were only about 6 carpenters employed. Even when the number of slaves increased to 49 in January 1865, there were only 5 carpenters, 3 blacksmiths, and 1 fireman; all the rest were laborers and helpers.

No slave worked more than twenty-four days during the thirty-one-day pay period of March 1863, and in July both free men and slaves worked no more than twenty-five days. Throughout the war, most white workers were full time and put in between twenty-five and twenty-seven work days per monthly pay period. A few worked less than that. Some of the white laborers and night watchmen worked thirty or thirty-one days per pay period. If slaves worked overtime, their masters were paid an overtime allowance. Such was increasingly the case in mid-1864, reflecting either fewer workers or a heavier workload, and the practice continued in 1865.[13]

The pay scale varied widely between jobs for all workers. In 1863 a free black fireman made between $1.00 and $1.25 a day, whereas a slave fireman was paid between 75 and 96.25 cents a day. A free black carpenter made $1.25 a day, whereas a slave carpenter was paid between 50 cents and $1.50 a day. A slave blacksmith made $1.15–1.25 a day. A free black gun stocker made $1.50 a day, and a free black pattern maker made $2.00 per day. Slave laborers were paid only 46.25 cents per day. In contrast, white workers made substantially more. Machinists were paid between $1.50 and $5.00 a day, with most being paid $3.50 to $4.00. Foremen were paid $4.00 to $5.50 a day. Carpenters made $2.00 a day, a wheelwright made $3.00 a day, and store keepers and clerks made about $2.60 per day. By late 1864 most of the hourly wage rates had doubled. The highest-paid salaried employee was Lopez, whose initial salary was $3,000 per year. In July 1863, the highest-paid hourly employee was the master machinist and soon-to-be general superintendent, J. Ralph Smith, who made $8.33 a day. Other high-wage earners were Julius C. Smith, the State Works' traveling agent, who made $50 a month in 1863 and $150 a month at the end of the war, and William Walton Smith, chief clerk, who made $66 a month in 1863 and $200 a month in early 1865.

Slaves appeared to have been well cared for with adequate housing and food. J. Ralph Smith, general superintendent of the State Works after August 1863, wrote to Colonel A. P. Calhoun, who lived in Pendleton, about one of his slaves, Simon, on October 18, 1864:

> I received your note a few days since. I have no objection to keeping your man "Simon," but as he appeared so homesick, and anxious to visit his home I thought you might exchange a younger fellow for him. Your boys seem very much satisfied (at least I have no complaints) but they beg me to write to you for their shoes, as the old ones are pretty well gone.
>
> Can you not let me have five or six more boys who can cut wood? I need a few more and would prefer them from you. I have applications from others, but have declined hiring until I hear from you, as boys who have worked together are more apt to agree than strangers. Since my last, I concluded to cut closer to the Works, and will require more hands in order to have wood on hand for next summer. If you can accommodate me I shall feel obliged.[14]

In another undated letter to Calhoun, Smith wrote: "Your boy Simon is very anxious to get a place here for his wife. If she is a good servant and a good washer I would like to have her, and could make her comfortable at my house. If she should not be the servant I want perhaps one of the other boys may have a wife that would suit. I want a woman with children."[15]

White Labor

Eleven payrolls exist for white employees, extending from July 15, 1863, to February 28, 1865.[16] The skilled labor pool was remarkably stable during that time period. Most of the men who resigned during that time were lesser skilled or unskilled labor. On any given month between July 1863 and February 1865, between sixty-four and seventy-five white men worked at the facility as it operated at its peak output. As the war progressed, the number of white workers slipped from the mid-seventies to the mid-sixties, but about seventy white men were working there in mid-March 1865.[17] Each department had a foreman, apprentices, and general helpers. The eight general areas of employment were the office and clerical staff, small arms (Morse) shop, gun-repair shop, carpentry shop, machine shop, foundry, blacksmith shop, and, finally, laborers and watchmen.

The office and clerical staff originally consisted of seven men plus the general superintendent, but for most of the war there were only five, and in February 1865, only four. When Lopez was general superintendent, his son, Moses, was employed in the office. Five other men worked in the office from the earliest payroll, dated July 15, 1863, until the last available payroll, dated February 28, 1865. Julius Clarence Smith, a thirty-three-year-old Charleston merchant, was the traveling agent for the State Works, tasked with locating and purchasing items necessary for its operations.[18] John A. Michel had done the original survey work at the site and was a draftsman in the office. Daniel S. Hart, Lopez's son-in-law, was the time keeper and a store keeper. William Walton Smith, a thirty-eight-year-old Charleston cotton factor, was the chief clerk.[19] E. E. Pritchard, who had been a private in the First Regiment, South Carolina Artillery (Militia), in September 1861, was a store keeper.

George T. Brooks was foreman of the small arms shop, which manufactured Morse's breech-loading carbine and inside-lock weapons. The men who worked there were all machinists and apprentices. Morse called Brooks his "master armorer," and he was still at the State Works in February 1865. In July 1863 there were twenty-three men in Morse's shop, but for the rest of the war there were only sixteen or seventeen. J. J. Mackey, a local gunsmith, was foreman of the gun-repair shop, also called the musket-repair shop. The men who worked there were gunsmiths and gun stockers. The gun-repair shop had only nine men in July 1863 but employed between fourteen and sixteen for the rest of the war. Mackey was still at the State Works in February 1865. In mid-July 1863, forty-seven-year-old John W. Sawner, a master carpenter from Charleston, was foreman of the relatively small carpentry shop, which performed both carpentry and wheelwright work.[20] Sawner was still there in February 1865. The carpentry shop had only two or three men for the entire existence of the State Works. E. T. Miller was foreman of the machine shop and remained there until February 1865. The machine shop started out with nine machinists

and held steady at that number until the summer of 1864, when it fell off to five by the end of the war. Phillip A. Mullane was originally foreman of the blacksmith shop, but he appears to have lost that job sometime prior to early 1864. The blacksmith shop had between ten and twelve men until it was combined with the foundry in June 1864. From that time to the end of the war, the two combined shops employed between fourteen and seventeen men. Robert McKay was foreman of the foundry, but he left the State Works before mid-June 1864. The foundry employed between six and nine men until mid-June 1864, when it was combined with the blacksmith shop under a single foreman. James Fraser was foreman of laborers and watchmen, and he, too, lasted until February 1865. There were between six and eight men in this department between 1863 and 1865.

In summary, no fewer than 240 men were employed at the State Works, and there were probably more than that because records for 1862 and early 1863 do not exist. One worker in the gun-repair shop, G. W. Taylor, gave a newspaper interview in 1884 and stated that between 300 and 400 men had been employed there.[21] See appendix 3 for a list of slave workers and appendix 4 for a list of white employees at the State Works.

CHAPTER 13

1863

The first six months of 1863 presented many challenges for Lopez as he attempted to accomplish the mission of the State Works. Hiring and maintaining adequate help was a perpetual problem, financial concerns persisted, and disagreements over his management of the State Works ultimately resulted in Lopez's resignation as general superintendent that summer. He also faced personal problems. His twenty-seven-year-old invalid son, David, died on September 11, 1862, and was buried at Coming Street Cemetery in Charleston.[1] In addition, another son, John, was unhappy at the State Works and was preparing to depart for better opportunities on active duty with the military.

John Hinton Lopez, a lawyer by profession, served in the Palmetto Guard in 1861 and possibly early 1862, and like his brother, Moses, joined his father at the State Works sometime in 1862. By early 1863 John wanted to leave the State Works and enter active duty, and his father tried to assist him in finding a spot in the military. David wrote to the secretary of war, J. A. Seddon, on February 25, 1863, requesting a position for John in the Signal Corps:

> John H. Lopez, who was a member of the Palmetto Guard at the storming of Fort Sumter, and who continued in after [sic] service was detailed by request of [the] governor of [the] state of So Ca for duty at the State Works where he has been usefully employed up to this time. He now desires to enter the service, and respectfully asks that the time employed in the State Service may not redound to his disadvantage— which it will if compelled to enter the ranks on the road to promotion which was then open to him now lost by said service. I respectfully ask for a position in the Signal Service which is the only one now open to him upon which he feels fully qualified having been educated at the Citadel Academy in So Ca. He respectfully refers to Col. James Chesnut Jr. and Hon. I. W. Hayne.[2]

David also enlisted the assistance of Isaac W. Hayne, who was attorney general of South Carolina and had been a member of the Executive Council. Hayne wrote to Colonel James Chesnut Jr. on February 28, 1863, requesting an appointment for John in the Signal Corps:

My Dear Chesnut

Mr. David Lopez has a son Mr. John H. Lopez who desires a place in the Signal Corps. From the qualities of the family and my knowledge of this young man I have no doubt he would be eminently qualified for service in that line. He has been in the army and was withdrawn to serve in the State Works and now wishes to return to more active duty. He prefers the Signal Corps to a place as a private in ordinary service and nothing else is now open to him. Any assistance you can give I will esteem a favor to myself.[3]

Unfortunately for John, there was no vacancy, and he did not receive an appointment in the Signal Corps. Instead, he left the State Works and joined the Palmetto Guard Artillery, Company A, of the South Carolina Siege Train, as a private on March 18, 1863, mustering in for the duration of the war. His mechanical skills were put to good use as an artificer and carpenter for the company. On January 12, 1864, John was granted a twenty-day furlough to visit his severely ill wife, who was living in a house they had purchased at 506 Augusta Street in Greenville. He was transferred to the Second Engineer Regiment on September 17, 1864. In a letter recommending his detachment, he was described as a "mechanic of the first order," "a very practical and useful man," and "an energetic and skilled workman."[4]

The State Works remained active in early 1863, repairing and altering small arms, and fabricating articles. Some of the articles were turned over to the State Arsenal in Columbia and others to the ordnance officer of the state.

In February 1863, the legislature passed a special appropriation of fifteen thousand dollars for the construction of one thousand Morse carbines at the State Works, and the workers in Morse's shop began to manufacture components. By the end of September 1863, Morse's workmen completed the first batch of one hundred carbines, at an expense of six thousand dollars to the state.[5] The legislature's appropriation of February 1863 ultimately achieved its goal, but not until early 1865.

Prior to this time, Lopez manufactured at least two of Morse's carbines at the request of Governor Pickens and the Executive Council. Based on Lopez's description of these as patterns, they were probably preproduction demonstration models, but we do not know if they were stamped with a serial number. A third carbine was also made about that time and sent to Richmond; likewise, we do not know if it was stamped with a serial number. Many Morse researchers have suggested, incorrectly, that as many as 200–300 of the carbines were made in Atlanta before the manufacturing process was moved to Greenville. The *Intelligencer* of Atlanta reported on December 13, 1862, that one Morse carbine was made by H. Marshall and Company in Atlanta. Some authors suggest that a serial number stamped using a large die, or font, indicates an Atlanta-made carbine and that small-die serial numbers were made in Greenville, but there is no substantiation for this theory. Most likely, only one Morse carbine was made in Atlanta; there are several arguments against the large-scale manufacture of the Morse carbine there. The strongest is that the South Carolina legislature ordered 1,000 carbines in January 1863, and records from the State Works show that 1,000 were produced by early 1865. In early 1865 the legislature ordered another 1,500, and carbine production continued at the State Works. Based on the fact that more than 1,000 were made at the State Works and that the highest surviving

serial number is 1,032, it is extremely unlikely that there was a significant production run in Atlanta. Another argument is that Morse and his machinery were not in Atlanta long enough to make any carbines in large numbers. Morse's July 6, 1863, letter to Tupper mentions his salary in Tennessee but none in Atlanta, arguing against his being seriously involved in Atlanta. The machinery was evacuated from Nashville to Atlanta in February 1862 and forwarded to Greenville about May, not enough time to set it up and get it into operation in Atlanta. Indirect evidence that all carbines were manufactured in Greenville except the single prototype is that the State Works manufactured or outsourced 1,079 carbine slings between late 1863 and early 1865, a number that corresponds closely to the full production run of carbines. Further, if carbines were being made in large numbers in Atlanta in late 1862, the South Carolina legislature could have simply evaluated one of those and would not have needed Lopez to make one for them to examine.

The South Carolina state auditor, James Tupper, requested an inventory and appraisal of the State Works, which was done on April 1, 1863. Included in this assessment were buildings, machinery, tools, material, stock, stores, and provisions of all kinds. The aggregate value of the property was $283,464.57.

A list of items that Lopez purchased for use at the State Works during the second quarter of 1863 survives.[6] Lopez called this an "abstract of items purchased," and it reflects the wide variety of items necessary for operating the facility, including raw materials and chemicals, office supplies, and daily living items such as food and cloth. Raw materials purchased included twelve thousand pounds of cast iron, one thousand pounds of bar iron, and eleven thousand pounds of castings. Lopez also required a variety of wood types, including oak, walnut, ash, and poplar, as well as hickory to be used as handles for the machines, such as trip hammers. Most sources state that the stock of Morse's carbine was made of butternut, which is in the walnut family. The necessary length of time to cure wood for use as a gun stock was considered to be three years, a luxury that neither Lopez nor Morse enjoyed.[7]

Lopez also required several chemicals, including soda, borax, muric acid, copperas, and ammonia. To operate the facility he had to buy charcoal at ten cents a bushel and pine wood at one dollar per cord. Lopez had apparently nearly completed his stock of tools and machinery by mid-1863. His purchases of such things during this time were minimal, but by the middle of the third quarter this would change as he bought a dozen files, some hooks and staples, rope, treble blocks, and castings for a screw machine. Indicating that there was still some construction going on at the site is the fact that Lopez bought fire bricks and fire clay, shingles, pine lumber, and window glass.[8] He obviously had to purchase food for the workers, including 572 pounds of fresh beef, 313 pounds of bacon, corn, flour, lard, and fodder for animals. Office and personal items included hairbrushes, envelopes, paper, ink, and crucibles. Lopez's most interesting purchase was a Sharps breech-loading rifle. Why he needed it is open to speculation. Possibly he wanted it to compare its material or construction technique with that of Morse's carbine.

A list of items fabricated at the State Works during the second quarter of 1863 also survives.[9] Lopez divided the list into two general areas: (1) those items manufactured by the "Ordnance service in all its branches" and (2) the "Manufacture of Morse's breech loading carbine."[10] The latter list was very short; only two machines dedicated to the production of Morse's carbine—a rolling machine for making cartridge cases and a punch

machine for punching cone seals—were made in Morse's shop between April and June 1863. Lopez's list documents that no Morse carbines had been mass produced at the State Works through the end of June 1863, but activity was progressing toward that goal. Morse personally owned a considerable number of the small milling tools used in making his breech-loading carbines, and on May 6 Lopez contracted with Morse to pay him three hundred dollars for the use of the tools to be used in the manufacture of one thousand carbines. The payment actually took place on July 21.[11] Fourteen months had passed since Lopez negotiated the use of the Nashville machinery from Governor Harris, and only two or three demonstration examples of Morse's carbine had been made at the State Works by mid-1863.

Lopez's list of items manufactured by the "Ordnance service in all its branches" in April, May, and June 1863 was impressive. Numerous items, especially wheeled vehicles, were manufactured at the State Works during that time. One four-horse wagon, two six-pounder field carriages and limbers, seven sponges and rammer staffs for six-pounders, and three gun-worm rammers for six-pounders were made. One four-pounder field carriage and limber, one sponge and rammer staff for a four-pounder, and one gun-worm rammer for a four-pounder were also made. Other articles were one caisson and limber, six maneuvering hand spikes, and six spare poles for field gun carriages. Sixteen field gun carriage wheels, ten ammunition chests, 131.5 pounds of railroad spikes, and 750 feet of railroad were also made. Completing the list were three hand lathes, three milling machines, and a four-spindle drill press. Of note are the items not produced at the State Works. No shot, shell, or shoulder arms had been made, and no cannon cast during the second quarter of 1863.

Finally in the summer of 1863 the State Works began to fulfill its original mission when it began to manufacture several new items. On July 11 the State Works began some work for the Confederate government for which it was to be reimbursed. This output included casting shot and shell, fabricating wrought-iron bolts, and making castings for Columbiad carriages.[12] Also included were the manufacture of gun carriages and various machines necessary for the armory, and the construction of parts for Morse's carbine. Except for the prototype carbines that Lopez sent to the legislature in late 1862, to Bonham in January 1863, and to Richmond in late 1862 or early 1863, this is the first documentation that the carbine was being made in Greenville.

Lopez took the train to Charleston to visit Bonham on official business on July 27, but its exact nature is unknown. On July 28 Lopez purchased 102 bayonets from Elijah Hall, who was a sixty-eight-year-old blacksmith from Lexington District. It is unclear if Lopez intended to use the bayonets on weapons being repaired or manufactured at the State Works or if he was planning to melt them down and recast them. Lopez paid Samuel Adams for services rendered in Richmond on behalf of the State Works on July 31. Adams examined a casting-gun screw machine and supervised packing and shipping it to Lopez in Greenville. Also, Adams bought tracing cloth and drawings for a double drop hammer, a screw-cutting machine, a profiling machine, and a dozen crucibles. It is not known if these machines were intended for use in making Morse's carbine or for other purposes. Lopez's total expenses for the month for July 1863 were $14,621.54.[13]

Lopez provided a detailed accounting of his receipts and expenses between January 1 and August 19, 1863. The single largest source of receipts was $84,000 provided by

the South Carolina Treasury Department. The next single largest source of income was $6,000 gained from the sale of provisions to the workmen. Lopez sold miscellaneous unneeded pieces of stock and machinery that generated about $2,000, and, that combined with $5,600 cash on hand, gave the State Works a net income of $97,858 for the first eight and a half months of 1863. Lopez's disbursements for the same time period were $95,578, leaving him with a cash balance of $2,339. His largest expenditures were $46,000 for payroll and salaries and $26,000 for stock. He spent $7,000 on provisions, $7,000 on building materials, $1,000 on fuel and coal, and $3,000 on incidental expenses.[14]

PLATE 1 A one-of-a-kind presentation cased set made by James J. Mackey at the South Carolina State Works in Greenville in 1864. The receiver is a standard Type III Morse carbine, but, unlike the military production carbines, this example has three interchangeable barrels. Governor Milledge L. Bonham purchased it for $345 shortly after leaving office in December 1864. Courtesy of McKissick Museum, University of South Carolina.

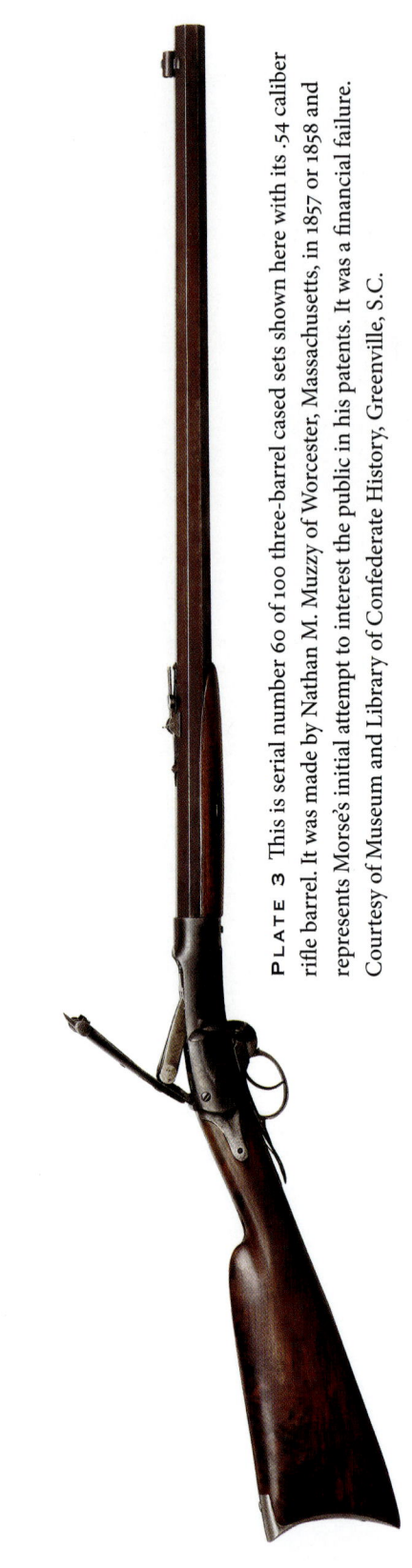

PLATE 2 Prototype of a Morse sporting rifle is nearly identical to the patent prototype and was most likely made in late 1855 or early 1856 by Daniel Searles, a Baton Rouge gunsmith. Courtesy of Buffalo Bill Center of the West, Cody, Wyoming. Gift of Olin Corporation, Winchester Arms Collection, 1988.8.1510.

PLATE 3 This is serial number 60 of 100 three-barrel cased sets shown here with its .54 caliber rifle barrel. It was made by Nathan M. Muzzy of Worcester, Massachusetts, in 1857 or 1858 and represents Morse's initial attempt to interest the public in his patents. It was a financial failure. Courtesy of Museum and Library of Confederate History, Greenville, S.C.

PLATE 4 David Lopez Jr. (1809–1884) was a respected Charleston contractor and pillar of the city's Jewish community. Lopez built the South Carolina State Military Works and was its first general superintendent, from February 1862 to July 1863. Portrait (oil on canvas) of David Lopez Jr. by Isabel Cohen Doud, 1938. Courtesy of Kahal Kadosh Beth Elohim Congregational Records, Special Collections, College of Charleston Library.

PLATE 5 Morse's prototype carbine, most likely made by H. Marshall and Company, an Atlanta sword maker, in late 1862. It does not have a serial number and is on display at the Greensboro Historical Museum in Greensboro, N.C. Courtesy of the Greensboro Historical Museum Collection.

PLATE 6 *From top to bottom:* serial number 135, a Type I that was converted to a Type II; serial number 715, a Type III; serial number 865, a Type III; rifle from the George Morse–designed, Nathan Muzzy–built three-barrel cased set serial number 60 of 100. Courtesy of Museum and Library of Confederate History, Greenville, S.C.

PLATE 7 Morse's carbines were issued to selected South Carolina mounted troops with this cartridge pouch that held twelve cartridges. A surviving example at the Atlanta History Center shows two cartridge pouches mounted on a cotton web belt, but the men of Percival's Aiken Mounted Infantry were issued one cartridge pouch per man in November 1864. The carbine also came with a sling, cartridges, bullet molds, loading tools, chargers, and wad setters. Courtesy of Museum and Library of Confederate History, Greenville, S.C.

PLATES 8 AND 8.1 Two views of a Type III Morse carbine, serial number 715, made at the South Carolina State Military Works in Greenville in 1864. Courtesy of Museum and Library of Confederate History, Greenville, S.C.

CHAPTER 14

Lopez's Resignation and Successor

Lopez resigned from his position as general superintendent of the State Works in late June or early July 1863. No letter of resignation is known to exist, and his exact reasons are unknown, but preserved documents show that tension between Lopez and his superiors was brewing in the spring of 1863. Lopez and State Auditor James Tupper disagreed about Tupper's accounting practices, and Bonham implemented major policy changes that aggravated Lopez. In addition, Morse did not respect Lopez's ability to manage the State Works, a situation that certainly created additional problems.

Clues to the reasons for Lopez's resignation are imbedded in several communications. Lopez wrote to Tupper on April 8, 1863, in an irritated tone implying that he was not receiving operating funds in a timely manner. Lopez reminded Tupper that on March 16 he made a requisition on the funds of the auditor's office to be able to meet the liabilities of the State Works for the quarter ending March 31. As of April 8, Lopez had received neither the money nor acknowledgment of Tupper's receipt of the original request. Lopez wrote, "As the time has passed for payment of many of the accounts, I beg to draw your attention again to the subject."[1] Letters between the two men crossed in the mail. Tupper had written to Lopez on April 7, returning Lopez's March 16 requisition on the grounds that Bonham required that it comply with Army Regulations. When Lopez received Tupper's letter from April 7, he replied on April 9:

> I beg you will say to his Excellency and be yourself assured that my great desire is to conform in that particular as in everything else appertaining to this department to every wish or order he [Bonham] will issue and if I do not readily fall into the formula of the "Army Regulations," it is because I have no military man and this department has heretofore never been put upon that footing. Since it is the wish of his Excellency I will do all I can to keep the accounts and make requisitions as he directs. To do this suddenly is impossible, yet it shall be done so soon as the nature of the case will admit. To do it more readily I propose to close all our accounts to the 1st instant

[April 1, 1863] and to begin from that time in the manner you indicate. As I informed you through my son [Moses E. Lopez] our clerical department would require increasing. I have employed an additional clerk, who entered upon his duties, and we are now busily engaged in placing the books in accordance with "Army Regulations." Yet I must draw your attention to the fact that these "Army Regulations" are observed with regularity in an old government on a peace footing (or even in war) with the system thoroughly perfected and all the departments working harmoniously but it cannot be done with that strict requirement laid down at a time like this, with no markets to buy in, or rather with no materials in the markets and no fixed value to base an estimate on. I may today make estimate for an article and before the funds are received, the article will have advanced one hundred percent, and this is of no uncommon occurrence.[2]

Lopez understood that one of the governor's new requirements was that he set up separate accounts—such as payroll, fuel, and railroad freight expenses—and that he make an estimate of items and funds required for each upcoming quarter. Lopez went on to give Tupper a hypothetical example of the problem as he saw it, and he asked Tupper for his recommendations on how to approach solving it. Lopez asked if, in the event a certain account expended all of its funds for any given quarter, would he be allowed to use funds from another account with a positive balance to support the underfunded account. Lopez wrote, "As I before observed the advanced estimate as per Army Regulations are based upon a regular state of things with a market of standard prices."[3] Lopez asked if a necessary item suddenly came on the market but was not currently budgeted for, was he allowed to purchase it using funds allocated for items in other accounts. He wrote: "These questions I ask for my future guidance, heretofore I was directed to exercise that discretion and have frequently when I heard of articles needed, even at distant points, sent and purchased, thus keeping up our stock rather than get out and be forced to go on the market to pay exorbitant rates. In this way I have collected valuable stock which if not replenished when opportunity offers will not be replaced when exhausted at any but extravagant rates, if at all."[4]

Lopez finished his lengthy letter of April 9 to Tupper saying that he was withdrawing his quarterly requisition of April 8 and submitting a new one for payrolls only so the workmen could be paid on time. Lopez asked Tupper to lay the matter before the governor and respond at his earliest convenience.

Lopez sent Tupper copies of his "Abstract of Disbursements," "Accounts Current," and "Statement of Receipts and Expenditures" for the first quarter of 1863 on April 23, and all seemed well.[5] Renewed friction arose when Lopez wrote to Tupper on April 27. Tupper had sent a check for six thousand dollars to Lopez on April 23, along with duplicate receipts for Lopez's signature. Lopez received the letter on April 25. The duplicate receipts were dated April 20, and Lopez requested in his letter of the twenty-seventh that Tupper change the date on the receipts to the twenty-third for two reasons. The first was so that the dates on both the receipts and the check would correspond, leaving no room for doubt to be cast if the issue were to come up in the future. The second reason was that Lopez's workmen had been due their monthly payment on April 15, and they were embarrassed and had complained about it to him because they had always been paid on time in

the past. Because their payment was delayed until the twenty-third, Lopez was concerned that, if they learned that the receipts were dated April 20, it might appear that Lopez had held up their payment for three additional days.[6]

The issue spilled over into May, and Lopez's tone became more caustic as he continued to disagree with Tupper's practices. On May 6, Lopez wrote to Tupper, who had suggested that, in the future, Lopez change the dates on receipts to conform with the dates on the checks. Lopez said that he did not feel at liberty to alter the date or amount on any document that had already been filled out and sent to him for signature. He did, however, suggest that Tupper send him blank forms so that he could make the receipts "conform to the facts, but you will see that a change of date or any figure is an awkward thing and I had supposed you would yourself have objected to it."[7] In his last known communication with Tupper, on May 8, Lopez enclosed his requisition for funds to cover the payroll for the month ending May 15, and his tone seemed to be somewhat less acerbic. Lopez wrote that his intention in submitting the requisition was to avoid a delay in paying the workmen, which had caused so much dissatisfaction among them in mid-April. Lopez said that he would be in Columbia on May 11 and would "call on you with the statements required by His Excellency the Governor, and hope to make such arrangements as to conform to all the prescribed forms of your office in reference to Estimates and Requisitions and Accounts of these Works."[8] Apparently Lopez could not tolerate his differences with Tupper any longer, and, after having implemented Bonham's new requirements for only one quarter, he resigned as general superintendent of the State Works at the end of the second quarter of 1863, certainly no later than July 6.

Morse was relieved when Lopez resigned. After working together for a more than a year, Morse, in his typical unabashed manner, made it clear in a July 6 letter to Tupper that he did not hold Lopez in high esteem for his ability to manufacture Morse's firearms. Morse was accustomed to working with men at the national, and even international, level, including superintendents of U.S. national armories, governors, secretaries of state, and representatives of foreign governments. Morse felt misled by and superior to Lopez, but was only slightly happier with his successor, J. Ralph Smith. He complained that his position under Smith was essentially unchanged, and he preferred that either he or an "outsider" be appointed as Lopez's replacement. Morse's self-serving letter to Tupper on July 6 reads:

> From what I can understand of the arrangements in contemplation here, I cannot but feel some dissatisfaction, and deem it best to express it at once. When I made my contract with Mr. Lopez as agent of this State, for the transfer of the Tennessee State Armory machines, tools, stock, etc. etc. it was agreed that I should be superintendent of the Small Arms Dept., having my master armorer with me together with many of my men, all of whom were to be under me, as they had been at Nashville. I had my draughtsman and patternmaker, and and [*sic*] expected to control my own business.
>
> When I found that I could not have such controls under Mr. Lopez, I demurred, and after a time, he has resigned. Now that Mr. Smith is appointed, I do not find that my condition is changed from what it was under the former supt., except that I have a man whom I esteem to deal with. Still, with him I am constrained, as I was before. If a bureau officer were appointed, as I suggested, the more he might know of the interior,

or working part of the business, the better, altho such knowledge is not essential, but still I asked to be the interior officer, as by taste, and inclination, the position would suit me; but to remain in the position in which Mr. Lopez leaves me, let who will be Genl Supt, I cannot consent to. I have more experience and skill, in all the work that is to be carried on, than any other available man. I have never before held any subordinate position, and those which I have held for many years at home, are far above any that the state has here to bestow. It cannot be said that my conduct towards this state has been of a mercenary character, nor that I have not always had her interests at heart. It was through my exertions that the state obtained nearly all of the small arms machinery, stock, tools, etc., etc., which she now possesses. She is now manufacturing my inventions, and I have made no trouble about remuneration for anything which I have done, and have rather held back, than otherwise, when the Governor advised me to present my claims.

At my own suggestions, I reduced my salary from $3500 per annum, as it was in Tennessee, to $3000. Had an outsider been appointed, my amour propre, could not have been affected, but to be thus overslaughed, is somewhat galling. All that I ask now is that my position in these works should be made such that I can hold it without the loss of my self respect, or the consideration of my friends. I am fully aware, that I stand next in line of promotion, in case an outsider was not to be appointed, as I had suggested. I think that you have received unfavorable impressions of me, or my management of business, which I could clear up, if I had an opportunity. Be so kind as to lay this letter, before His Excellency Gov Bonham.[9]

The Bacon Incident

Lopez resigned sometime before July 6, 1863, but he remained on the job for another six weeks. His last day as general superintendent was August 19.[10] The remarkable accomplishments of his eighteen-month tenure at the State Works were marred by an unfortunate event involving Lopez, Lopez's son Moses, and employee E. A. LeBlanc that took place just a few days before Lopez's last day of work at the facility. LeBlanc was a machinist who worked in Morse's small arms shop. On the afternoon of August 14, David and Moses Lopez were working in the front office when LeBlanc entered to request a refund on a fifteen-pound shoulder of bacon he had purchased at the commissary on July 25. Because of the ensuing incident, LeBlanc submitted a written complaint to Governor Bonham later the same day. Bonham delegated the handling of the affair to Tupper, who forwarded it for investigation to J. Ralph Smith, Lopez's successor as general superintendent. Affidavits concerning the event from the following five men survive: E. A. LeBlanc, E. E. Pritchard, Phillip A. Mullane, Daniel S. Hart, and William Walton Smith. A sixth witness was John Love, one of the day watchmen, and he was either not asked for a statement or it was not preserved with the others. J. Ralph Smith took the affidavits, investigated the bacon incident, and reported back to Tupper on August 21. The statements of all five men agree in general terms concerning the facts of the incident, but they disagree on some specific details.[11]

LeBlanc pleaded for justice in his complaint to the governor on August 14. He stated that he was a workman at the State Works who had purchased the bacon at the commissary, and he accused Lopez of swindling him. LeBlanc claimed that he had protested that the bacon was bad when it was issued to him on July 25, but that the store keeper told him to try it and bring it back if it did not suit him. LeBlanc wrote that he tried it but could not eat it because it was so musty and rotten, and he took it back to the commissary for a refund. LeBlanc's report gives the impression that he returned the bacon a few days before the incident with Lopez on August 14. LeBlanc said that Lopez refused to take it back, stating that it was good enough for anyone to eat and that LeBlanc must keep it. LeBlanc refused and left the bacon with the store keeper, who hung it in the storeroom. According to LeBlanc, within a few days it had fallen apart due to maggots and rot. W. W. Smith would later disagree with this conclusion. Leblanc wrote to the governor that, because he was unable to afford the financial loss, he went to Lopez's office on August 14 and asked in a gentlemanly manner for another piece of bacon. LeBlanc wrote that, as soon as he stepped into the office, Lopez turned in his chair and asked in an abrupt manner,

> What! do you want. Says I, Mr. Lopez I come to see if you will change my meat or not charge me for it. Says he, no damn you I won't! and told me to clear out. I said I came to have my rights and I thought that I ought to be satisfied as he had changed meat for others he ought to change it for me and that I did not like to be swindled by anyone I worked hard for a pitiful $3.00 per day to support my family. He ran to me to strike me. His son Moses grasped a double barrel shotgun and threatened to shoot me had it not been for the foreman blacksmith [Phillip Mullane] that jumped in between me and the muzzle. They called me all the Sons of B—hs that they could lay their tongues to and David Lopez distinctly told me before Wm. Walton Smith, Mr. Hart, Mr. Love, Mr. Pritchard, and Mr. Mullane that he intended to swindle me out of the $15.00 for what he called my impudence. He also caused my name to be stricken off the pay roll and discharged me from the State Works.[12]

William Walton Smith's statement was the most thorough of the four witnesses who gave affidavits. Smith was a thirty-six-year-old office clerk who had been a cotton factor in Charleston before the war and was one of the few salaried employees at the State Works making sixty-six dollars a month. His statement was undated, but it was included in J. Ralph Smith's papers forwarded to Tupper on August 21. W. W. Smith wrote:

> On the afternoon of the 14th August 1863 Mr. E. A. LeBlanc, an employee at the State Military Works, came into the office of Mr. Lopez, Genl. Supt., and inquired of him if he was going to charge him with that meat (referring to a piece of bacon he got from the commissary on the 25th July 1863 and which he said was not good.) Mr. Lopez replied that if it was charged on the books it would stand so. LeBlanc says, you intend to swindle me. Mr. Lopez relied, yes, I intend to swindle you, you D—n Son of a B—, clear out of this office. Mr. Lopez then ordered Mr. Hart the Time Keeper to strike Mr. LeBlanc's name off the rolls. LeBlanc then said, I ain't the first one you have swindled. Mr. Lopez then walked towards the office door, LeBlanc standing just outside on the Piazza. Mr. Lopez looking as if he wanted to strike. At this moment Moses Lopez stepped to the corner of the office where stood a double barrel gun loaded, and

took it up, saying, you D—n Son of a B— I'll blow your brains out, but did not point the gun at LeBlanc. Mr. D. Lopez stepped back and took the gun out of Moses Lopez's hands, saying at the same time, put that down, and placed the gun back in its original corner. LeBlanc then walked down the steps of the Piazza into the yard walking towards his workshop repeating that he was swindled. Mr. Lopez then called him a D—n Son of a B—. LeBlanc said you are a bigger Son of a B—. LeBlanc then went back to his shop.[13]

Phillip A. Mullane, foreman of the blacksmith shop, was also personally involved in the incident. He gave an undated affidavit that was also enclosed in J. Ralph Smith's report to Tupper dated August 21. His account closely agreed with W. W. Smith's. Mullane wrote:

I the undersigned went to the store room to get a modle [*sic*] for some work for my shop and was a coming out when I heard loud words in the office near by. [T]he 1st was from Mr. Lopez saying leave my office clear out and ordered his [LeBlanc's] name struck off of the payroll. LeBlanc says Mr. Lopez do you intend to swindle me out of that $15.00. Lopez replies that (in these words) I intend to swindle you out of that $15.00. LeBlanc says that shows what you are a damd [*sic*] swindler. Lopez jumpt [*sic*] two steps as if to strike him, with clencht [*sic*] fist and hesitated, saying that you want me to strike you that is all you want damn you. Simultaneously Moses Lopez jumpt over a platform desk saying kill the damn son of a bitch when I jumped in between David Lopez and Moses catching LeBlanc by the arm and advising him to leave and not have any more fuss. Lopez ordered him from the yard for a damn son of a bitch. LeBlanc replied that he was a bigger one and offered to fight him any way they wished and left for Morse's shop. That is all I know of the affair.[14]

Daniel S. Hart, David Lopez's son-in-law, worked as the time keeper and store keeper at the State Works. His testimony reads:

On the afternoon of August 14th LeBlanc entered Mr. Lopez's office and in a very abrupt manner asks Mr. Lopez if he intended charging him with a piece of bacon he had refused to keep. Mr. Lopez replied certainly, whereupon LeBlanc called him a swindler and made some other remarks which I cannot call to memory. Mr. Lopez ordered him out of his office, and threatened to discharge him if he did not return to his work immediately. LeBlanc then used the most indelicate language, which caused Mr. Lopez to curse him, and proceed to put him out of the office by force. LeBlanc appeared as if he intended to strike Mr. Lopez as he advanced to put him out of the office, and then Mr. Lopez' son [Moses] said, placing his hand on a gun that was by him "if you do, I'll blow your brains out." Some few words passed between them and the affair ended.[15]

Edward E. Pritchard was an assistant store keeper, but his testimony added very little to understanding the affair. In his statement of August 20, Pritchard wrote that he was at his desk in the storeroom during the incident in the office and knew nothing of his own knowledge. He heard the argument and went to the door to see Lopez standing in the piazza and LeBlanc in the yard. Pritchard could not distinctly hear their words, but did state that LeBlanc appeared to be insolent.

J. Ralph Smith wrote to Tupper on August 21 that he had handled the matter. He determined that the bacon was pronounced sound and sweet by several of the best men at the State Works when it had been issued to LeBlanc. He personally examined the bacon after the incident and wrote, "Being in need of provisions for the negroes, I examined this (rotten) bacon, and found two-thirds perfectly sound, and the balance preferred (by the negroes) to beef."[16] Several of the employees were anxious to get the bacon *after* the incident occurred, and Smith reissued the entire piece after he settled with LeBlanc. Smith disposed of the matter to LeBlanc's satisfaction by paying him for the bacon but terminated his employment, though Lopez had already terminated Leblanc on August 14 during the incident. Smith said that, if LeBlanc had kept quiet until Smith was in charge, he would have paid LeBlanc for the bacon and kept him as an employee. Interestingly, payroll records show that LeBlanc was back at work no later than January 15, 1864, and remained at the State Works until February 1865.

Lopez signed pay vouchers on August 19 for himself and his son, Moses, for work performed between May 25 and August 19, indicating that his tenure as general superintendent permanently ended at that time. His contributions were significant. Lopez and Gist were the driving forces behind the creation of the State Works. Gist had the political connections, but it was Lopez who had the practical knowledge and personal experience to get the program under way. Starting with no land, very little machinery, and a few employees based a hundred miles away, Lopez helped Gist analyze and choose the site, and he supervised clearing the ground, hiring more employees, purchasing, borrowing or making all the necessary machines and tools, and getting the operation up and running. The State Works was quite an extensive enterprise by the end of Lopez's tenure as superintendent. Most notably, it was Lopez who was responsible for brokering the deal between governors Pickens and Harris that brought George Morse and his machinery to the Greenville site, and it was Morse's novel breech-loading carbine that became the lasting legacy of the State Works. Within two months of Lopez's resignation, the State Works completed one hundred carbines and continued to do so at an average rate of about two per workday. Without the State Works, Morse's brass-frame carbine probably would not have been manufactured in significant numbers, and without Morse's carbine, the State Works would be remembered only for its production of long-forgotten war materiel.

After David resigned, he and Moses returned to Charleston. Ironically, David continued to generate some income from the State Works after that. Seven of his slaves—Kit, Barber, George, Marcus, Jacob, Jim, and John—continued to work there, and in November they earned Lopez about sixty dollars. It is unclear what Lopez did for the rest of the war. Some researchers write that he returned to Charleston and helped build the semi-submersible torpedo boats called "Davids." It is possible that Lopez worked on the construction of these boats since Francis D. Lee had some involvement with them, and Lopez knew Lee, having worked with him on Institute Hall in the mid-1850s. The only evidence for this is in postwar references, but no primary sources exist.[17] Moses rejoined his old comrades-in-arms when he enlisted for the duration of the war as a private in Company A of the South Carolina Siege Train on August 28, 1863. Known as the Palmetto Guard Artillery and also called the Buist Light Artillery, Company A had been formed largely from members of the Palmetto Guard militia company on February 28, 1862. By June 20, 1864, he was placed on detached service with the Engineer Department

and remained there for the rest of the war. When the war ended, Moses was with the Engineering Department at Alston, South Carolina, and he was paroled at Chester, South Carolina, on May 5, 1865.[18]

James Ralph Smith, a forty-one-year-old accomplished Charleston machinist, probably knew Lopez before the war and replaced him as general superintendent on August 19, 1863.[19] Smith was born in Charleston on July 26, 1822, and all extant records associated with the State Works show his name as "J. Ralph Smith," indicating that he went by "Ralph" and not "James."

Smith's function at the State Works prior to August 19 was that of master machinist, a position of leadership and responsibility. His name appeared at the top of a roster of employees in July 1863, when he was the highest-paid non-salaried employee at the State Works, earning a daily wage of $8.33.[20] This figure translated to about $2,500 annually based on the typical twenty-four or twenty-five days worked each monthly pay period. In comparison, both Lopez and Smith's annual salary as general superintendent was $3,000. After the war, Smith returned to Charleston and resumed his old trade. He lived at 7 Pitt Street and, throughout the late 1870s, was a partner in a machinist company, Smith and Valk. Smith was fifty-seven years old when he died from typhoid fever at his home on February 20, 1880, and he was buried at Charleston's Magnolia Cemetery.[21]

The political maneuvering continued after Smith took over Lopez's job in August 1863. Though Morse respected Smith more than Lopez, he felt just as constrained under Smith's leadership. Morse wrote to Tupper on August 20, 1863: "After mature reflection, I think that I can propose a plan, which will if adopted, relieve you of all the embarrassment suggested by our conversation of last evening. I therefore request an interview with you at your earliest convenience before you again visit the works, that I may explain myself in detail and give you as much time as possible for its consideration. Please name the hour when I shall call upon you."[22] Morse's exact plans are unclear, but perhaps he was proposing that he be paid for the state's use of his inventions, a request that he would formally make in December.

Meanwhile, Smith continued to manage the State Works after mid-August 1863. Many of his purchases were routine, such as crucibles, axles, old brass and iron, yarn, osnaburgs, charcoal, fodder, mutton, beef, rice, and salt. He also implemented some of the new machinery to fabricate items that the State Works had never produced. About two months after he took over management of the facility, he issued a statement reviewing all of the articles repaired and fabricated at the State Works from its establishment until October 1, 1863. Items repaired were: 2,480 muskets, 600 rifles, 190 carbines, 145 pistols, 1,564 bayonets, and 24 double-barreled guns. Smith recapitulated all of the items that had been manufactured at the temporary facilities on the State House grounds in Columbia as well as Lopez's list from his "Second Quarter 1863" report of the same nature. He also listed the new items made since July 1, 1863: a drop hammer, eight cast-iron pintle plates, eight wrought-iron pintles, and thirty-two Brook's rifle bolts.[23] The bolt was an armorpiercing, solid cylindrical projectile with a blunt or flat nose designed to reduce the chance of a ricochet, and they were often referred to in contemporary accounts as "bolts."[24]

Except for a few prototypes made in late 1862 and early 1863, no carbines of Morse's design had been completed by early September 1863. Henry Newton Reid, a private in Company D, Orr's Regiment of Rifles, was on detached duty at the State Works when

he wrote to Thomas D. Belotte, a thirty-eight-year-old farmer in Anderson District, on September 3, 1863. Belotte was a lieutenant, probably in the South Carolina Militia, and he was trying to obtain Morse's carbines for his company. Reid told Belotte that he must apply to Major W. G. Eason, state ordnance officer, in Columbia for the carbines. He also wrote that none had been finished, but that one thousand were almost ready to be assembled, and he anticipated that two hundred or more would be ready for service within three or four weeks. Reid wrote:

> I would be very glad for you to have the guns as I consider them the best gun in the *world*. I have never seen or heard of there [*sic*] equal. The following is a description as near as I can give though I will not give the particulars of construction as you could not understand unless you had the gun before you. The weight of carbine 5½ pounds[,] length of barrel 22 ins[,] caliber ½ in. The ball is a solid conical slug. The barrels are rifled with the Enfield groove and the range (is the same as the Enfield rifle) when they have an elevating site [*sic*] though site that we are putting on is stationary and to shoot to the center at 200 yards.
>
> The carbine can be loaded and shot 40 times in 5 minutes with ease, this *I have seen done several times* giving ample time to take good aim. It is very simple [and] can be loaded by any child of ten years of age. A man can load and shoot it with one hand as it requires but one hand to hold and load at the same time while you can hold your horse with the other. (It is got up expressly for cavalry service.) It is *perfectly* waterproof, it may be loaded and put under water for six or eight days after which it will shoot off the charge just as though it had been loaded but a minute before. It may be held under the water to load without interfering in any way, it never misses fire but always goes off.
>
> This is about all that I can tell you of the gun. I have seen some shooting done by Col. Morse (the inventor) that if I was to tell you it would look so incredible that you would not believe it. Give my respects to all of my old acquaintances and neighbors that is [*sic*] in your company.
>
> P.S. Col. Morse has one of his carbines that was made at the Springfield Armory as a modle [*sic*] that he will sell for five hundred dollars. The caliber is the same as the Colts Navy Pistol. It is a very finely finished Gun (or you might say an extravagantly finished Gun). It has a nice case and everything complete—would Capt. [A. P.] Calhoun like to buy it.
>
> Col. Morse wishes someone to buy it that would not let such a gun pass by unnoticed but to give it a fair trial and then say to the world what it is.[25]

Tupper documented that Morse's shop had been manufacturing the components of the weapon since July 1863 and completed the first hundred Morse carbines, along with their special ammunition, at a cost of $6,500, by October 1. Two of the carbines had been shipped to the Columbia Arsenal by September 30.[26] Another three hundred were being assembled from their various components and were slated to be finished by the end of the year.[27] In contrast, Smith's list from early October 1863 is oddly silent about the completion of any Morse carbines by October 1, possibly because he considered Morse's shop to be an entity separate from the State Works in general.

Mass production of Morse's carbine from 1863 to 1865

The first 100 carbines	Completed between September 3, 1863, and October 1, 1863
Approximately 698 additional models	Manufactured between October 1, 1863, and October 1, 1864
Total of 798 in the State Arsenal in Columbia	In storage as of September 30, 1864
234 models (highest known serial number is 1032)	Manufactured between October 1, 1864, and the end of operations in March or April 1865

NOTE: Documented production rates varied between 100 per month in September 1863, 33 per month in January–March 1864, and 67 per month in April–June 1864.

CHAPTER 15

Morse's Brass-frame Carbine

Morse's brass-frame carbine was neither his most complicated nor his simplest design. Clearly deriving its heritage from both Morse's firearms patents and his experience in improving his designs throughout the late 1850s, the carbine combined Morse's best features to make his most efficient military weapon. Its compact design and combination of brass and wood make the carbine a handsome firearm. It incorporates Morse's signature design of a breech cover, which exposes the open breech when lifted up and back. It employs the modifications that Munck made to Morse's U.S. Patent 20503 in late 1858 and early 1859, combining the sliding breech piece of Patent 15995 and the solid breech block of Patent 20503. No known patent, either United States or Confederate States, exists for Morse's brass-frame carbine. (See color plate 6.)

As in earlier patents, a firing pin mechanism runs through the carbine's breech block. Also like earlier designs, lifting the case cover causes "nippers" to automatically extract the spent cartridge, thus emptying the chamber. The spent cartridge fell out or was withdrawn from the top of the firearm, unlike the Muzzy-made Morse design, which allowed the case to fall out the bottom. Unlike the operation of some of Morse's previous weapons, the act of lifting the breech cover did not automatically cock the carbine's hammer, which had to be manually cocked. The carbine's hammer was what Morse termed a "guard lock" and was located within the frame, unlike his earlier "side lock" hammers. In some of Morse's previous patents, the breech cover had a spring-catch to hold it in place when closed; the earliest carbines did not, but later models did. (See color plate 7.)

Though exact measurements vary between individual models, the carbine weighs only six and a half pounds, the barrel on most survivors is about twenty inches, and the overall length is about forty inches.[1] It has a finger rest behind the trigger guard, which likely also functioned as an attachment point for a cavalry sling. On two survivors at the Greensboro Historical Museum, the finger rest has been expanded, whether through extensive use or intentionally to accept the clip of a sling. Leather pouches, each carrying twelve cartridges, were made specifically for Morse's carbines. A surviving example at the Atlanta History Center shows two cartridge pouches mounted on a cotton web belt, but the men of Percival's Aiken Mounted Infantry were issued one cartridge pouch per man

in November 1864. Carbines were shipped in wooden boxes containing twenty each. One such box survives at the Greensboro Historical Museum.

We do not know exactly how many Morse carbines were made, but it was approximately 1,040. In addition to at least 8, and maybe as many as 12, unnumbered prototypes, early models, and demonstration pieces, the State Works manufactured no fewer than 1,032 carbines, that being the highest known serial number.[2] Several prototypes were made, including 5 at Springfield Armory in 1860, of which some had side locks and others had guard locks. It is the author's opinion that these were of a similar design to the brass-frame carbines made in Greenville. In addition, there were other prototypes and early models. One prototype was made in Atlanta in late 1862, and it did not have a serial number. It is probably the same carbine that is on display at the Greensboro Historical Museum in Greensboro, North Carolina. Its previous owner was noted Civil War arms collector and author Dr. John M. Murphy, who referred to it as the "Atlanta Carbine." Two carbines that Lopez made in Greenville and took to Columbia in late 1862 and early 1863 were likely prototypes but might have been stamped with an early serial number. Another is the carbine that Morse sent to Richmond for evaluation in late 1862 or early 1863, which might also have been given a serial number. Two known demonstration models were made—the Bonham cased set made in Greenville in 1864, and the brass-frame carbine mated to a shotgun barrel at the Atlanta History Center. Neither has a serial number. Finally, a carbine was made in Greenville and sent as a gift to John Hunt Morgan, but it is not clear if this one had a serial number.

Seven hundred ninety-eight carbines were in storage at the State Arsenal in Columbia on September 30, 1864. Because the State Works' rate of carbine production varied between 33 and 67 per month, it is entirely conceivable that the total number reached 1,032 between October 1, 1864, and the end of production in March or April 1865. (See appendix 2.)

The only known existing prototype is at the Greensboro Historical Museum. There are several differences between it and the Type I production model. The floor plate of the prototype frame is made of wood, and on the production models it is made of brass. Other variations on the prototype include the lack of a storage place for the cleaning jag in the buttplate, a brass front sight, a larger, iron rear sight, and a .54 caliber barrel with seven lands and grooves.

Type I

The earliest production carbine was later designated Type I after improvements were made to the latch and sliding breech block and termed Types II and III. It is assumed that Type I carbines began with serial number 1, and the highest *known* serial number of a Type I is 257. The primary difference between Type I and Type II is that the breech cover on the Type I does not have an external latch, or catch, to hold it in place, allowing it to spontaneously fall open if the hammer is half-cocked and the carbine pointed up at more than a 45-degree angle. The breech cover is locked down only when the hammer is in the fired position. The latching mechanism on the Type I is internal, consisting of a movable iron rod within the breech cover. When the hammer is dropped, an iron rod attached to

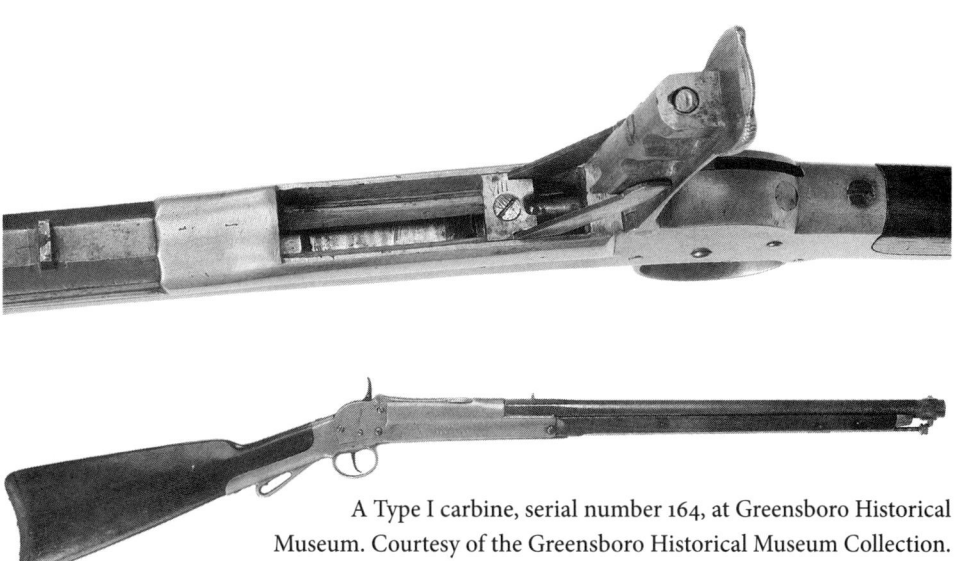

A Type I carbine, serial number 164, at Greensboro Historical Museum. Courtesy of the Greensboro Historical Museum Collection.

the hammer and in line with the single rod in the breech cover presses the iron rod in the breech cover forward, which in turn enters the sliding breech block, both locking it and pressing the firing pin forward. Cocking the hammer causes the locking mechanism to disengage and permits the breech cover to be raised.

Externally, the Type I can be distinguished by the absence of any iron latch, or catch, on the foremost aspect of the breech cover. The Type I breech cover is raised by grasping its knurled brass, forward edges, which extend slightly outward, and lifting it. On some existing Type I carbines, the brass breech blocks have two or three milled-out sections on each side of the breech block. This is an inconsistent finding, and its purpose has been postulated as being for weight savings and better control of the breech cover when the weapon was fired. Some Type I carbines were converted to Type II, as discussed below.

Type II

To correct the problem of unwanted opening of the breech cover on Type I carbines, an iron, spring-loaded catch, or plate, was incorporated into Types II and III, and was retrofitted to some Type I carbines. The Type II retained the sliding brass breech block but added a large-head screw threaded into its top. The breech cover was redesigned by adding an external iron plate with knurled sides that was attached to the front top of the breech cover. The plate was attached to a movable, spring-loaded iron rod which was added to the breech cover just above and parallel to the existing iron rod. The rod could be moved back about a quarter-inch by retracting the iron plate. As the new iron rod moved forward by the automatic action of the spring, its flanged forward end fit snugly under the head of the screw thus securely closing the breech even when the gun was not being fired.

The Type II modification dates from March 1864. Morse wrote to Bonham on March 9, saying that he had recently learned from J. Ralph Smith that the governor wished to

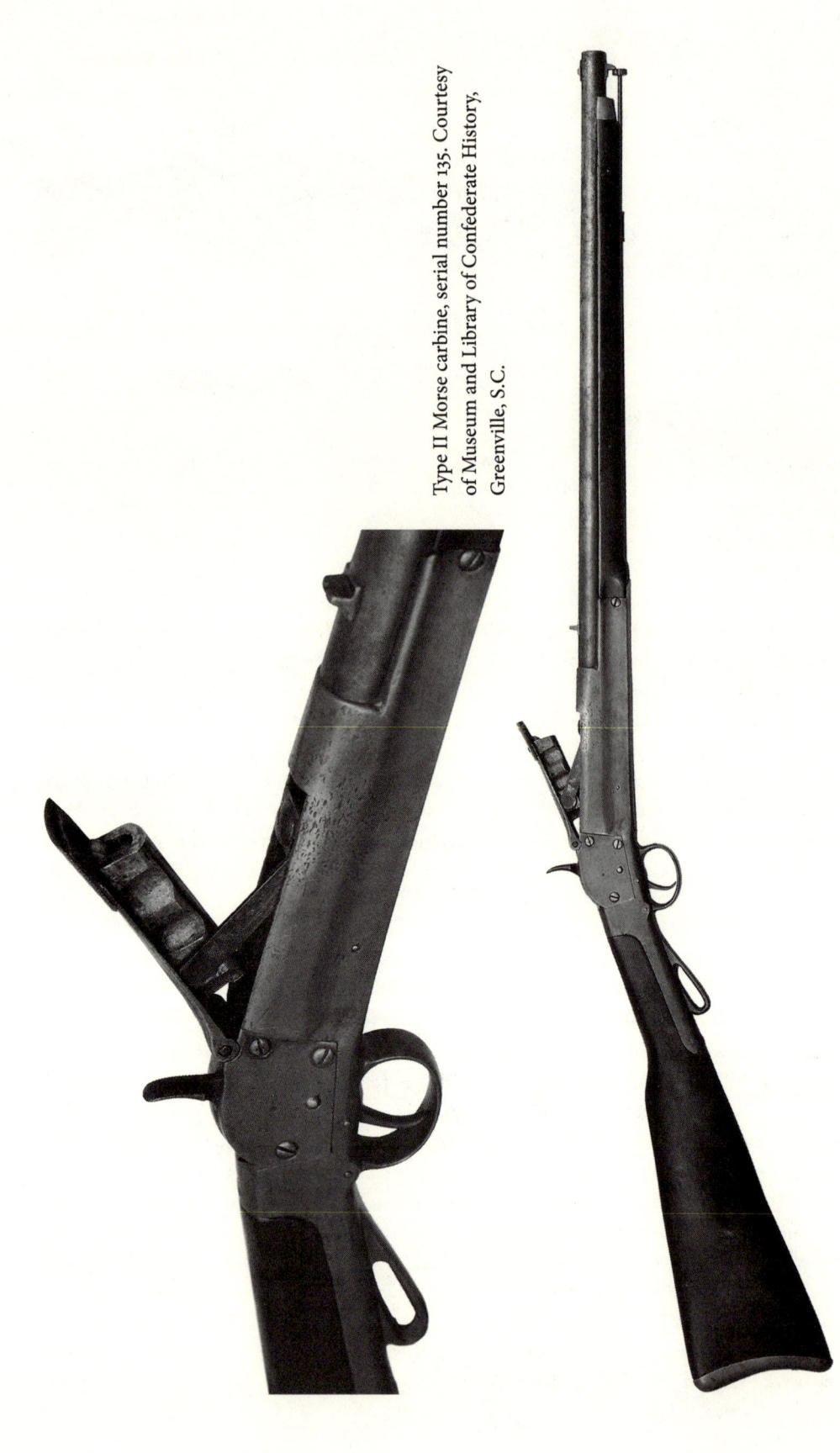

Type II Morse carbine, serial number 135. Courtesy of Museum and Library of Confederate History, Greenville, S.C.

have all of the completed carbines shipped at once from Greenville to the State Arsenal in Columbia. Morse wrote, "I have perfected, and will in a day or two have ready for your inspection, an improvement upon them, which in my judgment, should be applied to all before they leave. During the week, one model can be forwarded to you, with my views in reference to it, and in the meantime, I hope that the arms can remain here."[3]

In the same March 9 letter, Morse discussed making a gift of a carbine to Brigadier General John Hunt Morgan. Earlier Morse had left a carbine in a holster in Bonham's office and wrote that he had decided to give it to Morgan as a gift. Morse had also left two rifles and some pistols in a box at the State Arsenal in Columbia and intended to give them to Morgan as well. Morse asked Bonham to assist him in having the box shipped to Morgan but to return the carbine to Greenville so that he could apply the improvements, presumably converting it from Type I to Type II.[4] A note attached to the letter states that the carbine was shipped to Morgan on March 24, 1864, but we do not know its ultimate fate.

The conversions certainly took place in Greenville, probably throughout early 1864. In a likely reference to these conversions, a process that would have required the carbines that had already been shipped to the State Arsenal in Columbia to be sent back to Greenville, J. Ralph Smith wrote to Bonham on June 1, 1864, "I received the carbines at [railroad] cars yesterday and brought them up [to the State Works]; will have them attended to."[5]

Type III

Incorporating further improvements to the breech block, the Type III carbine was the final version of Morse's design. The large-headed screw and brass sliding breech block on the Type II were replaced by an iron, or steel, sliding breech block, which is a large oval rounded bar with a deep slot cut into its rear surface.[6] The flanged rod mechanism was retained on the Type III, but it was modified such that the forward flange is broader and flatter and fits into the slot of the iron breech block. Morse's final modification made the Type III superior to both preceding types because of its strength in locking the breech. (See color plate 8.)[7]

Much can be learned from examination of surviving examples. All of Morse's brass-frame carbines were made at the State Works in Greenville, with the exception of one made in Atlanta in late 1862. Almost all are .50 caliber, but there are a few exceptions. Five prototypes were made at Springfield in 1860, and we know that two were to be in the .54 caliber of the Harpers Ferry rifle and two in the .44 caliber of Colt's Army pistol. In 1863 Reid described a carbine that he said was made in Springfield in the caliber of Colt's Navy revolver, which was .36 caliber, but it was probably actually .44 caliber and was probably one of the five prototypes made there in 1860. The ultimate fate of the Springfield carbines is unknown. The prototype made in Atlanta in late 1862 was .54 caliber and is the carbine with no serial number that Dr. Murphy called the "Atlanta carbine," now owned by the Greensboro Historical Museum.[8] Dr. Murphy also observed that serial number 867 was a .58 caliber. Carbine serial number 180 in the Virginia Historical Society's Maryland-Steuart Collection is possibly a .52 caliber, and serial number 465 at the

Milwaukee Public Museum is felt to be .54 caliber. Serial number 790 has been altered to fire the .56-56 Spencer rimfire cartridge.

The carbines made in Greenville were manufactured using a rifled, steel barrel that probably threaded into the brass receiver, but it was permanently fixed in position. In contrast, the Muzzy-built Morse cased set was designed to accept threaded barrels of different calibers, which could be changed depending on the desires of the shooter. Unlike any of Morse's other firearms, the carbine was made from a solid brass frame, probably because brass was easier for semiskilled laborers to cast and machine.

On almost all of the known carbines, the barrel is round, but there are a few exceptions. Carbines with serial numbers 2 and 164 at the Greensboro Historical Museum have an octagonal barrel for the first 2.25 inches. Carbine serial number 180 at the Virginia Historical Society's Maryland-Steuart Collection also has an octagonal barrel for the first 2.5 inches. Finally, serial number 759 has a partially octagonal barrel according to Dr. John Murphy.[9] The inconsistency probably reflects the fact that the gunsmiths at Morse's small arms shop used whatever barrels were available to them at the time.

Almost all of the carbine barrels are rifled with three lands and grooves. The prototype at the Greensboro Historical Museum has seven lands and grooves, as does serial number 2 at the same museum. One unusual smoothbore carbine, serial number 161, is mentioned by Dr. Murphy.[10] Dr. Murphy also described the carbine's sights: "On most Morse carbines the front sight is a small brass block-blade inletted into a small, oblong iron block which in turn is inletted into the barrel just behind the muzzle. Rear sights are generally fixed-and-notched, but occasionally a specimen will have a small folding-leaf rear sight. Other variations in sights will be found. Carbine serial number 161 has the small brass block-blade front sight inletted directly into the barrel, and the front sight of another specimen has both blocks made of iron."[11]

The forestocks and buttstocks of most Morse carbines were of butternut, which is also called white walnut. At least one carbine has maple stocks, and the cased set at the South Carolina State Museum has cherry forestocks on all three barrels and a walnut buttstock. In 1863 Lopez purchased a variety of wood types, including oak, walnut, ash, poplar, and hickory. It is possible that his preference would have been some type of walnut since that was the traditional wood for gunstocks, but that he used whatever was available if walnut was scarce.

An iron cleaning rod, sometimes mistaken for a ramrod, is retained within the forestock and held in place by a stud fixed to the bottom side of the front end of the barrel. The rod is threaded to accept a brass cleaning jag stored in the brass buttplate.

On production models, the serial number is stamped on the brass receiver in front of the trigger guard, on the underside of the rear part of the breech cover, and on the wrist of the buttstock normally covered by the brass frame. Dr. Murphy described one exception, serial number 857, on which the number is stamped on the side of the barrel near the breech.[12] Several sources also state that the serial numbers appear on the inside of the buttstock and other places, and Dr. Murphy states that "serial numbers are stamped on most parts" of the carbine.[13] Serial number 1 is referred to in a standard text by Hill and Anthony; its location is unknown. Serial number 2 is at the Greensboro Historical Museum. Number 1013 is at the same museum, and 1021 or 1023 is shown in Dr. Murphy's book.[14] Dr. Sutherland writes that serial number 1032 is the highest known to exist, but its

Type III Morse carbine. Serial number 865. Courtesy of Museum and Library of Confederate History, Greenville, S.C.

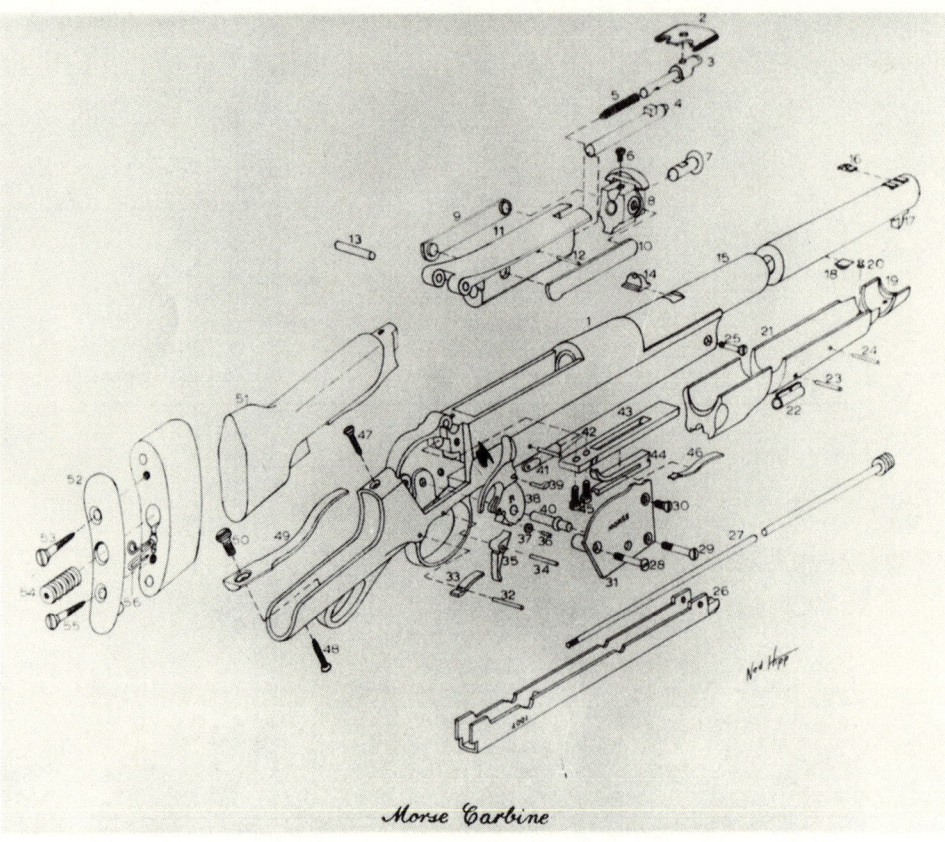

An "exploded" drawing of the Morse carbine prepared by Dr. H. L. Sutherland, Union, South Carolina, which clearly illustrates the complex construction of Morse's unique carbine. From Wray-Morse File 031. Courtesy of Atlanta History Center.

1. Brass receiver.
2. Breech lock latch (at top of diagram—number "2" not showing).
3. Breech lock slide.
4. Intermediate firing pin. Note three-piece firing pin.
5. Breech lock spring.
6. Front firing-pin screw.
7. Front firing pin.
8. Sliding breech block.
9. Left-side toggle.
10. Right-side toggle.
11. Pivoting breech block.
12. Intermediate firing-pin retainer pin.
13. Breech-block pivot pin.

14. Dovetailed rear sight.
15. Barrel.
16. Dovetailed front sight.
17. Cleaning-rod stop.
18. Front dovetailed forearm mount (rear dovetailed forearm mount not shown in diagram).
19. Brass forearm cap.
20. Forearm-cap retaining pin.
21. Forearm—wood.
22. Cleaning-rod thimble.
23. Cleaning-rod thimble retaining pin.
24. Front-forearm retaining pin (rear-forearm retaining pin not shown in diagram).

25. Front-floor plate retaining screw.
26. Brass floor plate.
27. Cleaning rod.
28. Rear-side plate screw.
29. Front-side plate screw.
30. Top-side plate screw.
31. Brass side plate.
32. Trigger spring pin.
33. Trigger spring.
34. Trigger pin.
35. Trigger.
36. Hammer-spring roller pin.
37. Hammer-spring roller.
38. Hammer.
39. Hammer and rear firing-pin connector pin.
40. Hammer pivot pin.

current location is unknown.[15] The serial numbers are stamped using two different-size dies, or fonts. Generally speaking, earlier production models are stamped with larger-size dies and later models, with smaller dies, but the application is inconsistent. Dr. Murphy writes that serial number 357 was the highest number he had personally seen with the larger-size die. On serial number 116 at the Upcountry History Museum in Greenville, both of the numerals "1" are in small die size, and the numeral "6" is in a large die size. Some sources record incorrectly that carbines with the larger-size dies were manufactured in Atlanta and those with the smaller size dies in Greenville.

In addition to serial numbers, various other markings and inscriptions appear on some surviving Morse carbines. Roman numeral "VII" appears on a dug Morse carbine frame from the Bentonville battlefield and now at the Greensboro Historical Museum. Roman numeral "V.I." appears on the breech block of serial number 180 at the Virginia Historical Society's Maryland-Steuart Collection. Serial number 643 in the E. Berkley Bowie Firearms Collection at Fort McHenry National Monument has "R" on one side of the serial number behind the trigger and "X" on the other side. Numbers 893 and 912 are also marked with an "R." A fourth carbine stamped with "R" in front of the hinge on the receiver is number 1013 at the Greensboro Historical Museum. Dr. Murphy states that such letters, Roman numerals, and dots "appear on most internal components."[16] They probably reflect the individual gunsmith who made the part.

Eleven surviving carbines are marked with "Morse," usually on the right side of the brass frame. They appear in a cluster of carbines manufactured in late 1864 or early 1865 and include 938, 953, 955, 956, 966, 978, 988, 1007, 1013, 1025, and 1032. The likely explanation for the appearance of Morse's name stamped on the side of the carbines is that the Committee on the Military recommended in late December 1864 that he be paid fifteen dollars for each carbine made at the State Works, and it was about that time that the stamps began to appear. Serial number 164 at the Greensboro Historical Museum has "South Carolina" engraved or stamped on the frame, which was probably a postwar addition. Number 288, also at the Greensboro Historical Museum, is marked "J Davis" on the top of the frame. Perhaps the most elaborately engraved carbine is number 676 at the Greensboro Historical Museum. The engraving is "Mar. 1, 1865, Cos. G and K Mass. Inf. by a night march surprised, near St. Stephens, S.C., a squad of Georgia cavalry who had followed Sherman's army from Chattanooga to Savannah. This carbine was captured in the fight by Capt. Chas. C. Soule."

In summary, Morse's carbine was an excellent firearm. Ahead of its time when first made in the late 1850s, it was soon surpassed as the war spawned rapid technological development in breechloaders. As a breech-loading weapon, it could be fired much more rapidly than a muzzle loader and with as much accuracy. It was highly portable and weather resistant. Its cartridge was time-consuming to manufacture and required strategic materials, but it was reloadable, theoretically offsetting

41. Rear-firing pin.
42. Ejector-slide stop pin.
43. Ejector slide.
44. Ejector.
45. Two screws connecting ejector slide (43) to breech slide block (8).
46. Ejector spring.
47. Top stock screw.
48. Bottom stock screw.
49. Hammer spring.
50. Hammer-spring screw.
51. Stock—wood.
52. Brass butt plate.
53. Top butt plate screw.
54. Brass cleaning jag.
55. Bottom butt plate screw.
56. Jag-retainer spring.

those drawbacks. The carbine did, however, have some flaws. First, only two screws hold the buttstock to the brass frame. Second, the buttstock is very thin in width. Third, the breech cover is weak at its attachment point, where it tended to crack under heavy use. The major shortcoming to Morse's carbine was that only about a thousand were produced, far too few to have an impact on the outcome of the war. At a time when large numbers of easily manufactured muzzle loaders were being made, Morse's carbine was a few critical years ahead of its time. By the time the war started, the U.S. government decided to employ breechloaders for selected troops, but it chose to use older designs like Burnside's for immediate use in the field and to develop some of the newer repeating breechloaders as the war progressed. Though borrowing heavily on some of Morse's concepts, the repeating breechloaders effectively made Morse's single-shot design obsolete by the time it was only a few years old.

CHAPTER 16

Sale of the State Works

The South Carolina legislature critically examined the financial status of the State Works in 1863 and, only a year after it had been built, made the determination to sell it. Relative to the potential sale, the issue of ownership of the machinery and tools that Lopez acquired from the Tennessee State Armory came up once again on February 3, 1863, when the legislature passed a resolution instructing the state auditor, James Tupper, to look into the precise terms of the original agreement with Governor Harris of Tennessee and to report back to Governor Bonham. The resolution stated that, if the price had been fixed at the time the equipment was transferred from Atlanta in the spring of 1862, Bonham was to pay that amount from the Military Contingent Fund. On the other hand, if the value of the machinery was to be determined at the time of payment, that is, at the present time or sometime in the future, the governor was to refer the matter back to the House of Representatives.[1] The Senate concurred with the resolution the next day.[2]

Tupper communicated with Harris, who replied by letter: "I have no authority or disposition to sell the machinery and tools which I turned over to the Governor of South Carolina in the spring of 1862. The Legislature of Tennessee will determine, when it meets, what disposition shall be made of the machinery and tools."[3] Tupper reviewed the history of the transaction and correctly concluded that the equipment had not been sold by the state of Tennessee to the state of South Carolina. Further, though an inventory had been taken in the spring of 1862, Lopez told Tupper by letter in early 1863 that an appraisement of the machinery and tools was never made. Additionally, the state of South Carolina had given a receipt to the governor of Tennessee in 1862 and agreed to fully account to Tennessee in a future settlement for everything it received in the arrangement. Tupper concluded: "It seems to be clear, that it remains with the State of Tennessee to determine hereafter whether a sale will be made, and upon what terms, for the machinery and tools accounted for by the State, as they were received from Governor Harris."[4] This conclusion made it difficult for Tupper to balance his books, knowing neither exactly how much the machinery was worth nor when the state might have to pay for it.

To facilitate Tupper's investigation, Bonham wrote to Harris on September 21, 1863, to inquire whether any valuation had been placed on the machinery and tools. Bonham mistakenly told Harris that Lopez had performed an appraisement at the time the

equipment was transferred to South Carolina's control, but Lopez had actually conducted an inventory, not an appraisal.[5] Harris's response is not preserved, but it is clear from Tupper's report issued within the next few weeks that no value had been assigned to the Nashville machinery and tools when Harris signed them over to South Carolina. In spite of the unresolved issues of ownership, the General Assembly wanted to sell the State Works as early as September 1863.

William S. Downer, superintendent of the Confederate States Armory in Richmond, was familiar with both the machinery at the State Works and Morse's carbine, and he made a trip to Greenville to inspect the State Works in late September or early October 1863. Bonham probably invited him to conduct the inspection in hopes that the Confederate government would buy the State Works. Downer reported his negative findings to Josiah Gorgas, Chief of Confederate Ordnance, on October 5:

> During my absence I visited the State Works of South Carolina at Greenville and find that in my opinion they are totally unpracticable for our purpose. My reasons for this opinion are as follows:
>
> First, the buildings are erected in an unsubstantial manner and are so constructed as to require different motive power for each shop thereby entailing great waste of fuel, wear and tear of machinery etc.
>
> Second, the State has given Mr. G. W. Morse an order for the construction of one thousand of his breech-loading carbines which they are desirous of having finished and which I would judge by present appearances will require six months to complete occupying all the power of the finishing shops.
>
> Third, the distance from sources of supply of coal and iron. These have to be transported [from the mines] to Columbia and thence over the Greenville Road, one hundred and forty six miles.
>
> Fourth, the great waste of power incident to the place. They have now four or five steam engines erected, of an aggregate power more than double what is required to run the machinery in operation. Yet Mr. Morse informs me he will require another forty horsepower engine to run his grindstones and polishing wheels. My opinion founded on the practical experience I have had is that these works carried on as they are now would prove ruinous to any private individual without unlimited capital in less than six months; carried on by the State, they will add unnecessarily to the burden of the war without producing any adequate results and as I said would be entirely unpracticable for our purposes having an eye to economy and efficiency. I take this opportunity of saying, though, I experienced every courtesy and assistance at the hands of Governor Bonham of South Carolina.[6]

Bonham made an inspection and fact-finding visit to the State Works on October 12.[7] J. Ralph Smith was not there during most of the governor's visit, and William Walton Smith, chief clerk, provided the governor with information relative to the capacity of the shops as well as a written statement of the amount of coke and coal on hand. J. Ralph Smith was able to see Bonham that evening, but time did not allow them to discuss the many matters that Smith had hoped to. Smith wrote to Tupper on October 16, asking if he had made a decision about disposing of some of the coal and coke, but warned Tupper that he would probably need all he had on hand because he was preparing to forge

rifled bolts and other heavy work as soon as the furnaces were prepared.[8] As a result of the visit, Bonham determined that there was enough coke on hand to offer some to the Confederate government.[9] Probably unaware of Downer's negative report dated October 5, Bonham wrote to Secretary of War J. A. Seddon on October 15 offering to sell the State Works to the Confederate government.[10] He wrote, "The legislature authorized me to exercise my discretion at transferring to [the] Confederate Government the State Foundry and Armory at Greenville, and the saltpeter plantation at Columbia. You can have them and the lead mines at a sum sufficient to cover costs and charges, which is less than half it would cost now to put them up. One thousand Morse carbines [are] now in the course of construction to be finished and reserved to the State according to a contract authorized by the Legislature."[11] Seddon responded by telegraph that he would give Bonham an answer within a few days.[12]

Tupper made his first annual report to the legislature about that time. Included in it was his assessment of the State Works, which bore a more favorable financial outlook than that held by the General Assembly. From its inception through October 1, 1863, the State Works cost the state $252,546.03. The Pickens administration was responsible for slightly more than half, $133,028.08, between March and mid-December 1862, and Bonham spent the remainder, $119,517.95, from late December 1862 to the end of September 1863. Tupper offered the following summary of expenses: machinery and tools, $31,839.57; building materials, $20,675.02; stock, $66,543.37; provisions, $11,718.53; labor and services, $102,089.72; railroad freight, $7,912.89; incidental expenses, $8,523.89; purchase of small arms, $1,251.00; and fuel, $1,992.04.

Tupper also reported that the value of work done at the State Works from its creation through July 11, 1863, was $67,513.00.[13] Though the report was made in late 1863, the auditor did not have an estimate for the work between July 11 and October 1. The breakdown on production was as follows:

Value of work done up to August 13, 1862: $22,893
Arms repaired and altered from August 13, 1862, to April 1, 1863: $19,386
Articles fabricated during April, May, and June 1863: $9,470
Articles manufactured between April 1, 1863, and July 11, 1863: $15,764

In addition, Tupper placed a value of $283,464.57 on the property at the State Works as of April 1, 1863. The property value combined with the value of the work produced to date should have given the state a positive account of about $98,521. Unfortunately, there were two significant financial problems that affected the balance sheet. First, South Carolina had not yet paid for the Nashville machinery, and neither its rightful owner nor its value had been determined. Second, the production of the State Works was not keeping up with its ongoing costs of operation.

Tupper discussed each of these issues in his report. He admitted to having no accurate appraisal of the value of the machinery, tools, and stock that Lopez brought from Atlanta to Greenville in early 1862. Tupper also acknowledged that failing to account for the State Works possessing articles that the State had not yet paid for would have a significant detrimental effect on the appraisal. Concerning the output of the State Works, Tupper felt that part of the problem was a "seeming unprofitableness" up to that time based on startup costs, which were one-time expenses of erecting the buildings and the

manufacture and purchase of machinery and tools.[14] Another issue harming the overall financial picture was that the State Works utilized steam power, which required coal. The ongoing cost of transporting both coal and iron ore by rail over long distances was a financial drain on the facility. In an effort to paint some of these problems in a more positive light, Tupper wrote: "If these more serious obstacles to a remunerative employment by the State can be overcome, I am satisfied that they will be by the skill and industry of the accomplished machinist [J. Ralph Smith] now at the head of this establishment."[15]

Dr. John Leconte, professor of physics and chemistry at the South Carolina College, wrote to Tupper on November 4, asking him to induce the governor to transfer the nitre beds (saltpeter plantation) and lead mines to the Confederate government but not the State Works, and he expressed astonishment that Seddon had not accepted the State Works.[16] His reasoning in not offering the State Works for sale was that the facility might be of some use to the state, while the nitre beds were useless without a powder works and the lead mines would be a "dead expense." Seddon's response to Bonham's offer was that the Confederate government would purchase the saltpeter plantation at Columbia and the state's interest in the Cameron lead mine in Spartanburg District, which it ultimately did, but he followed Downer's advice and declined to take over the State Works.[17]

Bonham addressed the Senate on November 23, informing it of Seddon's offer but recommending that it be rejected because he wanted to sell all three enterprises. Bonham felt that the lead mine was unprofitable and should be abandoned. Though the saltpeter plantation was potentially profitable, he felt that selling those two smaller operations and not the larger State Works was unwise. Having failed to sell the State Works, Bonham was determined to keep it in operation, and he made the following recommendations:

> The advantages of the State Works do not appear to be, so far, commensurate with the expenditures. It is believed, however, that they will in future be profitable, if the expenses of procuring coal and coke, and transporting iron, do not prevent it. It is a grave question for your consideration, in the present state of the country, whether it would not be better now, to change the locality of the works, placing them at Columbia, on the canal, or near the iron mines, at some place where the machinery may be run by water. And here I recommend to your consideration the question of how far it may not be desirable for the State to do something towards establishing more direct communication with the Deep River [North Carolina] coal mines. If these difficulties could be overcome, the information I have received on a recent visit to the Works, from the intelligent Superintendent and the heads of the different departments, satisfies me they will be profitable. It is a matter of great importance to the State to have it, in future, in her own power to manufacture the implements of war which she may need.
>
> In accordance with your resolution, I contracted for the manufacture of one thousand Morse's carbines. A part of them have been completed, and I regard them the best cavalry weapon in use. The compensation to Col. Morse has not been determined upon, and I recommend this matter to your consideration, as something more than the mere value of the article manufactured, it may be thought proper to allow.[18]

The General Assembly apparently did not completely share Bonham's outlook. They preferred to sell the State Works at a favorable price, but, failing to do so, to keep it in

operation. Unable to interest the Confederate government in buying it, the General Assembly passed a resolution on December 16, authorizing Bonham to sell the facility at
public auction, "If in your opinion it could be advantageously effected."[19]

Knowing that Seddon was already interested in buying other facilities in the state,
Bonham persisted in trying to sell the State Works to the Confederate government. He
wrote to Seddon on December 17 that the legislature had renewed his authority to take
that action and proposed making the transfer of ownership of the saltpeter plantation
and lead mine at that time. In a renewed effort to convince Seddon that the Confederate government should also take over the State Works, Bonham wrote: "I will take this
occasion to suggest further that the Confed. Govt. having control of all transportation
making it very difficult for me to procure coal for the "State Works" would do well to take
these works, as they can be made very valuable if kept supplied with coal."[20] Bonham told
Seddon that the State Works was casting shot and shell for the Confederate government
and could be prepared to cast cannon. Bonham's opinion was that the Confederacy, with
its control of transportation facilities, could do a better job than either the state or a private company, but he was not successful in changing Seddon's mind, and ownership of
the State Works did not change hands at that time.

Interest from the private sector arose about the same time when a group of private
citizens representing a company, probably Kalmia Mills, approached Bonham wanting to
purchase the State Works and possibly other state-owned assets as well. Bonham referred
the matter to the legislature, but nothing came of it for many months.

After the proposed sale to the Confederate government fell through in late 1863, the
issue lay dormant for nearly a year. In late 1864 Bonham revived the idea when he recon-
sidered the same three potential buyers: the Confederate government, Kalmia Mills, and
sale at public auction. Bonham ordered Tupper to carry out an inventory and appraisal
of the land, buildings, machinery, tools, materials, and other state-owned property at the
Greenville facility. Tupper delegated the task to J. Ralph Smith. Tupper and Smith were
instructed to exclude all the finished work and those items in the course of completion for
the state, as well as all machinery and tools owned by the state of Tennessee and the Confederate government. Smith provided an approximate estimate to Tupper about October
1, 1864, and Tupper reported to Bonham an estimated value of the entire State Works as
$853,265.25.[21]

The board of directors of Kalmia Mills was interested in buying the State Works in
late October 1864 and designated its president, Benjamin Franklin Evans, a partner in
Walker, Evans, and Cogswell Printers, to represent them in negotiations with the state.
Evans reviewed the inventory and appraisal and wrote to Tupper on October 26 that he
felt the appraisal was equal to or beyond the highest retail-market price. After offering
Tupper several examples with which he was knowledgeable, Evans concluded that the
entire appraisal was largely overestimated. He suggested that the appraiser made mis-
takes in some categories such as the value of bricks at $1.50 each instead of the current
market value of 3 cents each and the inclusion of worthless pikes in the appraisal. Evans
wrote that he was sincere in his wish to purchase the State Works but could not recommend to the board of directors of Kalmia Mills that they do so at the suggested price.
He recommended that the state fix a reasonable cash price, and he would act on it at
once.[22]

Tupper wrote to Bonham that Evans's valuation was only $5,000 less than the appraisal, and he explained that the two parties were actually not very far apart on the estimated value of the State Works. Tupper disagreed with Evans's assessment and felt that Smith was correct when he included 1,254 firebricks that Evans had failed to notice. Tupper concluded that Smith had overvalued the property by as much as Evans had undervalued it and that Smith's estimate was "intended to be used, only as the basis of an intelligent judgment as to the probable market value of each article, from which a price for the whole might be arrived at."[23] Tupper told Bonham that he provided Evans with an appraisal estimate but not a price at which the state would sell. Ultimately Evans and Tupper were unable to agree on terms, but Evans remained interested in the property.

Tupper wrote to Bonham on October 27, the day after Evans refused to buy the property, recommending that, if the governor chose to sell the State Works, it should be sold in bulk at public auction. He further recommended that advertisements be placed in newspapers in Charleston, Columbia, Greenville, and Richmond announcing that the auction would take place in fifteen days. He also recommended that the sale be in cash and that the buyer agree to complete all arms and munitions currently in progress. The buyer must also agree to properly store all machinery and other property belonging to the state of Tennessee and the Confederate States. Finally, Tupper recommended to Bonham that the sale not take place until either Kalmia Mills or another party had the opportunity to make an offer of three-fourths of the appraised value. At that point, Evans's offer would be considered a bid in the auction to be accepted if it were the high bid.[24] On October 29, Bonham agreed with all of Tupper's recommendations and instructed him to sell the State Works at public auction under the terms and conditions that he had recommended and to report the results.[25] Within a few days, advertisements appeared in the newspapers in Richmond, Charleston, Columbia, and Greenville that the State Works would be sold at auction in Columbia on November 15.[26] An advertisement in the *Daily South Carolinian* read: "Sale of the State Military Works at Greenville, South Carolina. To be sold at public auction on the 15th day of November, 1864, at 12 M all the lands, buildings, machinery, tools, materials, and other property of the State by order of the Governor."[27]

Opposition to the sale was immediate. Tupper wrote, "Upon the appearance of this notice of sale a number of communications were referred to this office from parties whose opinion on this subject were entitled to respect."[28] Both George Morse and J. Ralph Smith weighed in, arguing against the sale.

One complaint arose from an unexpected quarter, an attorney representing Vardry McBee's heirs. The sale notice precipitated a legal challenge to the ownership of the land, but it apparently dissipated as quickly as it arose. Tupper rendered his opinion on this new development on November 7:

> Respectfully returned to his Excellency the Governor with the remarks, that it cannot be reasonably supposed that the state would have expended the large sum which these works cost upon lands that were subject to revert in the event of a change (from unforeseen circumstances) of the use originally contemplated by the state and the donor. It is not however certain that either party contemplated a continuance of these works beyond the condition of things which induced their establishment. No consideration at that time of the supposed advantages of Greenville as a city, would

have justified the location of these works upon land for which any claim moral or legal might be justly set up in the contingency of the works being discontinued or transferred to other parties by the state. An *absolute* title to the state was an indispensable condition to the location of these works upon the land in question. And it was a very important consideration to the state, in view of the use to which they were to be applied, whether that title was obtained by purchase or by gift.

To act in this case upon the views of the learned attorney of the heirs of the donor might subject the state to considerable pecuniary loss and encourage claims in opposition to the legal title to lands for which I am not aware that there exists any precedent in the transactions of states or individuals.[29]

Though the auction was imminent, Bonham had not completely given up on the idea of selling the State Works to the Confederate government. On November 5 he informed George A. Trenholm, Confederate secretary of the treasury, of the upcoming auction. Bonham placed a reserve on the auction when he told Trenholm that he would not sell the State Works at auction, "unless some assurance is given that the Confederate Govt. will become a bidder at a price to cover the cost and expenses of these works to the state, and that other parties will offer at least 50 per cent over the sum of the said cost and expenses."[30] Bonham wrote that the state's cost up to that time was $450,000, that the cost of moving the machinery and tools to Columbia was $50,000, and that the appraised value was $896,500. Bonham might have padded the price a bit. In his end-of-the-year annual report, Tupper had determined that the state had spent only $400,000 on the State Works as of October 1, 1864.[31] Bonham told Trenholm, "The Confederate Government would thus secure for $500,000 this valuable property at an unexceptionable city of about one half the cost of building and furnishing an armory and foundry of this character in any other way."[32] Bonham also estimated that the sale of the land and such portions of the buildings that could not be moved would adequately cover the costs of moving the machinery and tools to Columbia. He also reminded Trenholm that some of the machinery at the State Works was owned by the state of Tennessee and the Confederate government and, though not included in the sale, would still belong to the government if it purchased the State Works. Trenholm sent either an unfavorable response or none at all.

Morse wrote to A. C. Garlington, adjutant and inspector general of South Carolina, on November 7, objecting to the auction and recommending that the focus of the State Works shift to the manufacture of domestic products while continuing to make his carbine. Protecting his personal interests, Morse argued for a plan that would ensure both ongoing production of his carbine at the State Works and wider exposure of his breech-loading musket, which had never been manufactured in Greenville, to the Confederate government:

> I regret that these works are to be sold by the state, altho I do not think that my personal interests would be materially affected by it. If the state would continue them in operation, and manufacture all kinds of agricultural implements, (and we are well prepared to do it) as well as repair such as muskets and at the same time make as many carbines as we could put out, the works would pay well for the investment and the state would not only have a good armory, but would supply a very important

want of her citizens, at a moment when it seems almost impossible for them to be furnished with what they want otherwise. We could barter for produce at old prices, and I have no doubt could furnish the State Commissary with most of the articles required by him. Labor was never so cheap for provisions, as at this moment. We have the machinery and skill, to make anything, and could turn it to good account in the manufacture of cotton, spinning and weaving machinery, or a thousand other things so much needed at the present moment. I have been urging this matter for some time, and only mention it to you, as I intended writing you upon another subject, as follows.

It seems a pity that the manufacture of the carbines should be discontinued, even if the works are sold. I own a considerable number of the small tools used in making them, while the state owns the balance. I wish to sell those belonging to me, and have them go with those owned by the state, to the confederate government, and have them continue the work. You have had the carbine in use during the last summer and can now give a very good opinion of its merits. An expression of your opinion, if favorable, would have considerable influence in Richmond, and consequently would aid in making the necessary arrangements. I am aware that the President does not like breech-loading guns, but still I know that he considered mine the best of them all. If Genl Chesnut felt authorized to add his influence to yours, I have little doubt that my object could be affected. Of course I could not ask favors of this kind, of either you, or Genl Chesnut, for the advancement of my personal interest, but if the public service would thusly be advanced, neither you, nor the Genl would object. All I should ask, would be a fair appraised value of the tools, and the right to manufacture the carbines at a fixed price for each, so that no patent rights would be purchased, except as they were used. I have my breech-loading musket here, to which I wish to add the improvement of the catch, like those on the carbines. I then wish to send it to Genl Robert E. Lee, and ask its trial by any soldier in his army, as the best weapon in the world for the defense of the trenches, and beaches, and fortifications. You are aware that buck shot or balls could be used at pleasure with great rapidity, and with fixed bayonet. I have all of the tools for this alteration of muskets, and the work could progress rapidly. If Genl Lee had a few men armed with these guns in the trenches, no amount of force could take them. Please give me your opinion of this project, as well as of the others referred to, and send an order to Mr. Smith to have the musket prepared. All it wants is the catch and loaded cartridges.[33]

J. Ralph Smith wrote to Bonham on November 12, only three days before the proposed auction, repeating Morse's theme of making the State Works profitable by converting it to the manufacture of agricultural implements:

As the Legislature of the State will shortly convene and will make disposition of these works, I beg leave to offer for your consideration, a few suggestions—Your Excellency is aware of the difficulties attending the purchasing of iron, and the impracticality of paying the market price, and working for the government. The government purchases [iron] at less than $1.00 pr. lb. [per pound]. These works have been offered it (at the mills) at $2.00 [per pound], consequently cannot furnish work unless at a loss to the State. The bar iron used for carbine barrels etc., is made here from scraps

and at one third the cost of same quality at the iron mills, but it cannot be made in sufficient quantities for general use unless expensive preparations are made and more workmen employed.

The government exempts (or details) over one hundred thousand planters & farmers to raise supplies for the army. Most of them are greatly in need of agricultural implements of every kind of which they cannot purchase. These implements can be made here, and can be readily exchanged for corn, bacon, and other produce[,] would yield the State a fair profit, and need not interfere with the manufacture and repairs of arms. The State may be able to make arrangements to obtain material from the iron mills in sufficient quantity to manufacture these implements, and much may be purchased in the country at prices which cannot be paid to do work for the government.

It is important to have planters at work, to feed our armies, it is equally important that these planters should be supplied with necessary implements, and if the State will supply these, the planters will derive material aid, and the "State Works" will be made remunerative. The works cannot purchase material at *market* prices, and work for the government; as the *cost* of *raw* material is more than government pays for finished work, but they *can* manufacture agricultural implements.

If the State appropriates funds for the manufacture of arms, she can hardly expect to have the *arms* and the *funds* also unless she will do other work and receive for it market prices. If she will dispose of the arms (to anyone who may wish to purchase) she can realize sufficient profit to make the "works" self sufficient. There has been done in the past year work sufficient (exclusive of carbines and cartridges) to reimburse the treasury to say nothing of small tools necessary to the establishment. (A list of this work is in [the] hands of [the] state auditor).

In addressing your Excellency on this subject, I am influenced by no selfish motive (as I have offers of more lucrative employment) but I should regret to see the valuable machines of this establishment, scattered through the country, when they might be made so conducive to the general good, if applied to the purpose I have mentioned.[34]

Finally, Tupper himself offered his opinion to the governor. He agreed with Morse and Smith's views and told the governor so, and he went so far as to present several arguments for not selling the State Works. He felt that the value of present currency and the fact that the State held a large surplus of Confederate Treasury notes rendered the sale of such valuable property "at this juncture impolitic and unnecessary."[35] Tupper wrote that he had been in favor of selling the State Works until the views expressed by Smith and Morse caused him to change his mind. Tupper called the two men "experienced machinists" who convinced him that the State Works could be operated without a loss to the state. He saw the major problem as being the high cost of obtaining iron and coal. However, Tupper wrote that the same challenges had been met and overcome by private enterprises whose resources and facilities were far below those of the State Works. The primary difference, as Tupper saw it, was that private enterprises had increased their charges for finished products to correspond with cost increases in labor and materials, whereas the State Works had not. Finally, Tupper added that the manufacture of Morse's carbines was profitable. The current market value of one carbine was $400, making the seven hundred carbines and appurtenances manufactured thus far worth $280,000, a figure that was

one-third of the present value of the State Works and more than two-thirds of the sum spent there thus far.[36]

After Bonham considered the opinions of Morse, Smith, and Tupper, and had taken into consideration the legislature's plan to meet very soon, he postponed the November 15 auction and referred the issue to the legislature's Committee on the Military.[37] Bonham addressed the Senate and House of Representatives on November 28, acknowledging his new belief that the State Works, heretofore a money loser, could be made profitable.[38] Citing the increasing lack of coal, Bonham reiterated his recommendation that the works be moved in stages to Columbia. His reasons were: (1) the works could utilize water power from the Columbia Canal; (2) they would be closer to iron mines; (3) they would be more secure within the confines of Columbia; and (4) they would be more likely to produce a profit in Columbia. Calling Morse's carbine "the best weapon for mounted infantry yet invented," Bonham earnestly recommended that the state manufacture as many as the works could turn out and sell them "to the citizens of the country" for the defense of their homes.[39] Bonham's hope was that, by combining a move to Columbia with the sale of large numbers of Morse's carbines and the manufacture and sale of cannon shell and agricultural implements, the State Works could show a profit, but he concluded his recommendations by conceding that these changes could make the State Works profitable even if the facility remained in Greenville.

It was about this time that Morse presented the cased set now on display at the South Carolina State Museum as a gift to Bonham. The timing of Morse's offer certainly appears be either his way of thanking Bonham for cancelling the auction or a bribe to keep the State Works open, and Bonham, perceiving it as such, refused to receive it as a gift. A few weeks later ex-governor Bonham bought the beautiful set for $345.

Bonham left office on December 18, 1864, and the new governor, A. G. Magrath, entered office the same day. The Committee on the Military made its report on December 24, breathing new life into the State Works. Though the committee considered the geographical location of the State Works "unfortunate," it did not recommend its removal or sale.[40] The committee agreed that the State Works could be made profitable if it diversified its efforts and began to manufacture agricultural and other domestic implements. The committee specifically recommended that the State Works continue operations for the year 1865 and that the superintendent be instructed to devote all the time and labor practicable to the manufacture of other articles and implements of precise necessity. The products of the State Works were to be sold and bartered under Magrath's direction. In addition, the committee recommended that another fifteen hundred Morse carbines be made and five hundred muskets be converted to breechloaders using Morse's patent and that Morse be paid fifteen dollars for each carbine completed and each musket converted as remuneration for the use of his patent.[41] Morse wrote in 1866 that Bonham ordered an additional five hundred carbines to be made, and "all of the work which [his] tools had to perform upon them, was completed, when we were stopped by the [Federal cavalry] raid. Supt. Smith then allowed me $150 for their use upon this last five hundred, as they had been considerably injured." It is highly unlikely that an additional five hundred carbines were completed between early 1865 and the end of the war. Morse might have been stretching the truth, or he might have intended to say that the *components* of five hundred carbines were made. It was about this time, late 1864 or early 1865, that the name

"MORSE" was stamped on the right side of the brass frame, as seen on several surviving carbines. There is no known documentation that any muskets were converted to Morse's breech-loading patent at the State Works, and there are no known survivors.

In December 1866 the General Assembly withdrew an undrawn appropriation of $15,500, which had been approved in 1863 and 1864 for the construction of Morse's carbine and the use of his patent.[42] And so, the efforts of the State Government to sell the State Works during the war ended, only to be resumed soon after the war was over.

CHAPTER 17

Late 1863 and 1864

The state of South Carolina paid for Morse's carbines and issued them to state troops. Edward Mortimer Boykin's Squadron was the first unit known to have been issued Morse's carbine. The two-company squadron was created in November 1862 and received its carbines sometime between September 1863 and January 19, 1864. At that time, Boykin's two companies were the only mounted infantry in state service.[1] Governor Bonham ordered Boykin and his men to Greenville on August 22, 1863, but they probably did not arrive there until October. The squadron participated in a scouting expedition to western North Carolina and eastern Tennessee from November 3 to 17, 1863, and saw action in eastern Tennessee in early December. Boykin's Squadron was called into Confederate service in early 1864, and he was notified on January 19 that his men were to return their Morse carbines and be rearmed with Enfield rifles.[2] Both companies of the squadron mustered in Confederate service no later than February or March 1864. Company A, under Boykin's personal command, became Company K of the newly created Seventh Regiment, South Carolina Cavalry, in early February. Company B, commanded by W. N. G. Rodgers, remained independent until December 1864 when it was attached to the Nineteenth Battalion, South Carolina Cavalry.[3] Morse stated in a letter in November 1864 that his carbines had been in use during the last summer. Since they were not available for issue until September 1863, he must have been referring to the summer of 1864, but it is not clear if it was Rodgers's Company or some other that was issued Morse's carbines at that time.

Morse wrote to the Senate and House of Representatives on December 1, 1863, requesting consideration of reimbursement for the use of his patents:

> In compliance with your orders, we are manufacturing at these works, one thousand of my breech-loading carbines, a part of which are completed, and the remainder well under way. In your appropriation for this purpose, it does not appear that you made any provision for remunerating me for the use of my patents, neither was I consulted upon the subject, previous to the passage of the law ordering them to be made. The peculiar model upon which these arms are being manufactured, was made here at my own cost, the authorities of the state here, being unwilling to incur even

this preliminary expense in the matter, altho it was my proposition to turn a lot of old hunting rifles, which we had on hand, into this kind of an arm, which has been done. The United States Government paid me five ($5) dollars each, for the right to alter two thousand arms to my plans, and were in treaty with me for the right to alter one hundred thousand (100,000) more, when the war broke out, which of course put an end to the transaction, altho they offered me every inducement to remain, and go on with the work. We are also manufacturing a new musket lock [the inside lock] of my invention, for which I shall have the right to claim some consideration, as it is admitted to be vastly superior to the old lock, and made with less than one half of the machinery required in the manufacture of a gun lock of the old style, and at about one half its cost. I never doubted that justice would be done me in these matters, and the object of this communication is to call your attention to the subject, with a hope, that after due investigation, and consideration, you would make me such remuneration, as in your wisdom may seem proper, and just.[4]

Morse's petition was referred to the House of Representative's Committee on the Military, which passed a resolution on December 5, 1863, to pay Morse five thousand dollars for the use of his patents in constructing the breech-loading carbine and inside-lock musket.[5] The House sent the resolution to the Senate on December 7.[6] Morse wrote in 1866 that the state agreed to reimburse him for the use of his patents at a rate of five dollars each for one thousand carbines, and he was to be paid at the completion of each batch of one hundred. He claimed that he was never paid for the last batch.[7]

Morse had many positive attributes, one of which was self-confidence. He wrote to Bonham on March 9, 1864, that the Confederate cavalry from East Tennessee had been in Greenville recently and brought with them some captured weapons. Morse had been able to personally examine "all of the lately invented Yankee carbines, several patterns of which have been made since we were in Washington, and I am very confident, that no one of them is as good as the one we are making. I am quite sure that my arm is still superior to any which they have in use, and this seemed to be the opinion of those who were using the Yankee guns."[8]

The State Works conducted business with several firms in Greenville District. Gower, Cox, Markley & Co. manufactured wagons and carriages, and it also made gun slings for the State Works. J. Ralph Smith paid Gower, Cox, and Markley to stitch 102 gun slings on December 28. He paid the same firm to make another 77 gun slings and one hundred cartridge belts and boxes on January 29 and February 5, 1864.[9] Altogether, Smith paid Gower, Cox, and Markley $361.20 in late 1863 and early 1864. It is not specified that these items were for the Morse carbine, but the number of gun slings approximately corresponds to the number of carbines produced thus far, and we know that slings were issued with the carbines. By early 1865 the State Works had paid for 1,079 gun slings, another number that roughly approximates the total number of carbines produced. No sling made specifically for the Morse carbine is known to exist, and the material of which the slings were made is also not known. It might have been either leather or cotton webbing. The carbine was issued with one or two leather cartridge pouches that carried twelve rounds each, and the belt of one surviving cartridge pouch at the Atlanta History Center was made of cotton webbing.[10]

In late October 1863 the citizens of Greenville became alarmed when a Federal raiding party engaged North Carolina troops twenty-five miles west of Asheville at Warm Springs, now known as Hot Springs. The citizens of Greenville feared that an invasion from across the mountains into the Upcountry of South Carolina could destroy the town, its factories, mills, and even the State Works. A delegation appealed to Bonham, who ordered Edward M. Boykin's Squadron of State Troops to the Upcountry. Beauregard also responded by ordering Captain Bachman's Light Artillery Battery and Colonel J. H. Williams's Fifth Regiment of South Carolina State Troops (Six-Months) there as well.[11] By mid-February 1864 the threat had subsided, but the concern was very real, as the populace would learn a year later. As part of his report summarizing the event, Major John D. Ashmore, commander of the Post of Greenville, requested fifty additional artillerymen to guard the mountain passes above Greenville. He stated that two artillery pieces were already at the State Works, and only the men were required.[12] Other than a two-pounder brass cannon known to have been issued to the State Works in the summer of 1862, Ashmore's statement is the only discussion of cannon being there, leading some researchers to speculate that they had been cast there, though there is no evidence of such in the records of the State Works.

The State Military Works remained a facility with a dual purpose in mid-1864, producing general war-related goods and Morse's carbine, but not yet making domestic products. J. Ralph Smith generated a report of all articles purchased and fabricated between April and June 1864. The purchases included routine items such as food, fuel, fodder, and raw materials.[13] Just as Lopez had done in his report of items fabricated in the second quarter of 1863, Smith broke down the report of fabricated items into two parts. The first part represented the ordnance service and all its branches, and the second part covered a separate department consisting solely of Morse's breech-loading carbine and cartridges.

Smith wrote to Tupper on March 2 to say that the State Works had built six milling machines, only three of which were being used. Smith proposed exchanging two of these machines with the Confederate shops in Columbia for a compound planer and vertical drill press.[14] On March 15, Smith completed an inventory and appraisal of all the tools and machines received from Nashville in 1862. His list was actually taken from an inventory of the State Works made on April 1, 1863, and that, in turn, had been taken from Lopez's inventory when the items were shipped from Atlanta in the summer of 1862. Smith noted that a few objects were missing and might have been consumed or omitted. He also wrote that those items belonging to the Confederate government were appropriately marked on the list. In spite of the minor discrepancies, Smith felt that he was able to comply with Tupper's primary objective, which remained unstated but was probably an effort to assess how much equipment at the State Works belonged to Tennessee or the Confederate government and how much to the state of South Carolina. Acknowledging that he would act only on orders from the governor, Smith asked Tupper if he should "deliver" the tools solely upon the order of the chief of ordnance or should he wait on the endorsement of the governor of Tennessee. As it turned out, this exchange was moot; all of the equipment remained in Greenville.[15]

In mid-April 1864, Smith was confronted with an issue that had caused problems between Lopez and Tupper a year earlier. Tupper was late in delivering the April payroll check to Smith. Indicating that Smith had a better relationship with the state auditor than

had Lopez, Smith wrote to Tupper on April 14, informing him that the check was uncharacteristically late. Smith suggested that the check might have been "miscarried" and was merely informing Tupper of the event. Expressing hope that the delay was due to causes beyond Tupper's control, Smith was nevertheless forced to delay paying the employees, and he politely requested that Tupper advise him about the status of the check.[16] The tone of Smith's letter is substantially different from the frustration displayed by Lopez in the spring of 1863.

During the second quarter of 1864 Morse produced one hundred carbines, 6,660 cartridges, and twenty "moulds for cartridge metal," probably meaning molds for cartridge casings, not bullets. The ordnance service manufactured 340 ten-inch mortar shells, eighty muskets, and sixty rifles of Morse's inside-lock design. Reflecting the diversity of its capabilities, the State Works also produced a four-spindle drilling machine, a wire drawing machine, and even a fire engine during the second quarter of 1864. It is unclear if the fire engine was intended to be used at the State Works or elsewhere. The Palmetto Fire Engine Company of Charleston presented an old India rubber hose to the State Works in June, probably to be used on the new engine.[17]

One of Morse's character traits was his inclination to continually develop improvements to his breech-loading firearms and cartridges. His only known endeavor in improving a muzzle-loading firearm was called the inside-lock conversion and was an alteration of the lock mechanism of an existing muzzle loader. It was also the only firearm Morse designed in which he did not utilize one of his patents. Because the inside lock did not use a lock plate, it required less metal for the lock. In addition, the lock mechanism was located inside the body of the firearm, making it quicker and cheaper to manufacture and, because less wood was removed from the stock, it resulted in a stronger weapon. Two oval brass side plates were located externally, one on each side of the stock. A square shaft ran between the two side plates, and to it was attached the hammer. The lock mechanism was located centrally within the stock.[18] Lewis described the mechanism: "Its unusual and simple lock mechanism is recessed into the stock under the trigger guard. A lock frame made of one piece of metal holds main and trigger springs, while a square lateral shaft carries the hammer. Besides its simplicity, this lock made a far stronger stock possible, as a minimum of wood was cut away."[19] Edwards wrote, "The whole outfit was a 'package deal' that could have been fabricated in enormous quantity at a properly equipped shop and shipped out all over the nation to musket and carbine makers, if anybody had been foresighted enough to see its merits."[20]

The State Works manufactured 80 muskets and 60 rifles using the inside-lock design. At least 16 are still in existence. The Greensboro Historical Museum owns 3. One is a short rifle that has no serial number stamped on it and has the numerals "1842" on the barrel, indicating that it was most likely from a U.S. model 1842 musket. Another is serial number 50, which is marked "Morse's lock, State Works, Greenville, S.C.," and the third is serial number 104, which is a musket marked "1863, Morse's lock, State Works, Greenville, S.C." The Atlanta History Center owns both number 70 and another with no serial number, possibly a prototype. The South Carolina State Museum owns number 12, and the South Carolina Confederate Relic Room displays number 19, remnants of an inside lock. Springfield Armory Museum owns number 93; Virginia Historical Society, number 110; and American History Museum at the Smithsonian Institution, number 118.

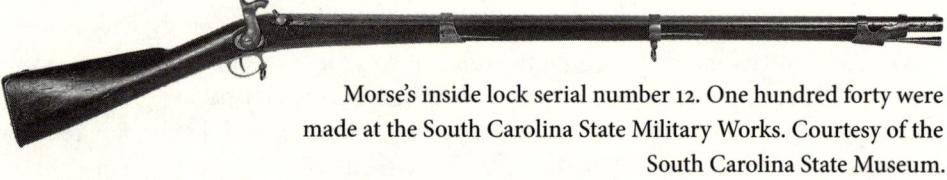

Morse's inside lock serial number 12. One hundred forty were made at the South Carolina State Military Works. Courtesy of the South Carolina State Museum.

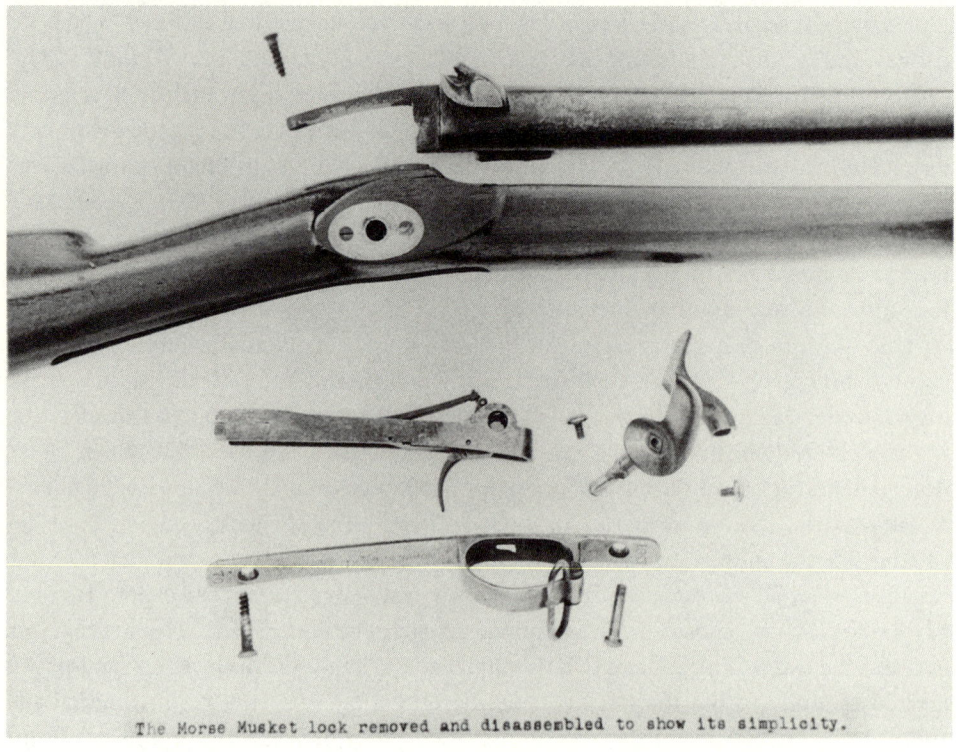

The Morse Musket lock removed and disassembled to show its simplicity.

Components of Morse's inside lock. From Wray-Morse File 047. Courtesy of Atlanta History Center.

Serial numbers 5, 6, and 48 are privately owned, and numbers 122 and 129 are discussed in the literature and are also probably privately owned. Finally, Dr. Sutherland wrote that number 163 is the highest recorded serial number for an inside lock, but this statement conflicts with the fact that only 140 were manufactured.[21]

Smith's accounting ledger for the third quarter of 1864 showed a balance of $1,446 at the beginning of the quarter and $7,922 at the end.[22] The State Works' income for that time was $69,365. The auditor's office provided the State Works with its single largest source of income—$20,164 in July, $13,565 in August, and $19,092 in September. Profits from the sale of provisions to workmen were next, amounting to nearly $11,000. Miscellaneous income came from altering and casting 288 pieces for the Batesville Manufacturing Company, a manufacturer of cotton yarn on Rocky Creek twelve miles east of Greenville at Batesville. Additional miscellaneous income came from casting brass boxes for Lester and Brothers, a manufacturer of cotton cloth on Enoree River at Pelham, and casting spindles for Weaver's Factory, a manufacturer of cotton cloth. It also appears that the State

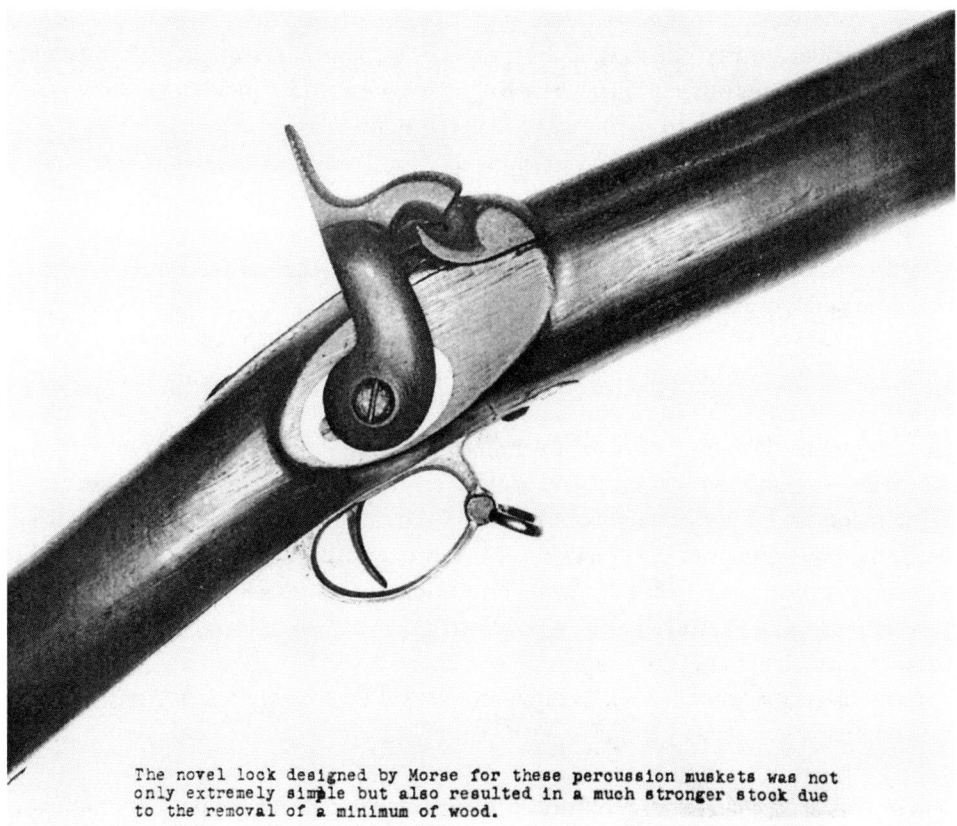

The novel lock designed by Morse for these percussion muskets was not only extremely simple but also resulted in a much stronger stock due to the removal of a minimum of wood.

Morse's inside lock. From Wray-Morse File 047. Courtesy of Atlanta History Center.

Works repaired small arms for private citizens and charged them for the work, but Smith reported an income of only $66.50 from this endeavor during the third quarter.

Smith's report of articles purchased during the third quarter of 1864 includes the typical items such as iron, brass, charcoal, walnut lumber, food for man and beast, clothing, and office supplies. He also issued a report for raw materials expended and consumed during the third quarter.[23] All of the raw materials were metals—old cast iron being the majority, but also bar iron, scrap iron, bar steel, sheet steel, spring steel, old brass, brass castings, copper, zinc, tin, bell metal, and other miscellaneous metals. The articles manufactured during that time included Morse's breech-loading carbines, ordnance stores, brass, and bar iron. He made no mention of casting any cannon. He also made no mention of manufacturing any muskets or rifles, indicating that the eighty muskets and sixty rifles already produced represented the extent of the production of Morse's inside lock design at the State Works. The State Works also manufactured and repaired machines and tools, and repaired and altered small arms. Smith also showed a heading for "making artillery work" probably meaning some type of accessory for artillery since only 166 pounds of bar iron and forty-nine pounds of brass were used for this purpose.

In the abstract of articles fabricated at the State Works during the third quarter of 1864, Smith deviated from the previous format and included Morse's carbines and cartridges with the report on the Ordnance Service and all its branches. The State Works

made 200 carbines and 6,800 cartridges in the third quarter, a rate of about 67 per month. It also produced thirty-one axes, eight syrup kettles, four Dutch ovens, fifteen skillets, four frying pans, four wagon wheels, one rolling mill, thirty-four molds for casting metal, and 6,976 pounds of bar iron. In August 1864 the State Works paid for railroad shipping expenses on ten boxes of arms and in September for five boxes of guns and one-quarter keg of powder.[24]

Though Morse's carbine went into service in late 1863 or early 1864 with E. M. Boykin's Squadron and, according to Morse, remained in use throughout the summer of 1864, new companies organized in the fall of 1864 were also issued the carbine.[25] Shortly after the evacuation of Atlanta, Governor Bonham called on every man not already in service to "mount and equip himself" for the defense of the state. His declaration was published in the *Charleston Daily Courier* on September 7, and he noted that the state would issue all such mounted troops a carbine of the most improved kind.[26] The governor agreed to accept any company with at least sixty-eight men on the roster. He planned to organize the new companies into battalions and regiments, but there were never enough to do so.

In the late summer of 1864, William F. Percival, an Aiken physician, began to enlist exempt men from Barnwell, Edgefield, and Lexington Districts for state service in one such new company, called the Aiken Mounted Infantry.[27] By September 10, some men were on the roster but not the requisite sixty-eight rank and file. The South Carolina adjutant and inspector general, A. C. Garlington, wrote to Percival on the tenth and told him that Bonham appreciated his efforts, wished him well, and would accept his company for service and furnish it with arms of the most improved kind when he had sixty-eight men on the roster. By October 6, Percival had enough men, and the company was accepted pending muster and inspection. Garlington promised Percival that his men would be issued an approved carbine *after* they mustered in service. Bonham appointed Percival captain of the company on October 22.[28] On November 16, Garlington's office ordered the following to be issued to Percival's Company: seventy-five Morse's carbines, seventy-five cartridge pouches and belts, seventy-five slings, two thousand cartridges (about twenty-six or twenty-seven per man), four bullet molds, four loading tools, four chargers, and four wadsetters.[29] The next day he ordered Percival's company into the field at Hamburg, South Carolina, where they were to receive the arms. From Hamburg they would be in a good position to guard the ferries and crossing places on the Savannah River, operating downriver to within twelve miles of Savannah, Georgia, as Sherman marched on that city. It is likely that Percival's men carried Morse's carbine into combat as his company was present during the Battle of Aiken, South Carolina, on February 11, 1865.

Percival's company was in camp near Mathew's Bluff in Barnwell District, South Carolina, on January 7, 1865, when he extolled the virtues of the carbine in a letter that was published in the *Edgefield Advertiser* on January 25. Percival wrote, "The Company now numbers about sixty members, and is armed with Morse's Patent Breech Loading Carbine—one of the most effective and convenient weapons ever used by mounted men, weighing only eight pounds, and can be loaded and discharged five or eight times in a minute."[30] Percival urged South Carolinians not to allow Sherman to advance without at least making an effort to stop him. He referred to his company and another of mounted infantry commanded by Captain Alfred J. Fredrick, which he felt "should be filled instantly and others formed. General Garlington desires those who can mount themselves

Altman Andrew Seigler and his wife, Cecelia Nancy Courtney Seigler. Seigler is the author's great grandfather. He served in Percival's Company, which was issued Morse's carbine in late 1864. Photograph in the author's file.

to join mounted Companies in preference to the infantry, as they can be made far more efficient in watching the movements of Yankees and retarding their progress."[31]

Several of the new mounted infantry companies were being raised for state service at about the same time as Percival's. When mustered in service, they were armed with Morse's carbine.[32] One was Frederick's Orangeburg Company of State Cavalry, also called the Orangeburg Mounted Men, to which Garlington referred when he noted that it mustered in service at Orangeburg on October 25, 1864.[33] On November 13, Garlington ordered both Frederick and Percival's companies, as well as the Battalion of State Cadets, to be ready to go into the field on short notice.[34] Frederick's Company was on active duty at Mathews Bluff at the same time Percival's Company was at Sisters Ferry in January 1865.[35] Another such company was raised by William Robinson Jones and was called both Jones' Mounted Infantry Company, South Carolina State Troops, and the Pendleton Mounted Infantry. It was organized in Pendleton on September 26, 1864, and went into service about December 3. A fourth company of mounted infantry was raised by forty-seven-year-old Joseph Josiah Brabham on January 24, 1865, in Barnwell District, and it went into service on February 11. Governor Magrath's correspondence in February and March 1865 is addressed only to Percival, Frederick, and Brabham, though Jones' Company existed until April.[36] The size of these four mounted companies varied from sixty to

Joseph Josiah Brabham (September 6, 1817–July 8, 1883) signed the South Carolina Ordinance of Secession in December 1860 and commanded a company that carried Morse's carbine in 1865. By permission of the family.

a maximum of a hundred men each.[37] This accounts for an issuance of between 240 and 400 Morse carbines. Two hundred ninety-eight carbines had been issued prior to September 30, 1864, but those issued to Percival, Brabham, Jones, and Frederick's companies constituted a separate group of carbines.

Were Morse's carbines issued to any other South Carolina troops? Law Professor Neil Alford postulated in a 1964 letter that W. P. Emanuel's Cavalry might have been issued Morse's carbine when it was in state service and then carried the firearm with it to Virginia. Alford's great-great-uncle Hugh Milton Stackhouse (1839–1936) was in Emanuel's Cavalry and told Alford prior to his death that he had carried a Morse carbine in the battle of Trevilian in Virginia in 1864. Emanuel's Cavalry was Company C, Twelfth Battalion, South Carolina Cavalry, until it became attached to the Fourth Regiment, South Carolina Cavalry, in December 1862. Emanuel's Cavalry left state service long before the first carbine was issued.[38] It is possible that Stackhouse had purchased a Muzzy-built Morse carbine before the war and carried it to Virginia as a personal weapon. Less likely is that Stackhouse somehow got his hands on an issued carbine and took it to Virginia. There are unconfirmed reports of Morse bullets being found in the vicinity of Stoney Creek, Virginia, where the Fourth, Fifth, and Sixth South Carolina Cavalry regiments had their camps, but, again, there is no documentation that Morse's carbine was ever issued to any of those regiments. Except for Boykin's brief excursion into North Carolina in late 1863, there is no evidence from existing records that any unit of South Carolina Cavalry or mounted infantry was issued Morse's carbine and later carried it in service out of the state. The brass frame of a Morse carbine was dug from the Bentonville battlefield and is on display at the Greensboro Historical Museum, but it is likely that it had been captured in Columbia a month earlier and was dropped by a Federal soldier on the battlefield. It is certainly possible that other South Carolina units were issued Morse's carbine, but documentation for such is lacking.

Tupper's Second Annual Report

Tupper made his second annual report to the legislature for the year ending October 1, 1864, in about November 1864.[39] The State Works received $236,000 during the fiscal year ending September 30, 1864. The single largest source of income, $175,000, came from the Executive Department through the state auditor's office. The sale of provisions and cloth to workmen produced nearly $28,000, and the Charleston Arsenal purchased Columbiad shells, mortar shells, and coke for almost $20,000. The remainder came from repairing

arms for citizens, the sale of machinery and tools to the Batesville Factory and to Lester and Brothers, and the sale of twenty wrought iron bolts to James M. Eason of Eason Brothers in Charleston. In addition, the State Works sold trunion plates, pintles and plates, and hardware bolts and washers to the Confederate States Ordnance Department. Also the State Works sold nails to the South Carolina Commissary Department, coke and iron to the Greenville and Columbia Railroad, and forty spindles to Weaver's Factory.

Expenditures, or disbursements, amounted to $228,000. The single largest category was payrolls and salaries, amounting to $104,000. Other categories were stock, provisions, fuel, coal, mules, and incidental expenses. Also, the State Works returned $33,000 to the auditor, giving the State Works almost $8,000 cash on hand at the end of September 1864.

By October 1, 1864, the state had spent a total of $434,345.31 on the State Works. About $252,000 of that was spent prior to October 1, 1863; about $175,000 from October 1, 1863, to October 1, 1864; and $6,000 for the construction of Morse's carbines. The State Works had done about $33,000 worth of work for the Confederate government, resulting in the actual cost to the state as of October 1, 1864, of $400,775.31.

Tupper enumerated all the items manufactured at the State Works after October 1, 1863, the value of which was $191,272. In addition to numerous small tools for the manufacture of arms and ammunition, the following had been made:

One steam hammer—$9,000

Two, four-spindle drilling machines—$8,000

Two punching and drawing machines—$7,000

One rolling mill for shot and shell—$2,000

Two polishing machines—$400

Two [railroad] trucks and one railroad
    car—$800

One screwing machine not completed—$2,500

Five axes—$1,100

One wagon and a set of extra wheels—$100

Patterns and gauges for shot and shell—$800

180 muskets—$18,000

57 pair of shot and shell flasks—$7,000

54 iron moulds for sheet brass—$2,000

23 artillery wheels without tires—$1,000

900 gun wipers and 900 wrench
    screwdrivers—$4,500

Iron work for Columbiad carriages—$43,500

Ordnance stores sent to Charleston—$35,522

Four milling machines—$16,000

Two rolling mills—$2,500

One wad punching machine—$500

One fire engine—$1,500

One double jig saw—$500

One foundry ladle—$1,000

25 water buckets—$250

One corn mill—$500

Eight syrup kettles—$6,000

60 rifles—$6,000

Bar iron made from scrap—$10,000

Ordnance stores on hand—$900

Tupper reported the production of Morse's carbine separately. Since October 1, 1863, the State Works had produced 700 Morse's carbines, 24,000 carbine cartridges, 48 bullet molds, and 48 tools. The original order was for 1,000 carbines. Tupper wrote in his report in October or November 1864 that an additional 300 carbines "are now being furnished, and of this number, 100 with their implements and 10,000 cartridges will be ready for delivery during the present month, leaving 200 to be completed under the order of the legislature for the manufacture at these Works of 1,000 of these arms for the use of the

state."[40] Thus, the total number of carbines produced at the State Works since mid-July 1863 was 800. This number is consistent with a "Return of Ordnance and Ordnance Stores at the State Arsenal, Columbia, for the year ending September 30, 1864," which showed that the arsenal had acquired 798 weapons labeled "Carbines, Morse BL [breech loading]" by that date.[41] The same return indicates that 298 had been issued, leaving 500 on hand. The same return reported that 80 muskets and 60 .54 caliber rifles, both of Morse's inside-lock design, had been received during the previous fiscal year and were still on hand on September 30, 1864. This number is in exact agreement with Smith's second-quarter report. In contrast, Tupper reported that *180* muskets and 60 rifles had been manufactured at the State Works in the fiscal year. It appears that either Tupper's number of 180 was incorrect or the State Works continued to produce muskets throughout the third and fourth quarters of the fiscal year.

Smith's accounting ledger for the fourth quarter of 1864 showed a balance of $7,900 going into the fourth quarter and $9,936 at the end of the quarter.[42] The State Works' income for that time was about $51,000. Tupper continued to provide the State Works with its single largest source of income—about $44,000 for the quarter. Profits from the sale of provisions to workmen were next, amounting to about $6,500. This is more than a 50 percent drop when compared to the third quarter of 1864, probably reflecting a much smaller workforce in early 1865. Miscellaneous income came from repairing small arms and the sale of one of Morse's breech-loading carbines to Governor Bonham.

CHAPTER 18

1865

The legislature elected a new governor, Judge A. G. Magrath, on December 18, 1864. In an effort to make the State Works more productive, Magrath assigned Colonel James Monroe Eason, whose rank stemmed from his role as the governor's aide-de-camp, to the State Works in February 1865.[1] Eason was born in Charleston on March 22, 1819. In 1846, James and his younger brother, Thomas, inherited Eason Iron Works, which had been founded in 1825 by their father, Robert. The company manufactured heavy machines,

Charleston machinist James Monroe Eason (1819–1887), was placed in charge of the State Works in February 1865. Courtesy of David West and Sonya Behling Eason of Jacksonville, Fla., owners of the 1859 oil portrait of James Monroe Eason in its original frame.

including steam engines, pumps, threshing machines, rice mills, sawmills, grist mills, sugar mills, and cotton presses. It also built a huge steam dredge that deepened Charleston harbor. In addition to operating the family business, Eason was elected to the South Carolina Legislature in 1860.

During the war, the business was called both James M. Eason & Brother and J. M. Eason and Company, and it enjoyed a very productive relationship filling government contracts. Eason was superintendent of construction of the state gunboat, CSS *Chicora*, in 1862. In 1863 the company rifled and banded two eight-inch Columbiads, which were then mounted at Fort Moultrie, and rifled at least four thirty-two-pounder banded guns and two twenty-four-pounders. In addition, the company produced a variety of ordnance, including rifled shell and bolts, hollow shot, and rifled conical shot, as well as carriages for Columbiads and other heavy artillery support equipment. It also performed tasks such as building anchors and chains for harbor obstructions and repairing various workboats. After the war, Eason continued the business and served in the South Carolina legislature from 1878 until 1880. Eason died at his home in Charleston on December 31, 1887, and was buried at Magnolia Cemetery.[2]

After Charleston and Columbia were evacuated in mid-February 1865, Eason's company was no longer able to function, releasing him for duty elsewhere. In a turn of events that justified the earliest proponents of the State Works, South Carolina had now become completely dependent on that facility for the domestic manufacture of its war materiel. Magrath wrote to Eason on February 23, five days after Charleston and Columbia had fallen:

> You will find enclosed an order appointing you one of the Executive Staff, and assigning you to duty at the State Works near Greenville.
>
> In the direction and universal supervision of these works, I desire that you would represent me.
>
> It is now of the greatest consequence that these works would be made contributory to the wants of the state and its citizens.
>
> Our arms have been lost and must be supplied as soon as possible. Our transportation because of the destruction of our railroads must be supplied with wagons. In all other respects, our wants are only to be supplied by these works.
>
> Hitherto they have not been productive; now they must be so. Our necessities are so great that there can be no excuse because of any difficulties. Our security requires that all difficulties should be overcome.
>
> Under your direction I expect and hope great success in the effort.[3]

Oddly enough, it was Eason's business in Charleston that fulfilled many of the functions that Gist and Lopez had hoped the State Works would be able to accomplish but never did. Also, Gist and Lopez's choice of a relatively protected location for the State Works was confirmed when Eason's business was occupied by Federal troops. Eason's new role at the State Works was somewhat murky. J. Ralph Smith did not relinquish his position as general superintendent even after Eason was appointed as the governor's personal representative with universal supervision at the facility. Magrath apparently chose his words carefully when he named Eason as Smith's superior but retained Smith in his former

capacity. Smith signed numerous documents as "general superintendent State Works" throughout March and as late as April 6, 1865.

Columbia was evacuated, and about one-third of the city was burned on February 17. For three days Federal troops destroyed public buildings, railroad tracks, locomotives, depots, arsenals, machine shops, warehouses, and much more.[4] In the arsenals were immense quantities of ammunition, which were dumped into the Congaree River. A large amount of ordnance, ordnance stores, and machinery was also destroyed. More than ten thousand serviceable muskets and rifles were reported to have been destroyed, and four hundred of them were recorded as "Morse rifles (South Carolina)."[5]

An exact accounting of the fate of Morse carbines does not exist, but we can deduce a close approximation. We know that 798 Morse carbines had been received at the State Arsenal by late September 1864, and 298 had been issued by September 30, 1864. In late 1864 and early 1865 about 300 carbines were issued to the seventy-five-man companies of Percival, Jones, Brabham, and Frederick, leaving about 200 in storage. New carbines were being manufactured and shipped to the arsenal throughout late 1864 and early 1865 at a rate of about 50 per month, placing the total number in storage at about 400 in February 1865. We know that 400 carbines were destroyed, or were reported to have been destroyed, by Sherman's troops in Columbia in February 1865. By mid-February 1865, the total number produced was about 1,000, 600 of which were issued and 400 destroyed or captured. We know that some carbines were manufactured in Greenville in March and April 1865 and that a total of about 1,032 were manufactured, giving a nearly full accounting for most of Morse's carbines.[6]

Magrath was still very much interested in keeping the State Works functional in March 1865. He wrote to the presidents and directors of the iron works in South Carolina on March 1, noting that the recent invasion of the state resulted in destruction of communication and transportation systems, many of which, especially railroads, could not be easily repaired. Magrath suggested that the state's commerce must revert back to wagon transportation. He wrote: "To supply this necessity and provide transportation, the State Works will need iron."[7] He also wrote that iron would be needed to replace the arms destroyed by Sherman's troops in Columbia: "These State Works at Greenville are the chief, perhaps the only, modes in which the great and present wants of the State can be relieved. But to your iron works must we look for the supplies which will enable the State Works to do their work."[8] Magrath concluded by saying that he hoped that "such arrangements can be made for the supply of iron from your works as to enable the State Works at Greenville to be actively employed in the production of those things which will contribute to relieving the necessities of the people, and diminish the general suffering of the state."[9]

Eason wrote to Magrath on March 2 to say that he had arrived at the State Works by train the day before. The letterhead he used was that of the Office of the General Superintendent of the State Works, though he did not technically hold that position. Eason wrote that he planned to go to work at once manufacturing percussion caps.[10]

Smith's accounting ledger for the first quarter of 1865 showed a balance of $9,900 going into the first quarter and $5,010 at the end of the quarter.[11] The State Works' income for that time was about $42,000. Tupper's office continued to provide the State Works

with its single largest source of income—about $35,000 for the first quarter. Profits from the sale of provisions to workmen were next, amounting to about $4,500. Miscellaneous income came primarily from the repair of small arms, which generated about $1,200.

Smith's expenses for the first quarter of 1865 were $46,641. The largest expense categories were payroll and salaries, $21,103, and provisions, $20,144. Lumber and building supplies, stock, charcoal, fuel, and incidental expenses made up the rest.[12] Smith's expenses in 1865 were for the usual items such as oak and pine cordwood, charcoal, old cast iron and brass, pork, lard, corn, ink, and fodder, and renting a post office box. He was still doing some construction work at the State Works when he bought 5,445 feet of pine lumber and hired a man named Self to do some masonry work. Smith got a raise in 1865; his annual salary went from $3,000 to $3,500.[13] Morse's annual salary, however, remained at $3,000.

Julius C. Smith, State Works traveling agent, left Columbia on February 13, only four days before Sherman entered the city, and continued to fulfill the duties of his office until at least March 1865.[14] From December 2, 1864, through March 14, 1865, his travels were limited to the still-operational Greenville and Columbia Railroad, including Pomaria, Allston, Newberry, Belton, and New Market near Greenwood where he purchased hogs and beef and shipped them back to Greenville.

Production continued at the State Works along the same general lines as late 1864. The State Works produced Morse's carbines, ammunition, and related equipment at least through the end of March 1865. It also stepped up its production of agricultural, mechanical, and domestic products, and continued to manufacture and repair machinery and tools, as well as brass castings and bar iron.[15]

Just as he had done previously, Smith continued to outsource some of the leather work in 1865. In late 1863 and early 1864 he had used Gower, Cox, and Markley. On December 16, 1864, Smith paid J. C. C. Turner $411 for making 548 carbine slings, and on February 24, 1865, he paid Sarah Taff $264 for making 352 carbine slings.[16] This made a total of 900 carbine slings, and if one assumes that the 179 gun slings made by Gower, Cox, and Markley in late 1863 were also for carbines, the total comes to 1,079, roughly approximating the total number of carbines made.[17]

In a letter that surely would have made Gist, Lopez, and Smith cringe, Eason wrote to Magrath on March 18, 1865, offering his analysis of the State Works and recommending sweeping changes after having been on site for only seventeen days:

> The past week has been spent at this place, in all of which time I hoped to be able to have an interview with your Excellency but your non arrival precluding the same, conclude to offer my opinion on paper, as to what is required to make these works be beneficial to the State at large. Outside of the manufacture of small arms, they can only be considered a neighbourhood jobbing shop, consequently doing but little good to the state. They have in all about one hundred and twenty men at work, while the machinery is capable of employing five times that number and can be only made self sustaining by a large increase, by self sustaining is meant to pay the interest on the present currency value, say one million dollars, the wear and tear, risk etc. and also yield a profit.
>
> Before the force of workmen can be increased, it will be necessary to erect some cabins adjacent to the works for the men to live in, for the village affords none. They

should be allowed rations and clothing as in the Army and Government workshops, which should be furnished by the proper state officers. Under any other system the men cannot sustain themselves for where their entire time is required at the works they cannot, in such a location either look up food or clothing[.] To give them enough in money to purchase at *retail* would appear extravagant and not serve the purpose, for men must be made cheerful to have the work done, they must be cared for, and shown that their labor is appreciated, for they are without the excitement of the army to cheer them on. My opinion is that all products of the works should be turned over to the proper issuing officer of the State, an invoice rendered at a price commensurate with the times, and they receipt for the same, issue or dispose of them at the invoice price, and in turn when they supply the works on requisition, let it be charged at cost price. There is no reason why these works should not be managed as an individual establishment, if the State will furnish the capital to make the purchases, the only objection is the present condition of affairs. The officers charged with gathering supplies can better regulate the whole than by having separate agents, who are coming in contact and competition for the same articles. The manufacturer should not be a Trader, Quarter Master, or Commissary, every man to his particular business when all have one common interest. The Treasurer, Quarter Master and Commissary should respect and comply with the requisitions made on them by your representative in charge of the works. With your acquiescence to all these suggestions, it will be some time before the whole could be completed, as before above stated accommodations must first be provided, before the men can be obtained. But with diligence and energy infused into all concerned, I could but hope that great good would result from the change of system, and increase of the works. If you concur with me in what I have advanced, and will make the order covering all, I will do my best to carry them out, and leave the result to demonstrate the wisdom of the course or if your Excellency should differ, I can wear the tittle [*sic*] you have been pleased to confer upon me, visit the works as often as necessary and allow things to proceed under the present system. I shall be pleased to hear from you at your earliest convenience.[18]

Greenville remained unscathed in April 1865, but the town was not completely in the clear yet. Colonel W. W. Perryman was commander of a company of soldiers at the post at Greenville. Anticipating a Federal raid, he notified Magrath on April 11 that he had ordered the men in his command under arms and that he could put six hundred men in the field if necessary. Perryman requested an immediate supply of ammunition and stated that he was cooperating with Colonel Martin.[19]

The State Works was still in business in late April. Thirty-five small arms, all in poor condition, had been discarded by the Army of Tennessee and were shipped from Newberry to Greenville for repair on the eighteenth.[20] The state militia, especially Perryman, was in desperate need of arms and ammunition. Instead of looking to his own State Works to fill the need, Magrath sent W. G. Eason, state ordnance officer, to Augusta, Georgia, to obtain arms and ammunition from Colonel Rains. Eason arrived at Augusta on April 14 to find out that Rains had no small arms or ammunition to loan or sell.[21]

Greenville was fortunate to have been spared exposure to Sherman's Army as it came through South Carolina in February 1865, but Federal cavalry did pass through the town

in early May. Fortunately, the war was essentially over and the troopers' mission was not to burn and pillage, but to capture Jefferson Davis.

Union General George Stoneman led six thousand cavalrymen from Tennessee into western North Carolina and southwestern Virginia on March 25, 1865, with orders to disrupt Confederate supply lines, destroy public works, and free prisoners at Salisbury, North Carolina. Robert E. Lee surrendered the Army of Northern Virginia at Appomattox on April 9, and Joseph E. Johnston surrendered the Army of Tennessee near Durham on April 26, effectively ending hostilities. Stoneman was in the vicinity of Asheville, North Carolina, when Johnston surrendered. Many of his troopers soon returned to Tennessee, but a cavalry division under Brigadier General Alvan C. Gillem was sent over the mountains into South Carolina to pursue Jefferson Davis, who had fled Richmond on April 3 and was moving in stages through southwest Virginia, North and South Carolina, and into Georgia by early May.

While Stoneman returned to Tennessee on April 19, he sent two brigades, one under the command of Brevet Brigadier General Simeon B. Brown (Eleventh and Twelfth Kentucky and Eleventh Michigan regiments) and the other under Colonel William J. Palmer (Fifteenth Pennsylvania, Twelfth Ohio, and Tenth Michigan regiments) to scout down the Catawba River towards Charlotte. Stoneman also ordered Colonel John K. Miller's Brigade (Eighth, Ninth, and Thirteenth Tennessee regiments) to Asheville.[22]

Stoneman ordered Miller and Brown's brigades in pursuit of Davis on April 27. Under Brown's overall command, the brigades were to cross the mountains near Flat Rock, North Carolina, follow the Saluda River down to Anderson and Belton, South Carolina, and march from there to Augusta, Georgia. Stoneman ordered Brown "to follow [Davis] to the end of the earth."[23] Palmer was to follow with his brigade. Brown and Miller's brigades left Asheville on April 28 and camped near Hendersonville that night. On April 29 they moved to Brevard and on the thirtieth they crossed into South Carolina at Saluda Gap, also called Jones Gap, and marched on to Pickensville.[24] The Twelfth Ohio of Palmer's command had already marched through Hendersonville and crossed into South Carolina at Saluda Gap on the twenty-eighth, but they were moving eastward into Spartanburg District as the other regiments in Palmer's Brigade—Fifteenth Pennsylvania and Tenth Michigan—paralleled them over the border in North Carolina. A smaller group of 150 troopers under the command of Major Joseph Lawson of the Eleventh Kentucky crossed the mountains and rode down Buncombe Road into Greenville on May 2.[25]

Brown's men fought two skirmishes in Anderson District on May 1, one near Piedmont and another near La France, and then rode into the town of Anderson later that day. They plundered Anderson on May 1 and 2 and finally moved on toward Georgia on the third.[26] It is possible that South Carolina troops fought with Morse's carbine in both skirmishes. Louise A. Vandiver, a historian from Anderson, South Carolina, wrote in 1928 that, during the fight four miles north of Anderson at La France (also called Pendleton Factory), a home-guard company of young boys from Pendleton under the command of a Captain Jones was engaged with Federal cavalry. Captain William R. Jones did command a company matching that description which was armed with Morse's carbine. Though Vandiver's account is plausible, it is anecdotal, and there is no supporting documentation.[27] Equally anecdotal is Vandiver's version of the skirmish near Piedmont,

about ten miles from Greenville and fifteen from Anderson. She records that A. D. Hoke commanded a company of home guards in Greenville. Hoke, a respected Greenville physician, had been captain of Company B, Second Regiment, South Carolina Volunteers, in 1861 and was severely wounded in the right arm at First Manassas. His health was so adversely affected by the wound that he resigned in May 1862. One of Hoke's lieutenants was W. P. Price, most likely William Pierce Price, who was a young Greenville lawyer and editor of the *Greenville Enterprise*.[28] According to Vandiver, Price "opened the state armory in Greenville and equipped the company with its arms" sometime on or before May 1. We do not know if these arms were Morse's carbines or some other firearms. Vandiver writes that Price and some armed civilians left Greenville before Federal troops entered on May 2 and fled toward Anderson. They caught up with the Arsenal Cadets near Piedmont and participated in a brief skirmish near there on May 1.[29]

Greenville residents had been hearing rumors of the Federal raid for several days. The Arsenal Cadets had been posted north of town near Marietta in late April, and on the night of April 30–May 1, they moved southward through town, surely informing the citizens that Federal raiders were on the way. Greenville's authorities asked the cadets' commander, John Peyre Thomas, not to make a stand in the town, hoping to avoid retaliatory destruction.[30] With Thomas's cadets marching on toward Anderson and Hoke's Company dispersed, Lawson's troopers met no opposition as they moved into town about mid-day on the second. Compared to Sherman's march through South Carolina, Lawson's men treated Greenville relatively well, generally ransacking the town and terrorizing many of its citizens, but leaving its structures intact.

G. W. Taylor, a gunsmith at the State Works, gave an interview to the *Enterprise and Mountaineer* in 1884. He was present when Federal raiders came through Greenville on May 2 and recalled that the State Works had been abandoned and all of the employees had scattered before Lawson's troopers entered the town. Taylor said, "Everything of any value had been removed, except much machinery and some guns, the latter being boxed up."[31] In an effort to hide the gunpowder from the Federal raiders, one of the last things Taylor did was to assist in placing a number of kegs of powder in a well on the Foundry premises, which had been specially prepared to receive them without putting them into the water. Taylor's recollection is consistent with Vandiver's account of W. P. Price taking some guns from the State Works before Lawson rode into town.

The following statement, attributed to Morse as having been written in 1866, also suggests that the portable contents of the State Works were stolen at the end of the war but that there were no carbines there at the time:

The few arms manufactured, were kept exclusively for State Service, and the Governor has always positively refused to allow them to go into Confederate service. Captain Boykin's company obtained about 100, and when mustered out of State service, these arms were all, by order of the Governor returned to me. These works have been employed most of the time in the manufacture of tools, and in assisting farmers, and have turned out in three years, 1000 guns which were deposited in "State Arsenal" at Columbia, where they were destroyed by General Sherman's Army. In the fall of 1864, the shops were advertised for sale, but at the suggestion of myself and others, the legislature determined to convert them to agricultural works for the purpose of supplying

the country with the necessary implements for farming . . . and continued in this field until raided by a mob, which carried off everything portable.[32]

It is certainly possible that any number of civilians, Confederate troops, and Federal cavalrymen entered the State Works and carried off weapons and tools in early May 1865. But so much in the way of machinery and tools remained there in October 1866, including numerous portable items such as three hundred dozen files and rasps, that Morse must have been incorrect in writing that a mob carried off everything portable.[33]

Major Lawson was forty-two years old at the time and was from Kentucky. He had been captain of Company G, Eleventh Regiment, Kentucky Cavalry, since November 1862, and was commissioned major in January 1865. He commanded a battalion in the regiment from that date, but was in charge of a group of detached men in early May 1865.[34] According to Taylor, Lawson rode into Greenville and first asked if there were bushwhackers in town. Upon hearing there were none, he assured the citizens of Greenville that his men would respect their property and persons. He even assigned individual cavalrymen to some houses to protect property and occupants.[35] Even so, a significant amount of plundering took place.

Lawson's troopers were primarily looking for arms and horses and behaved only slightly better than Brown's men in Anderson.[36] They entered private dwellings, ransacking some and terrorizing the dwellers, while leaving others alone. Mrs. Arthur M. Huger wrote that a cavalryman strode into her house, demanding to know if any firearms were there, but ignoring silver forks, spoons, and cups in plain view and assuring Huger that no one would be harmed.[37] Mrs. Caroline Howard Gilman, a refugee from Charleston, requested and received a Federal trooper to protect her home.[38]

Advancing into Greenville, the raiders marched to the Goodlett House on Main Street, which was occupied by a Confederate Wayside Hospital and a relief organization known as Soldiers Rest, managed by the Greenville Ladies Association in Aid of Volunteers of the Confederacy. The raiders proceeded to strip the rooms of Soldiers Rest of every article, leaving the society without the means of carrying on any further operations and effectively putting an end to the ladies' efforts to offer shelter, food, clothing, and sympathy to the soldiers.[39] Gilman wrote, "The Raiders, about two hundred in number, went to Main Street and opened the Commissary Stores, robbed the Bank, pillaged every article from the rooms of the Ladies' Association, and then proceeded to private houses and property. The refugees from the Coast had put their property in various closed chambers in empty stores three years ago, all over town. Everything was rifled. Books, costly plate, wines, pictures, bed linens, thrown into the street to be picked up by any passer-by. All the afternoon we saw white and black, laden with goods, passing by the house."[40]

Some of Lawson's men entered the home of Dr. James P. Boyce, who was a theologian at the Southern Baptist Seminary in Greenville, former chaplain of Greenville's Sixteenth Regiment, and a state legislator.[41] Boyce held the rank of lieutenant colonel as an aide-de-camp to Governor Magrath in late 1864, and as provost marshal of Columbia he claimed that he was the last Confederate to leave Columbia as Sherman's troops entered the city in February 1865. Lawson's men might have been aware of Boyce's history, but more likely they were attracted by his large house on the edge of town as well as rumors

that Boyce owned a large amount of plate, jewelry, and diamonds. They threatened him with drawn pistols in an effort to locate the valuables but were unsuccessful. Having plundered his house "in a wild search for precious things," they finally left with his horses, watch, and other personal items.[42] Gilman wrote that one trooper pointed a pistol at an eight-year-old named Forsyth, and the child "fell insensible," never recovering from the shock.[43]

Only one house, belonging to Captain Wesley Brooks, was set afire, but it was quickly extinguished. Lawson's men also robbed a bank, which was operating out of Hamlin Be-attie's store. Taylor recalls that it held eight thousand dollars in gold coins, but other ac-counts place the figure at thirty thousand in Confederate paper money. Lawson learned that a significant amount of wine was stored at David's Warehouse on Pendleton Street in West Greenville. Concerned that he would not be able to prevent his men from burning the town and abusing the people if they were drunk, Lawson ordered all of it poured out into the streets, and the gutters flowed with wine for some distance. Much other property stored in the warehouse was thrown out into the street and carried off by both white and black.[44] The warehouse was demolished, and large quantities of valuable articles in it were lost.[45]

The raiders also found their way to the railroad depot just off Main Street, where they destroyed thirty or forty bushels of salt. Taylor said that a large amount of tax-in-kind, such as cloth, belonging to the Richmond government was being stored at the depot. The raiders took it out and gave it to whoever wanted it. It is logical that Lawson's men would have ridden only a half-mile farther down the railroad track to examine the State Works. There is only a single, yet reliable, account that definitively places Lawson's men at the State Works. Mrs. Gilman requested and received a personal guard for her home, and a Lieutenant West responded by assigning a soldier to guard her house.[46] West later returned to Gilman's house at twilight with a stand of U.S. colors and, approaching the piazza, he said, "I hope you will excuse me madam, for bringing my flag, which I have captured in the foundry. I hope I don't hurt your feelings. . . . I should be very sorry to do so"[47] He then walked into the parlor, unrolled the flag and extended the flagstaff along the high, old-fashioned mantel, allowing the flag itself to spread out fully against the wall. Another soldier, named Simpson, arrived at the Gilman house for supper, carrying the cloak and watch of James P. Boyce. Simpson asked West how many Rebs he had bagged that day, and West relied, "None, but I captured a splendid number of guns, and a stand of colors at the foundry."[48]

Likely, only two people were killed during the raid on Greenville. Josiah Choice was a fifty-seven-year-old man who was killed by Lawson's troopers on May 2. One version of events is that Choice wounded a Federal cavalryman and was shot dead. Another is that he was shot when he threatened to kill anyone who took his horse.[49] He was buried at Springwood Cemetery.[50] Taylor recounted that a former slave who probably belonged to the estate of the late Vardry McBee shot at him from a distance of several hundred yards as he crossed an open field between Pendleton Street and the Gaillard School House. Taylor was not hit, and the next day Lawson's troopers caught up with the man in the woods near the State Works, where they shot and killed him "for impudence."[51] Lawson's "discretion, judgment, and rightmindedness" were credited with Greenville being spared

the disastrous fate of Columbia and several other South Carolina towns.[52] By the evening of May 2, Lawson and most of his men rode on toward Anderson.

Magrath formally suspended the functions of his office on May 22, 1865. He was arrested on May 25 and taken in an ambulance to Fort Pulaski, where he joined the secretary of treasury, George A. Trenholm. Magrath was released in December and returned to the practice of law in Charleston.

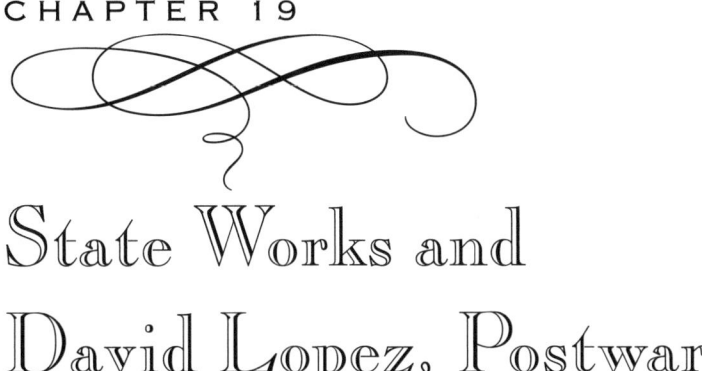

CHAPTER 19

State Works and David Lopez, Postwar

Another attempt to dispose of property and machinery at the State Works came up in November 1865 when the South Carolina Senate referred a resolution relative to the sale to the House Committee on the Military. The House concurred with the Senate, agreeing that the State Works was no longer of any practical use to the state and should be sold. The Senate recommended that a commission be responsible for selling the State Works, but the House felt that the better approach was to entrust James Tupper, who was still state auditor, to handle the transaction. The House recognized that because Tupper was intimately involved in previous efforts to sell the State Works he would be the best person for the job, and they recommended allowing him to either sell the State Works for cash or on limited credit, as he deemed best.[1] A Special Joint Committee of the General Assembly met on November 13, 1865, to resolve the differences between the House and Senate concerning the sale. They agreed that the sale of the property, along with all its machinery and materials, was essential to avoid further loss to the state and should be accomplished as soon as possible. The committee preferred a cash sale as long as the price was not seriously sacrificed, but it agreed to authorize a sales agent the discretion to extend limited credit to a buyer for those portions of the property that could not be sold for cash. If credit was extended, the title was not to be delivered until the purchase was made in its entirety, and the state reserved the right to resell the property if the buyer failed to comply with the terms. The committee recommended appointing four prominent Greenville men, John W. Stokes, C. J. Elford, G. F. Townes, and H. P. Hammet, as commissioners to oversee the sale.[2] Both houses of the General Assembly agreed on November 13 to authorize the commissioners "to sell the [State Works] at auction or private sale, as in their opinion may be most advantageously done for the State, for cash, or on such limited credit as they may believe advisable."[3]

The *Charleston Courier* reported on February 16, 1866, "The property at the State Works in Greenville had been seized by the United States authorities, and, unless the Secretary of the Treasury shall reconsider his recent decision, it will, we presume, be sold by

them before very long. This will be a serious loss to the State, and will serve to increase, of course, the burdens of taxation upon all our people."[4] The *Southern Enterprise* clarified on March 29, 1866: "We learn that the State Works at this place are to be turned over to Governor [James L.] Orr, provided their assessed value does not exceed the sum of sixty thousand dollars. All finished and unfinished arms and ammunition to be taken by the United States authorities."[5] Recognizing that the late Vardry McBee had donated twenty acres of land for the State Works, the General Assembly soon authorized a portion of the proceeds of the sale to be set aside as a token of appreciation to his estate.

On May 1, 1866, a Major Parker provided a list of stores at the "State Arsenal" in Greenville, and Major General A. B. Dyer, chief of ordnance, recommended to the secretary of war that Parker be allowed to sell them at public auction because the items on the list were of little value and were not of such character to justify sending them out of the state.[6] Parker's list does not survive, but another one does. After trying to sell the State Works since 1863, the General Assembly was finally successful in selling the remaining machinery in 1866 and the land in 1872. The commissioners appointed by the legislature to sell the land and machinery at the State Works announced that its contents would be put up for public auction beginning October 17, 1866.[7] In a sixteen-page catalogue, the commissioners made available to the public much of the machinery and tools that Lopez and Morse had so painstakingly acquired in 1862. Not all of it was auctioned, however. Some of the equipment from the State Works went to the new penitentiary authorized by the General Assembly to be built in Columbia in 1866. In late November, Governor Orr directed the commissioners to "reserve such machinery as might be usefully employed in the manufacture of wood and iron in the Penitentiary when fully in operation. The reservation was made, and there will be in the future only a trifling outlay needed for machinery to operate most branches of the manufactures of wood and iron."[8] In the fall of 1866 the General Assembly also authorized the former superintendent, J. Ralph Smith, to purchase some of the machinery. Smith had resumed his family business in Charleston after the war and was planning to use the machinery there. The South Carolina Senate passed a bill on December 20, 1866, authorizing the commissioners for the sale of the State Works to sell to George Morse, or any other purchaser, the personal property connected with the State Works that had not yet been sold, providing the sales price was equal to the appraised value.[9] Between Morse, Smith, the penitentiary, and the public auction, we must assume that most, or all, of the equipment in the State Works was removed from the property in late 1866.

The commissioners also wanted to sell the property. The *Mountaineer* reported on November 22, 1866, "The real estate of the State Works at this place has been sold to Dr. S. S. Marshall and other gentlemen associated with him, who intend converting it into a cotton factory, which will be put in operation as soon as the necessary machinery can be obtained."[10] Samuel Steene Marshall, M.D., was born on August 10, 1819, died on March 6, 1883, and was buried at Christ Church Episcopal in Greenville. Land records at the Greenville County courthouse show that Marshall did, in fact, purchase a twenty-acre piece of property in Greenville District on October 1, 1866, but it was located along Richland Creek and the Paris Mountain Turnpike, two miles away from the site of the State Works.[11] It is possible that Marshall also purchased the State Works site at about the same time, but a deed for such purchase cannot be located. In 1872, W. J. Whipper of Beaufort

om Mexico! New,
hionable perfume.
l or manufactured
ry it and be con-

ed Sweet Opoponax
red by E. T. Smith
aking a sensation
Is very delicate,
s on the handker-
Evening Bulletin.

w Perfume from
nionable Perfume

only elegant Per-
toilets, and never

the sweetest Ex-
rsedes all others.
ther.

ies, in their morn-
l gladness, when
oponax.

CELIBACY—An
Instruction for
ases and Abuses
powers, with sure
free of charge in
Address Dr. J.
Howard Associa-
Aug 15 3mo

EY SOAP.

t Soap, in such
made from the
nild and emol-
rantly scented,
al in its action
by all Druggists
s. March 28 1y

IAIR DYE.

t in the World.
ect HAIR DYE.
l Instantaneous.
splendid Black or
injuring the hair
ill effects of bad
sts. The genuine
helor. Also, RE-
CT OF MILLE-
and Beautifying
ATCHELOR.
New York.

H IN MEDICINE.—

the Legislature of South Carolina
to Sell the State Works.

EXTENSIVE SALE
OF
MACHINERY,
Materials, Tools, &c.,

AT public auction, at the State Military Works, Greenville, S. C., beginning on WEDNESDAY, 17th October, 1866, consisting of:

2 Steam Engines—one 25 and one 40-horse power.
Engine Lathes, Milling Machines, Vises.
Hand Lathes, Wood Turning Lathes.
Drilling Machines, Punching Machines.
Anvils, Grind-stones.
Trip Hammers and Forges, Steam Hammer.
Smiths' Tongs, Smiths' Hammers, Shafting.
Screw Cutting Machines.
Wire Drawing Machines.
Morticing Machines.
Bellows, Machine Blowers, Portable Benches.
All descriptions of Gun Machinery.
Wood and Iron Planing Machines.
Circular Saws.
Leather Belting, Rubber Belting.
Flasks, Patterns, Pouches, Wrenches, Drills.
Several hundred gross Iron and Brass Wood Screws.
300 doz. Files and Rasps.
2,500 lbs. Cast and Spring Steel.
14,000 lbs. Swedes and American Iron.
Sheet Rubber.
500 cords Pine Wood, Coke.
Brass and Iron Trimmings and Filings.
56 pair Window Sashes.
Corn Mills, Hand Screws, Wagon Wheels.
Planes, Augers, Braces and Bits.
Stoves and Pipes, Fire Brick.
And numberless other articles.

All these are in large numbers and quantities, and constitute a stock of Machinery, &c., such as is seldom offered.

A catalogue, of sixteen pages, has been printed, containing a general description of the Stock, which may be had on application by mail to the Secretary.

☞ Terms cash, or a note at sixty days, with approved security, bearing interest, at option of the purchaser.

For further information, apply to
C. J. ELFORD,
Sec'y to Commissioners State Works,
Sept 7 f6 Greenville, S. C.

South Carolina Railroads.

New
CLOT
FRENCH AND ENGI
CASSIMERE
HAT

THE subscribers h
will open THIS
voices, comprising
CLOTHING, ENGI
CLOTHS, CASSIMI
first importation to th
Our READY-MAD
pally from our own
made up in the lates
Our stock of mate
embraces every styl
to order at the shor
tomers in the countr
send their orders,
promptly attended t
We will also open
ment of HATS, of t
latest fashion.
We invite the at
friends in the city
feeling assured that
satisfaction to all
themselves.

R. & W. C.
Sept 19

Cocka
PARLOR
SOLD to the TR
any other Match
Sept 5 2mo CA

CALNAN &
DEA
GROCERIES,
Wines, Liqu
SELECT GOODS
never offered
COST.

An announcement of the public auction of the State Works on October 17, 1866, including an extensive list of machinery and tools remaining there eighteen months after it ceased operations. *Daily Phoenix* (Columbia, S.C.), October 5, 1866.

was a special commissioner for the Sinking Fund Commission of the State of South Carolina, which was established to sell off unproductive state assets. Whipper was authorized to sell the State Works property in 1872, implying that the state still owned it and that Marshall probably did not purchase it in 1866. Vardry McBee's son, Alexander, purchased the property at auction for $2,850 on March 4, 1872. The deed describes the property as having been conveyed to the state in August 1862 and does not list any other owners between that time and 1872, again making it unlikely that Marshall ever owned it.[12] Possibly the deal with Marshall and his fellow investors fell through in November 1866, and they chose instead to purchase a different piece of property.

Amidst of all this activity, Morse, who was still living in Greenville, wrote to one of the commissioners, G. F. Townes, on November 27, 1866, requesting that the state reimburse him $1,200. Morse claimed that the state owed him $400 in unpaid salary, $150 for the use of personal tools that he had hired out to the state, $150 for damage to his tools, and $500 for the use of his patents in the manufacture of the tenth batch of one hundred carbines at $5 each.[13]

Little is known about the fate of the buildings on the grounds of the State Works. About 1866 or 1867, Charles T. Hopkins purchased one of the storehouses there and planned to use it for a schoolhouse.[14] Hopkins was a former slave from the South Carolina Lowcountry who was "for many years a voluntary exhorter among his people" and "a meek, amiable, judicious, virtuous, godly man, zealous for the good of the freedmen."[15] Born about 1820, he obtained an education and by 1870 was a trial justice living in the Butler Township in Greenville and, later, a minister of the gospel. Hopkins was instrumental in working with the Freedmen's Bureau to establish an elementary school which met in 1866 at the Goodlett House, a former Confederate Hospital seized by the U.S. government after the war. When the Goodlett House was returned to its owners in 1866 or 1867, Hopkins raised $260 from Greenville's white citizens to purchase a new school building. He bought one of the storehouses from the State Works site and moved it two miles to a lot on Laurens Street, which he had leased from Randall Croft.

It is hard to know for certain which of Lopez's structures Hopkins purchased. Lopez built one store house measuring sixty by twenty-five feet, and another storage building measuring thirty-five by sixteen feet. Regardless of which building was moved, the fact that it was movable at all attests to the quality of the original construction. The building was remodeled, and the Freedmen's Bureau ultimately covered the remaining costs. The school eventually became the social center for freedmen in Greenville.[16] Some buildings at the site were probably dismantled, and the lumber used to make other structures and for firewood. Some of the material might have been used in 1884 to build Grace Cathedral for black Episcopalians on White Horse Road near Green Valley.[17] The State Works site was used for the Fourth Grand Annual Fair of the Greenville Agricultural and Mechanical Association of South Carolina in 1876.[18] Soldiers might have used the site as a campground during the Spanish-American War.[19] By 1910, all of the machinery and buildings had been removed from the site, and the sole surviving warehouse, which had been moved elsewhere and used for cotton storage, as a building-supply business, and as a livery stable, among other things, was destroyed that year.[20] Today, private homes and a school occupy the site.

After the war, Lopez remained in Charleston. By December 1865 he was a representative of a New York–based company that sold heavy machinery.[21] Lopez resumed his building enterprise in Charleston with his sons, John and Moses, no later than mid-1866, operating as David Lopez and Sons until 1869 when John left to create his own business.[22] In 1866 David was elected president of Charleston's Hebrew Orphan Society, an organization his father had founded in 1801. In 1870 Lopez was a builder, supporting an extended family living in his home. Three of his children, his sister, two grandchildren, and a daughter-in-law were living with him at the time.[23] Possibly seeking more secure income, Lopez became superintendent of a phosphate manufacturer, the Coosaw Mining Company, in 1870. Moses worked with his father at the six-hundred-employee company, where they developed new mining techniques and made the company extremely profitable. Lopez's success in the phosphate-mining industry helped make the state a world leader in that area and assisted in its recovery from Reconstruction. Lopez held the position until his death.[24] In 1880 he was living in Charleston with his widowed daughter, Priscilla, her children, and his sister, Sally.[25] The seventy-five-year-old Lopez died from congestion of the brain on April 21, 1884, while visiting Philadelphia. He was buried three days later at the plot he originally purchased for his first wife, Catherine. It was later incorporated into Kahal Kadosh Beth Elohim's Coming Street Cemetery in Charleston.[26] Moses E. Lopez was superintendent of the Coosaw Company from 1893 to 1899. He died from chronic nephritis in Atlanta on June 3, 1907.

CHAPTER 20

Morse's Postwar Patent Petitions and Lawsuits

Like many Southerners dealing with the war's aftermath, Morse struggled to carry on with his life. Captain William James, provost marshal in Greenville, administered the Oath of Allegiance to the United States to Morse on August 30, 1865. It said, "I, G. W. Morse, do solemnly swear in the presence of Almighty God, that I will henceforth faithfully support and defend the Constitution of the United States and the Union of the States thereunder, and that I will, in like manner, abide by and faithfully support all Laws and Proclamations which have been made during the existing Rebellion with reference to the emancipation of Slaves—'So help me God.'"[1]

Morse's wife had joined him in Greenville no later than December 5, 1864, when her name appeared on the membership roll of the Greenville Ladies Association in Aid of the Volunteers of the Confederacy.[2] During the war, Morse rented a house and five-acre piece of property on Garlington Street, now called Arlington Avenue. It was about a half-mile from the State Works and about a mile from the Greenville Court House.[3] Morse purchased the house and property from Mrs. M. M. O'Neal on September 1, 1865, for eleven hundred dollars, demonstrating his intention to continue living in Greenville.

The secretary of war ordered a board of officers to assemble in Washington, D.C., in March 1866 to consider the best breech-loading design for infantry and cavalry arms, as well as for alterations of existing muzzle loaders. No fewer than eighteen carbine models and twenty-two rifles and muskets were entered in the competition, but Morse did not submit a design to the board.[4]

The Morse family was still living in Greenville in 1870, but had probably moved to the opposite side of Pendleton Street.[5] Their home at that time was in Ward 5, which was bounded on the north by Reedy River, the east by Main and Pendleton streets, and the west and south by city limits, which extended in a one-mile radius from the courthouse. The 1870 U.S. Census showed George's age as fifty-four years old, but he was actually

fifty-eight. He was listed as a planter, meaning that he was still planting his land in Louisiana and deriving his income from that source. Morse testified in 1877 that the war had left him penniless, which was not completely accurate. Though the war had, in fact, cost him dearly, Morse was worth $25,000 in real estate and $300 in personal property in 1870, and his wife was worth $1,300 in real estate and $8,400 in personal property.[6] The Morse's 1870 net worth in today's dollars would be at least $500,000. Also living in the house in 1870 were their children Lelia and Bryan, as well as some servants and a twenty-three-year-old nephew, E. A. Sompayrac. At the same time, George and Marianne's first son, Peabody, was living in Natchitoches, Louisiana. The Morses would live in Greenville for only about two or three more years.

Morse's death certificate stated that he had lived in Washington, D.C., for nineteen years preceding his death in 1888, placing him there in 1869.[7] It is likely that he began shuttling between Greenville and Washington about 1869 as he launched a lifelong campaign to regain control over his patents and recoup lost money. Though his primary residence was still in Greenville, Morse visited his brother, Isaac, in Massachusetts on August 13, 1870, as they worked on a statement as part of a plea for renewal of his patents.[8] Morse stated in a petition to the U.S. Congress dated January 20, 1872, that he was living in Greenville, South Carolina.[9] At some point in 1872 he was in Jefferson, Texas, and later the same year he was in Natchitoches, Louisiana, though the exact extent and nature of his business in those places is not preserved.[10] It appears that Morse and his family moved permanently to Washington, D.C., sometime between mid-1872 and late 1873. A letter dated June 7, 1872, shows him to be living in Greenville, but he signed another letter on June 29 in Washington, D.C., and on November 27 of that year Morse wrote that he was temporarily living in Washington.[11] The Morse family had already moved to Washington when they sold their Greenville property to Mrs. Katherine M. Bollin for $1,200 on November 11, 1873.[12] Washington city directories initially listed Morse in 1874. George kept his family in Greenville for seven or eight years after the war because he was cash poor. Though his cotton crop and gin were burned at the end of the war, he attempted to plant his land in Louisiana after the war but failed, losing $8,000. Staying in Greenville was his best option until he could recover financially. His plan to do so lay in the renewal of his patents.

Reflecting his mental state as well as his fiscal condition, Morse wrote in an August 1870 appeal for the renewal of his patents, "I am now poor and broken down, while the gun makers who have taken advantage of my invention are rich and prosperous."[13] Overshadowing the rest of Morse's life was the sentiment that others had usurped his inventions, and he attacked them with as much vigor as he had displayed in the development of his patents in the late 1850s. About 1870 Morse began to fight back in an effort to regain control over his patents; the fight went on for the rest of his life and beyond. He launched a multipronged assault to regain his fame and fortune. First, he simultaneously sued the Winchester Repeating Arms Company and initiated efforts to renew his patents via the U.S. Patent Office and the U.S. Congress. Second, he appealed directly to the U.S. Congress for financial compensation, and third, he brought a series of lawsuits against the U.S. government.

Winchester Suit

About October 28, 1870, Morse claimed in the U.S. Circuit Court in the District of Connecticut that Winchester Repeating Arms Company had infringed on his patents. George and his brother, Isaac, entered into an agreement with Spencer H. Cole and Henry S. Lambert, both of Brooklyn, New York, on December 19, 1871, in which Cole and Lambert would receive one-half of all money received in any legal action involving infringement on Morse's patents.[14] The Morse Arms Manufacturing Company was formed on July 25, 1872, when Morse assigned his patents, licenses, and agreements to the new entity.[15] James A. Skilton, Cole, Lambert, and George Morse formed the new corporation, which was based in New York City.[16] Cole, Skilton, and Morse were listed as managing trustees for the first year, and Morse's patents represented its capital.[17] The stated purpose of the company was "for the manufacture and sale of breech-loading firearms and ammunition therefore under the general patents heretofore granted by the United States of America to George W. Morse."[18] The initial value of the company stock was seventy-five thousand dollars, divided into three thousand shares of twenty-five dollars each, and the corporation would last for twenty years, beginning August 1, 1872. In reality the purpose of the new company was not to manufacture firearms but rather to initiate lawsuits in order to regain some financial compensation for Morse.

On August 21, 1872, the Morse Arms Manufacturing Company and the Winchester Repeating Arms Company reached an agreement in which Winchester agreed that it had made and sold firearms that contained a substantial part of Morse's inventions. Winchester paid Morse ten thousand dollars in damages for patent infringement, and gave a promissory note to pay an additional fifteen thousand if certain conditions were met. Upon payment of the fifteen thousand, Morse would consider the matter to be fully settled and would discharge his company's claims. Claiming that it had been given false information, Winchester did not pay the fifteen-thousand-dollar note, and another lawsuit ensued.

The Morse Arms Manufacturing Company brought a new suit against Winchester on November 23, 1874, or 1875. Morse made two claims of infringement on his first firearm patent, 15995. First, he claimed that Winchester infringed upon his breech-block patent in which the front of the breech block comes into contact with the cartridge, which in turn contacts the rear of the barrel. Second, Morse claimed that Winchester infringed on his use of nippers to extract the spent cartridge. More than a decade later, in 1887, Winchester filed a cross-bill contesting the original agreement on the grounds that the ten-thousand-dollar payment was based on fraudulent papers and stating that Morse's two claims did not apply to all breechloaders and all extraction systems. Winchester sought reimbursement of the ten thousand dollars it had already paid Morse and cancellation of the fifteen-thousand-dollar note. A hearing was held on July 18, 1887, and Judge Shipman of the U.S. Circuit Court found that, instead of false information regarding the promissory note, a mistake had been made and dismissed both companies' claims, effectively concluding Morse's dealings with Winchester.[19] Of note is that V. D. Stockbridge testified in favor of Morse in this lawsuit, and he would do so several more times in Morse's suits

against the government.[20] Morse's early success against Winchester surely encouraged him to pursue similar matters with the U.S. Patent Office and the U.S. government.

Morse applied to the commissioner of patents for extensions on his first two 1856 patents, 15995 and 15996, in 1870, but both were rejected. Stockbridge, who worked at the Patent Office as an assistant examiner specializing in firearms in 1870, became a strong supporter of the originality of Morse inventions throughout the 1870s and 1880s. Stockbridge delivered his opinions on both patents to the commissioner on October 7, 1870. Concerning the firearm patent, 15995, Stockbridge did not oppose Morse's application for extension. He wrote that the part of Morse's invention covered within the scope of the claims was undoubtedly new and was still useful. Concerning the cartridge patent, 15996, Stockbridge wrote that the concept of a closed-cup metallic cartridge did not present any remarkable or decided advance in 1856, but that a cartridge combined with a primer as described by Morse was new.[21] Likewise, Stockbridge did not oppose Morse's application for renewal of 15996. Morse's lawyer, E. L. Sherman, wrote to Samuel A. Duncan, assistant commissioner of patents, on October 22, 1870, stating that he had received a letter that very morning saying that Duncan wished to have more information on Morse's loyalty during the war before making a determination about his patent extensions. Duncan gave Morse only five days, until October 27, to respond, but Sherman wrote that he did not think he could locate Morse in time to comply. Sherman thought Morse was in Boston with his brother, Isaac, and he immediately sent Isaac a message instructing him to have Morse respond with the requisite affidavit and forward it to Duncan at once. Sherman told Isaac that, if Morse could not be reached, that he (Isaac) should submit such an affidavit as he could swear to. Sherman wrote to Duncan, stating that Morse had always maintained that he was never in the "rebel army and I know he belongs to a very worthy and noble family in New England."[22] Sherman's statement was factually correct, but it was also an incomplete and misleading description of Morse's wartime activities. Sherman was leaving for Chicago the next day and planned to return on October 28, one day after Duncan's deadline.

Morse, who claimed in 1876 that Duncan had no legal right to make such a request, was away on business as a traveling agent for a Massachusetts-based company in late October 1870.[23] Because it was impossible to reach him before the October 27 deadline, Isaac complied with Sherman's instructions and wrote a supporting, albeit very inaccurate, affidavit on October 24: "I know that he [George] made every reasonable effort to derive benefit from his inventions but without success."[24] According to Isaac, all of George's efforts to promote his inventions had failed, and George had maintained neutrality during the war. Isaac wrote, "I have every reason to believe he kept his promise [not to raise his hand against the flag of the United States] and that he was not inclined and did not in any way aid in the rebellion, on the contrary I am confident that he resorted to various plans to keep out of the rebellion and that he never took up arms against our Government or in any way aided those who did."[25] Isaac was either ignorant of the facts relative to George's actions during the war years, misinformed by his brother, or intentionally misleading in his statements. Isaac went on to describe how George lost all of his slaves, for which Isaac rejoiced, and how his brother's cotton crop and gin had been destroyed by fire, and that he was left without financial means of support. According to Isaac, George lost his desire

to pursue any of his inventions after the war, but Isaac felt that both the U.S. government and various individuals continued to derive great benefit from them. According to Isaac, it was he who advised George to apply for an extension to his patents, and he who furnished George with the money with which to make the initial deposits.[26]

J. M. Thacher, acting commissioner of patents, wrote on November 29, 1872, that both extension applications were denied in 1870 mainly because Morse had failed to show due diligence, especially during the period covered by the late Civil War.[27] Though Isaac's misleading statements about George's wartime activities led to a temporary victory with the Patent Office in late 1872, both Isaac and George's misrepresentations ultimately came back to haunt George in 1879 when his final appeal to the U.S. Congress was denied in part because Congress recognized the disingenuous nature of their testimony and found that, "After the close of the war, by professions of loyalty, he [George] secured the passage of a bill by Congress authorizing the extension of his patents."[28]

After the failure in 1870 to obtain renewal of patents 15995 and 15996, Morse petitioned the U.S. Congress on January 20, 1872, to enact a bill that would authorize the commissioner of patents to extend the two.[29] The commissioner opposed the extension and accurately summarized Morse's activity between April 1860 and the end of the war.[30] Specifically, W. C. Dodge, attorney for opponents of the extension, wrote on May 22, 1872, that (1) Morse's invention was not new in 1856; (2) it was not useful and valuable to the public; (3) the extension would be injurious and detrimental to the public; (4) Morse had already been adequately remunerated; (5) Morse had not used due diligence; and (6) Morse had filed his patent with the archives of the Confederate government and used his invention for the benefit of the rebels against the U.S. government, thereby forfeiting any rights he might have for an extension.[31]

Repeating many of the claims in Isaac's 1870 affidavit, Morse's 1872 petition to Congress was filled with a mixture of fact and fiction. He accurately stated that he believed he was the original and first inventor of the improvements made in the patents' specifications. Further, he made the valid argument that his inventions were still valuable, and if extended, the patents would constitute "almost the only available property remaining, or within the possible reach of your petitioner."[32] Morse then provided a correct description of his prewar financial situation relative to the patents: "that your petitioner though he received royalties amounting to $13,000, under these said patents, and three subsequent patents, and six thousand five hundred dollars for an interest in foreign patents, covering all of said inventions, has failed to realize any net remuneration whatever from his said inventions, but that instead they have cost him a large portion of his income for four or five years, amounting probably, to six or seven thousand dollars annually, besides the constant labor of both head and hands while experimenting upon, perfecting and seeking to introduce his said inventions, during said four or five years, previous to the year 1861."[33]

At this point in the statement, Morse began to stretch the truth when he referred to his wartime activity. He wrote quite incorrectly that he "was unable to realize any remuneration during the war, because, though well known as a gun man by all the Confederate authorities who offered to adopt these said inventions, and urged by them to take charge of their manufacture, at the U.S. Armory, Fayetteville, North Carolina, he declined to do so."[34] Morse implied that he refused to support the Confederacy by turning down the appointment at the Fayetteville Armory, but there is no evidence that he was offered or

declined that position. He failed to mention that he was superintendent of the Tennessee State Armory from the summer of 1861 to the spring of 1862 at annual salary of thirty-five hundred dollars, and that he earned three thousand a year at the State Works from the summer of 1862 to the end of the war. He also failed to state that he most certainly did receive royalty payments from the state of South Carolina for its use of his inventions—both the brass-fame carbine and the inside-lock muskets. Further, Morse repeated similar claims in one of his postwar lawsuits. Morse stated under oath that he went south before hostilities started "partly to see what would be the upshot of things and as a man would go home out of a rain storm that was rising, and anyhow to place myself in a position that my property—my only real property I possessed in the world—could not be confiscated on account of my being absent from the country."[35] In the same deposition he denied holding any office under the Confederate government and denied giving any voluntary aid to the rebellion. Further he stated that he was offered a position in the Confederate Ordnance Office and refused, and reiterated that he was requested to make his arms at the Fayetteville Armory and refused.[36]

Morse also wrote in his 1872 petition to Congress that near the end of the war Confederate authorities burned nearly seventy-five thousand dollars' worth of cotton and a valuable gin house and buildings, depriving him of his entire available capital and preventing him from engaging in the development of his inventions from the spring of 1865 to 1872. He was left with his land but lost about eight thousand dollars attempting to plant it after the war. He failed to disclose in the 1872 documents that he and his wife were worth approximately thirty-five thousand dollars in 1870. Morse claimed that he was largely indebted and "hopes by means of these said inventions, which now seem to have become of some value, through the seed of your petitioner's own efforts and labors, planted long years since, to pay off his debts, and provide for the infirmities of advancing age."[37] Finally, Morse claimed that he had not received notice of the patents' impending expiration until after the patent commissioners had rendered their opinion not to extend them.

Morse signed a second affidavit dated about two weeks later, on February 1, 1872. Possibly feeling that he had not adequately explained himself in the first document, he dealt primarily with his loyalty to the Union in the second one, but in a very duplicitous manner. Morse's second affidavit is an astounding string of falsehoods. He said that he was opposed to the dissolution of the Union and was opposed to the movements of individuals who organized the Confederacy and to the movements of the Confederacy itself. He stated that he went south in 1861 solely for the purpose of protecting his property and family, and specifically not for the purpose of engaging in the rebellion. When he did so, he left four thousand dollars in personal property in the North which he directed his agent to invest in U.S. government bonds. He said that he did not enter the army or serve in any branch of service in the Confederacy, nor did he engage in any employment for or under the Confederacy. He wrote that he was urged to sell his cotton to the Confederate government but refused to do so. He swore that during the entire rebellion he studiously avoided all connection with the Confederate government. Morse testified that he embraced the opportunity to sign the oath of allegiance in 1865 and did so as soon as he could reach an official authorized to take his oath. He claimed to have kept the oath faithfully between 1865 and 1872, to be a true and loyal citizen of the United States, and he

expected to receive the benefits and immunities offered to such individuals. Morse closed by stating "that his grandfather fought in the Revolution to establish the independence of this Government and all his feelings were in favor of the Union of the States; and that no one could more deeply deplore the rebellion and the war for its support."[38]

The Forty-second Congress (March 4, 1871–March 4, 1873) referred a bill to its Committee on Patents, which agreed with all of Morse's assertions, including his claim that he remained in the South during the war and devoted himself to the protection of his family and personal property because the U.S. government was unable to protect it from confiscation. The committee noted that Morse in fact did apply for an extension to his patents within the time limit required by law. However, within six days of the expiration date, the patent commissioner notified Morse's attorney that Morse was required to submit an affidavit stating "whether facts and circumstances justified him in absenting himself from the north during the war."[39] The Congressional Patent Committee noted that Morse was absent at the time and did not receive the notification of the U.S. Patent Commission in time to comply with its requirements before the patents expired. The Committee on Patents determined that Morse was not negligent in his failure to provide an affidavit on time and recommended the passage of the bill. The committee introduced a bill (H.R. 1108) to the House of Representatives on January 22, 1872, titled "A Bill for the Relief of George W. Morse," which authorized the commissioner of patents to exercise jurisdiction over patents 15995 and 15996 and to grant or refuse Morse's petition for a seven-year extension based on his usual approach in such matters, and Congress passed it on March 11, 1872.[40] A minor threat arose in April 1872 when a bill was introduced to repeal the one passed on March 11. In response to this challenge, Morse's attorney, James A. Skilton, wrote on April 3, "We are willing and prepared to let the light shine to the bottom of our case. And when it is all presented it will make a story of inventive genius, life long struggle, hope deferred, claims ignored, ridiculed and traduced, together with world wide benefits confirmed upon mankind."[41] Apparently the April bill was defeated.

The patent commissioner vigorously opposed all extensions on the grounds that Morse was not the original inventor or discoverer of the inventions, but he received testimony throughout 1872, and as a result, extended three of Morse's five U.S. patents at various times that year.[42] First, the original firearm patent, 15995, was granted on October 28, 1856, expired October 28, 1870, and was approved for consideration of extension by Congress on March 11, 1872. The Patent Office granted a seven-year extension on November 29, 1872, with an expiration date of November 29, 1879.[43] The extension was granted most likely because of a petition and disclaimer Morse filed with the commissioner of patents on November 27, 1872, in which he reserved for himself and his legal heirs all rights, title, and interest in his inventions and in which he modified the original claim. Morse stated that he "had reason to believe that through inadvertence, accident, or mistake, too broad a construction may be placed upon" one clause of the patent, and he entered a disclaimer "to such a construction of said claim as would make it cover broadly all and every mechanism for performing the same functions without regard to substantial identity or equivalence of invention."[44] Morse's addition of three words, emphasized as follows, satisfied the patent commissioner: "Also the combination of *the* moveable parts *herein described,* or their equivalents, whereby I retract or deliver the gun of a cartridge, drop it, open and clear the way for the insertion of another cartridge, whether the previous cartridge was

fired or failed to fire, and cock the hammer automatically at one motion, substantially in the manner described."[45]

Second, in a decision that was particularly harmful to Morse in the long run, the patent commissioner did not extend the original cartridge patent, 15996, which was issued October 28, 1856, and expired on October 28, 1870. Morse correctly referred to the cartridge patent as "the very soul of the invention," and persisted in his efforts to get it renewed.[46] Because the statute of limitations on 15996 expired in 1878, Morse's claim to this patent was barred from his 1884 lawsuit.[47] Third, Morse's second U.S. firearm patent, 20503, had been issued on June 8, 1858. He filed a disclaimer as part of the renewal process on 20503 on June 7, 1872, stating that too broad a construction had been placed upon the clause describing the use of a lever for the purpose of extracting the cartridge case. Morse entered a disclaimer "to the use of said lever except when the same is applied directly to the cartridge case and is in fact the retractor of the said case."[48] The patent commissioner granted an extension for 20503 on June 7, 1872, and it expired on June 8, 1879. Fourth, another cartridge patent, 20727, had been issued June 29, 1858. Morse filed a disclaimer with the commissioner of patents on Patent 20727 on June 29, 1872, and was granted a seven-year extension the same day. The extension expired on June 29, 1879, but, because of insufficient proof, it was abandoned and excluded from Morse's 1884 suit. Fifth, Patent 20214 was not extended in 1872 for reasons that Morse said were obviously "an error of the then Commissioner of Patents, or for some other insufficient error."[49]

Encouraged by the renewal of three patents, Morse proceeded in his efforts to gain some degree of personal recognition, vindication, and remuneration. He presented a fifteen-page report, called a memorial, to the Forty-third U.S. Congress (March 4, 1873–March 4, 1875) in April 1874. His military title on the memorial header is shown as "Colonel," which he earned for his service in the prewar Louisiana Militia. When he presented another memorial to the Forty-fourth Congress only two years later, he did not use any military rank. The author has seen an occasional document, including his obituary, in which Morse is referred to as "colonel" but has never seen one which Morse signed as such, except the 1874 memorial.[50] The memorial was a highly technical, detailed document reviewing Morse's experience in dealing with water-control issues along the lower Mississippi River. He held strong opinions about the best approach to handling flood waters along the river and its tributaries, specifically the construction of levees and the use of low-lying swamp land as overflow areas in times of high water. It is not clear if this was an invited report or one spontaneously generated by Morse, but in it he makes specific proposals as if he were being considered for a job.[51]

In 1876 Morse presented another memorial, this time to the Forty-fourth Congress (March 4, 1875–March 4, 1877) "in regard to his claim as the inventor of the modern metallic cartridge system of fire-arms."[52] He repeated the appeal that he had made to Congress in 1872 for an extension of his cartridge patent, 15996, which had been rejected that year, but he also went much farther in 1876. The Committee on Patents of the Forty-fourth Congress, which ultimately agreed with him, initially described his claims as "apparently extravagant," but it felt the matter warranted thorough investigation.[53] Never lacking in self-confidence, Morse's opening sentence in the 1876 memorial read, "Your petitioner, the undersigned, respectfully represents to your honorable body that he was and is the original and first inventor of the modern metallic centre-fire cartridge system

of breech-loading fire-arms, which has within the last few years been adopted by all the civilized nations of the earth, making our American workshops the armories for all the world, by which great wealth has been acquired by our manufacturers and our country vastly enriched."[54]

Morse claimed that all the major arms makers in the world, including Allin/Springfield, Remington, Sharps, Winchester, Henry Martine, Snyder, Mauser, and others were all only different mechanical modifications of his inventions. He argued that his inventions were so advanced in the late 1850s that he was unable to obtain patents that would secure for him the full benefits of his inventions because Patent Office officials and practicing attorneys were insufficiently skilled in the art to properly comprehend his interests. He explained further that the government was slow to adopt breechloaders on a large scale because that would have required discarding huge stores of ammunition for muzzle loaders and that by the early 1870s his patents had expired.

Morse pleaded with Congress that he should have been granted extensions of the cartridge patents, 15996 and 20214, according to the Act of 1870 and that their unjust denial had caused him great and irreparable injury, which he claimed was twofold. First, he had been denied the ability to receive any remuneration from the application of his patents over the preceding years as they became more widely used. Second, his financial backers were discouraged from supporting his ongoing efforts to reclaim his patent rights. Morse wrote that he was poor and had received no reward for his labors, which had vastly benefited the whole world. He appealed to the Forty-fourth Congress to allow its Patent Committee to properly investigate the case and asked that such relief be "granted as may in their wisdom appear just and proper under all the circumstances."[55] To help Congress arrive at a figure, he stated that his expenditures exceeded his receipts by twenty thousand dollars, not including his loss of time. Morse concluded, "Some idea of the vast importance may be obtained, when it is considered that all the millions of soldiers in the world have been, or are now, obliged to throw aside their old weapons and adopt [breech loaders] . . . and a new system of army tactics must be adopted in consequence. That the new weapon, when opposed by the old, can tear down or build up Empires. That by its use we easily subdue all uncivilized nations, and introduce the railroad, the telegraph and the Bible."[56]

The sixty-four-year-old Morse's sentiments were clear and bore some merit in that he perfected the first practical metallic, center-fire, reloadable cartridge and an accompanying breech-loading firearm. One must judge if he was accurate in his analysis that his cartridge invention was a world-changing invention on the order of the things to which he compared it.

Among many supporting documents in the memorial, Morse presented an affidavit from a longtime supporter, V. D. Stockbridge. Stockbridge, who for many years had been chief examiner of the class known as "Firearms" in the U.S. Patent Office, was currently a member of the Appeal Board of the office and in 1876 was acknowledged to be the highest authority in such matters. Stockbridge reviewed the development of breechloaders and their cartridges, and wrote, "in this country nothing amounting to a breech-loading fire-arm in connection with a cartridge sufficiently portable, and at the same time waterproof, to be practically used, was known anterior to Morse's inventions, with the single exception of that described in a patent to Smith & Wesson, of February 14, 1854."[57] Twelve

years later, in 1888, Stockbridge testified that guns like the U.S. Springfield breech-loading rifle, Winchester, and Remington were all based on Morse's concept of the loose breech block.[58]

Stockbridge provided an even more detailed document in 1876 that strongly favored Morse. He enumerated what he considered to be Morse's achievements:

1. Morse was the first man in the world to make a breech loading gun in which the breech was moveable relative to the barrel and in which the relatively loose-fitting joint between the breech and barrel was made tight by a yielding, or elastic, cartridge.
2. Morse was the first man to make an extractor for withdrawing the cartridge, fired or not, from the gun.
3. Morse was the first man to combine a cartridge extractor with a moveable breech-block such that the extractor came from behind the cartridge and raised over then fell in front of the cartridge's flanged rear.
4. Morse was the first man in this country to demonstrate how to withdraw with certainty a cartridge from its seat in the barrel.
5. Morse was the first man, with the exception of an Englishman, A. E. L. Bellford, to make a cartridge extractor moveable relative to a moveable breech-block.
6. Morse was the first man to provide a tapering cartridge and a corresponding tapering seat within the firearm in connection with a means to withdraw the cartridge.
7. Morse was the first man to provide a practical means of any kind for mechanically inserting and withdrawing a cartridge entirely from its seat.
8. Morse was the first man, with the exception of Bellford, to combine a moveable breech-block with a firing pin.
9. Morse was the first man to make a flanged center-fire, outside-primed, yielding metallic cartridge adapted for use in an open-breech jointed gun, having a gun chamber larger than the cartridge case.
10. Morse was the first man to construct a properly-formed cartridge-chamber that would support the entire annular surface and rear end of a yielding metallic cartridge case.
11. Morse was the first man to use a flange-headed and loosely fitting metallic cartridge in connection with a chamber seat, and to depend on the expansion of the material of the cartridge to seal the joint when fired.[59]

Stockbridge went on to discuss his view of the development of the breechloader from the early 1860s to the mid-1870s. He wrote that in 1861, five or six years after Morse's first patent, private gunmakers began to embrace some of the most essential features of Morse's patents. Between 1861 and 1866 gun manufacturers adopted most, if not all, of Morse's improvements and marketed them all over the world. About 1866, ten years after Morse's first patent, the U.S. government began to manufacture firearms that incorporated all of the essentials of Morse's designs. Reviewing Morse's eleven achievements, Stockbridge wrote, "This may be said to be the beginning of the general manufacture in this country, and in the world of firearms embracing all of these essential characteristics."[60] Since 1866 almost all of the inventions relating to breech-loading firearms adopted the same plan. Stockbridge wrote that in 1876 "there is not a practical military small arm

in the world which is not made upon his plan."[61] Stockbridge acknowledged that many improvements had been made upon Morse's original designs, but that they all used the essentials of Morse's system. Finally, he concluded by saying that Morse's original patents had not covered all that he invented or that he claimed at the time.[62] Stockbridge summarized his opinion when he testified that "Morse invented all the essential features of the modern breech-loading system of firearms, in use in all parts of the world."[63]

Agreeing with Stockbridge, the 1876 Committee on Patents finally offered Morse some validation when it found in favor of his appeal, writing that there was no room left for doubt on the subject. The committee wrote, "the conclusion is irresistible that Morse is, indeed, the real inventor of the new system, which has turned our American workshops into armories for all the nations of the earth; thereby giving remunerative employment to a great number of our citizens, and largely adding to the wealth of our country by giving us a monopoly of the trade in fire-arms."[64] Further, the committee agreed that Morse's patent applications understated their significance because they covered neither the inventions nor the claims which were set up in his original specifications. The committee blamed this on three things: (1) Morse's inventions were so new that the Patent Office did not know what to allow him to claim; (2) Morse, himself, did not really comprehend what was really new; and (3) at the time there were very few competent patent attorneys. The committee wrote, "It is now admitted that he had a right to make such claims as would have prevented the construction of any one of the breech-loading arms now in use, without infringing his patents, if his claims had been drawn up so as to cover his inventions."[65] The committee also found that Morse had developed several devices which he had not patented but others had, and that they were currently being used in the manufacture of the U.S. Springfield rifle.

After the 1876 Committee on Patents specifically found that Morse's inventions were not known in the United States or abroad prior to his first patents in 1856, it recommended the passage of a bill which provided Morse with twenty-five thousand dollars for "full compensation for his services and expenses, and the cost of models furnished by him to the United States, in adapting his system of breech-loading fire-arms to the arms of the United States."[66] The recommendation alone should have offered some consolation to Morse, but, unfortunately, the bill did not pass.

Morse's memorial to the Forty-fourth Congress was presented *in toto* once again to the Forty-fifth Congress (March 4, 1877–March 4, 1879) on December 5, 1878, but by that time sentiment had turned against him.[67] Morse claimed once again that he was "the original inventor of the modern metallic-cartridge breech-loading system of fire-arms and ammunition adapted to their use"; he also submitted "his claim for compensation and expenses incurred for five years' services rendered to the United States in adapting his new system to its military arms, including many unpatented devices used by the government, and also for reimbursement for expenses incurred in the construction of several models, and for traveling expenses to and from the several United States armories and arsenals."[68]

Morse's memorial was referred to the Committee on Patents, which recommended its passage by the House. A Senate bill dated March 1, 1879, once again proposed appropriating twenty-five thousand dollars to Morse, but the same day, only three days before the end of the congressional session, the Committee on Patents reversed its opinion and

recommended that the Senate postpone the bill indefinitely.[69] For the first time in all of the congressional appeals, the committee discussed Morse's loyalty to the Confederacy and wrote that he "contends that he was employed by the States of Tennessee and South Carolina, and not by the Confederate Government. His patents appear to have been registered at Richmond, and could not have been thus registered had he not claimed to adhere to the Confederacy. Your committee can find no evidence of any acts of loyalty [to the Union] on his part during the war."[70] Disagreeing with Stockbridge, the committee also discussed Morse's lawsuit against the U.S. government and concluded, "But your committee are satisfied that there is very trifling, if any, originality in the Morse patent. They are satisfied by a careful examination of the patents which antedate it in this and foreign countries that it has not a single element which had not been anticipated by other inventions. They therefore report adversely to the bill, and recommend that it be indefinitely postponed."[71] This failure spelled the doom of Morse's appeals to the U.S. Congress for financial relief, though he would not give up for several more years.

In June 1880, George was a civil engineer living in Washington, D.C., with his wife, Marianne; his children, Lelia and Bryan; Bryan's wife, Hattie; and their son, George E. Morse. Morse appealed, once again, to the Forty-eighth Congress (March 4, 1883–March 4, 1885) on February 26, 1884, and its Committee on Patents again recommended the House pass a bill giving Morse financial relief.[72] Accompanying this appeal was an affidavit from the U.S. Army's chief of ordnance, Brigadier General Stephen Vincent Benet, who wrote on January 17, 1884, "In my opinion Mr. Morse fairly and justly deserves this much at the hands of Congress, and I strongly recommend the passage of the bill."[73] The secretary of war, Robert Todd Lincoln, concurred with Benet's opinion.[74] Though Morse clearly enjoyed some support within both the government and the military, he also had many detractors, and the bill met the same fate as its predecessors. Morse's great-granddaughter, Mary M. Kane, confirmed in 1963 that neither George nor Marianne received any reimbursement from the U.S. government.[75]

In addition to his unsuccessful appeals to the U.S. Congress for remuneration and his partially successful pleas to the U.S. Patent Office for extensions, Morse turned to the civil courts in an effort to attain financial compensation for the time and effort spent on his patents between 1855 and 1861. He was encouraged by the relatively quick initial settlement of his lawsuit against Winchester Repeating Arms Company, and he soon initiated a series of lawsuits against the U.S. government which dragged on until after his death in 1888. Morse's basic argument in all such cases is summed up by a statement presented in one. It declared that about 1866 the U.S. government began to manufacture military breechloaders which contained the features found in Morse's inventions. He described this class as

A fire-arm having a moveable breech-block, which was not intended to make a tight joint by contact of its parts by the rear end of the barrel and said breech-block, but which was to have its joint between them sealed through the medium of a yielding cartridge case and breech-block, having a firing-pin passing therethrough, a taper in the barrel, tapered in connection with a means for extraction, a centre-fire or centre-primed cartridge, an extractor which would throw the cartridge clear of the gun—all these were and are embraced in the arm made by the Government. This may be said

to be the beginning of the general manufacture in this country and in the world, of firearms embracing all of these characteristics.[76]

Lawsuit Case 10270, 1874–1875

Morse was living in New York City in June 1874 when he presented a formal claim of infringement on his patent rights to the War Department through his attorney, James A. Skilton. The War Department declined to consider the claim based on the grounds that it was not authorized to adjust what it considered to be unliquidated damages, that is, the claim of past unauthorized manufacture and use of Morse's patents.[77] The Morse Arms Manufacturing Company then brought a suit, case 10270, against the United States in the U.S. Court of Claims on December 9, 1874, claiming infringement on patents 15995, 20503, and 20727, the three which had been extended in 1872.[78] Numerous petitions and motions were filed on several dates, including February 9, 1875, and November 4, 1875, and depositions were taken in the U.S. Circuit Court in Springfield, Massachusetts, between November 13 and December 17, 1875.[79] George's brother, Isaac, was a consultant in the case, which would finally be dismissed twenty years later.[80]

Morse was a sixty-four-year-old civil engineer and inventor living in Washington, D.C., when he gave a deposition for the lawsuit in February 1876.[81] He claimed that alteration of old muzzle loaders had resumed at Springfield Armory in the winter of 1865–66 using his patents, and that Springfield Breech-loading Rifles were made in large numbers using his designs, tools, and techniques between 1866 and 1875.[82] His fundamental belief was that "All the breech-loaders they [the U.S. government] have made involve the principal of my invention fully, the only difference from my models being in the mechanical details or equivalents."[83]

The Morse Arms Manufacturing Company claimed damages of $650,000, based on $5 for each of the 130,000 breechloaders made at Springfield. The company later amended the claim to $876,675, contending that the United States paid about $80,000 and continued to pay about $20 daily for the use of what was known as the Preston patent, which postdated Morse's patent and, as he claimed, infringed on it. The Preston patent had been issued on February 5, 1867, to A. B. Ely of Newton, Massachusetts, and had passed through several owners since that time. By early 1873, the U.S. government had already paid $82,000 and was currently paying twenty-five cents per gun for the use of the patent.[84]

The assistant U.S. attorney general argued that the principles claimed by Morse were known to others and were patented long before Morse was granted his patents in 1856 and 1858. Specifically, he named Jean Samuel Pauly, who had patented similar designs in England in 1815 and 1816, along with several others.

Brigadier General Stephen Vincent Benet, chief of ordnance at the War Department, argued against Morse in 1875 but supported him in 1884. Benet rendered his opinion to the secretary of war, William W. Belknap, on March 6, 1875, summarizing his arguments against Morse's claims.[85] Benet wrote that Morse had two contracts with the U.S. government, one from September 13, 1858, and another in February 11, 1860. He said that the

rights granted by Morse to the government in each case were limited to a specific number of arms, and the government never completed the total number enumerated in either contract. Further, Benet denied Morse's claim to have "entered in a contract of license with the Government of the United States, through the Secretary of War, whereby the Government was licensed to make and use arms and ammunition under and according to the specifications in said letters for and during the life of the same."[86] Benet wrote that because Morse claimed five dollars for each of the approximately 130,000 breechloaders and ammunition for them produced by the U.S. government since 1865, his claim was essentially one of damages for infringement on his patent rights.

The defense argued that the U.S. government had not manufactured or altered a single breechloader using Morse's patents, other than the ones it had specifically contracted and paid for, during the time periods that the patents were in effect. The court agreed and dismissed the petition, but as of the end of December 1878, case 10270 remained undecided.

The legal issues lay dormant for six months until July 17, 1879, when Morse and his attorneys filed another case, George W. Morse *v.* the United States (#12065), just a month after the expiration of patents 20503 and 20727 and only four months before the expiration of 15995. Both sides subsequently filed petitions relative to the original case, #10270, and they were considered in the Court of Claims in December 1880. Morse's motion was that the U.S. Patent Office had agreed with the secretary of war that the government might use his invention and pay a reasonable compensation. Morse also claimed that licenses were given authorizing the manufacture of limited amounts at a fixed royalty. To support this, Morse's amended petition specifically cited his September 13, 1858, contract in which he was to be paid ten thousand dollars for altering two thousand muskets and the contract dated February 11, 1860, for one thousand carbines at three dollars each. Morse's motion also stated that, after the war, Congress authorized the manufacture of the arms according to the Springfield system, which included Morse's invention, and that a large number of arms had been made, but no royalty had been paid. The petition stated that new experiments were made after the war, resulting "in an order made July 26, 1866, under which the manufacture was made for which compensation is now demanded."[87] Morse further claimed that a board appointed by the secretary of war recommended that Congress adopt the Springfield system, which incorporated his inventions, for breech-loading muskets and carbines, and Congress agreed in 1873. Morse claimed that, based on his original contracts, the government continued to use his inventions and was bound to pay a reasonable royalty to him.

The next year, the court denied Morse's request for a rehearing and denied his motion, writing that the government had not infringed on his patent and that Morse was not the original inventor of the breech-loading system used by Springfield.[88] The case remained active when a motion filed by the U.S. attorney on September 20, 1880, was considered in March 1881.

Lawsuit 14340, 1884

The original suit, 10270, had not been settled when the Morse Arms Manufacturing Company filed a new motion on April 4, 1884, with a new attorney, Alfred Ely of New

York. The Morse Arms Manufacturing Company brought yet another suit, 14340, against the United States in the U.S. Court of Claims on April 12, 1884, based on the same three patents—15995, 20503, and 20727. The company claimed that Morse was "The lawful patentee of a device for sealing a breech joint of a breech-loading gun, purposely made open, by the expansion of a yielding metallic cartridge case, purposely made to fit loosely in its chamber."[89] The company claimed a royalty of $1 per gun for 15995 and 25 cents per gun for 20503, damages totaling $274,000 between July 10, 1875, and November 29, 1879. The two cases, 10270 and 14340, were soon merged.

Morse wrote to former governor Bonham on August 26, 1885: "My old friend, I am still alive and fighting for the establishment of my rights to the paternity of the modern breech-loading system of firearms. I have a case in the U.S. Court of Claims, in which I wish to prove that the State of S.C. paid me $5 royalty for the right to manufacture carbines under my patents. Lopez is dead, but he paid to me the sum stated with your sanction as Governor, and also paid me for superintending the work. Can you give me an affidavit touching the facts? Please tell me what has become of the gun made for you at the state works."[90]

Bonham did not reply, and Morse wrote to him again three months later in November, "Dear Sir, I have written to you several times and cannot conceive of any reason why I am unable to obtain an answer."[91] Apparently Bonham never answered; he died in 1890.

Virgil Stockbridge, Morse's longtime supporter and now a Washington, D.C., patent attorney, gave a lengthy deposition in July and November 1885 for case 14340, and again in 1887 as 10270 and 14340 were being tried as one case. The forty-eight-year-old Stockbridge had been a firearms instructor as a lieutenant in Company G, Second Regiment, District of Columbia Infantry, during the war. He had been with the U.S. Patent Office from 1868 to 1881, working most of that time as an examiner of firearms, and rising to assistant commissioner of patents by 1880. In 1881 he took up the practice of law as an expert in patent cases. He testified, "The Springfield rifle referred to in the petition infringes the Morse patent, because the use involves the assemblage of a gun barrel open at the breech, and a breech block carrying a firing-pin to be secured in place before firing, under such an arrangement as to leave an open space all the way around between the rear end of the barrel and the front face of the breech block."[92]

Because three of Morse's U.S. patents had either been denied extension or their statute of limitations had expired, the primary focus of the 1884 suit became patents 15995 and 20503, both firearms patents. Morse's claim for 20503 was based on the use of a lever to extract the spent cartridge case, and he claimed that this lever system was used in the Springfield gun. The court found that the concept of using a lever to extract a cartridge was not new in 1858, and, because Morse's specific lever design was not used in the Springfield gun, the court dismissed the petition. As far as Patent 15995 was concerned, the court dealt harshly with the late Morse, dismissing the petition on the grounds that Morse's claims were too broad. The court wrote in June 1892:

> The state of the art in 1856 was not such in our opinion as would permit a patent to issue covering so broad a claim as that made by plaintiff's counsel in this case. The advantages of loose breech mechanism were known when Morse applied for his patent; the metallic cartridge as an invention was old, although practical difficulty had

been encountered in manufacturing it; the use of a cartridge as a gas check was by no means a novel idea. It was not open to Morse in 1856 to patent every "device for sealing a breech joint of a breech-loading gun, purposely made open by the expansion of a yielding metallic cartridge case, purposely made to fit loosely in its chamber;" quite the reverse, this idea had been disturbing inventors for many years. Morse came to the patent office limited to some specific mechanical device which should solve the long-studied problem, and he must be held to the one described in his application. Morse was by no means the first in the field, and therefore he is not entitled to the very liberal construction which is here asked.[93]

All three cases involving either George W. Morse or the Morse Arms Manufacturing Company *v.* the United States, 10270, 12065, and 14340, were tried on May 23–26, 1892, four years after Morse's death and eighteen years after the first case was filed. The court issued its opinion on June 20 and made no finding of facts and entered no judgment on both case number 10270 and 12065. The court found in favor of the defendant on number 14340 and dismissed the case. The U.S. assistant attorney general requested that 10270 be dismissed in December 1895. As one last gasp, the Morse Arms Manufacturing Company appealed case 14340, but the U.S. Supreme Court finally dismissed it in the spring of 1896.[94] Defense attorneys for the United States summarized their position in a lengthy document, in which a single sentence suggests the real reason for Morse's ultimate legal defeat, "When he [Morse] went out of the Union, designing and scheming against its life, he abandoned all claims to its patents and their values."[95]

CHAPTER 21

Morse's Final Productivity

Whitney-Burgess-Morse lever-action magazine carbines were manufactured between 1878 and 1882. Courtesy of Buffalo Bill Center of the West, Cody, Wyoming. Gift of Olin Corporation, Winchester Arms Collection, 1988.8.836.

About three thousand Whitney-Burgess-Morse lever-action magazine carbines were manufactured between 1878 and 1882. Chambered for the .45-70 cartridge, the repeater was produced in three major variants—a sporting rifle, a military rifle or musket, and a carbine. In addition there were numerous variations within the three major types. The firearm, which came to be known as the Whitney-Burgess-Morse lever-action repeater, did not gain widespread public acceptance.

V. J. Whitney entered into a license agreement with Andrew Burgess in 1878 to use two of Burgess's patents to produce the firearm, which Whitney termed the Burgess Repeating Rifle. Whitney stamped Morse's name and first firearm patent number, 15995, on the gun, but Morse's personal role in this venture is not entirely clear, and his compensation, if any, is unknown. There are several theories, but all are conjecture. One is that Morse financed the first production run to bolster his ongoing lawsuit against the U.S. government, hoping to profit from the venture. Another is that Burgess credited Morse because he was concerned that Morse, who had already favorably settled his patent infringement case against Winchester, might win his nearly identical case against the U.S. government. Yet another theory is that Burgess paid Morse outright for the right to use Patent 15995.[1] Several surviving Whitney-Burgess-Morse firearms are known to exist, including serial number 170 at the Buffalo Bill Center of the West, and another at the Atlanta History Center.[2]

U.S. Patents 345,165, July 6, 1886, and 361,996, April 26, 1887

As his lawsuits plodded along in the mid-1880s, Morse was granted two new patents, one for improvements to the center-fire metallic cartridge and another for a handheld reloading tool. Neither was financially successful. Morse applied for a patent for "Improvements to Reloading Cartridges" in May 1883. It was rejected, amended, and finally issued on July 6, 1886, and it represented Morse's final attempt to make a better cartridge. Morse wrote that his new invention improved on his 1856 patents, 15995 and 15996, as well as other more recent patents but, interestingly, he did not specifically mention his other two, more recently issued, cartridge-related patents, 20214 and 20727. By the mid-1880s, the problem with cartridges was that they were prone to fracture inside the chamber of the firearm, making ejection difficult, if not impossible. Morse noted that in the older, shorter cartridges the entire casing would move rearward when the gun was fired. But the length of cartridges had expanded over the years, causing the forward section of the cartridge to become immobile as it expanded in the gun chamber when the weapon was discharged and resulting in a fracture of the casing as its rear part was forced backward. As a result, pieces of the cartridge failed to eject. He claimed to have corrected the problem by inventing a cartridge with a movable base in which the interior portion of the head of the cartridge moved backward or the head of a folding-head cartridge moved rearward.

Morse wrote to the Frankford Arsenal on June 2, 1885, to reestablish a relationship for the manufacture of his newest cartridge. At about the same time he was working on a new handheld reloading tool, which would be patented as a pocket cartridge loader, patent 361,996, on April 26, 1887. It was designed as a complete kit for capping, uncapping, cleaning, and loading cartridges. In addition it could also crimp the shell around the bullet and could expand and fit reinforcing cups on the base of the cartridge. In his characteristic manner, Morse wrote to Brigadier General Stephen Vincent Benet, chief of the Ordnance Department, in 1885, "In view of the fact that I am the inventor of the primed, expansive metallic cartridge system of breech-loading firearms, and that my long practical experience in this matter makes my opinions of value, I have the honor to submit these inventions to your Department, in the hope that my rude instruments may be duplicated and my inventions thoroughly tested by the Ordnance Department."[3]

As usual, the issuing of patents lagged behind research and development. Morse began working with the Frankford Arsenal on preliminary testing of the new cartridge in the summer of 1885. In typical fashion, he continued to write to Benet, espousing his favorable views of the testing results and the superiority of his invention. Tension arose between Morse and some officials at Frankford both because of his ongoing lawsuits against the U.S. government and his loyalty to the Confederacy twenty years earlier. Nevertheless, testing continued into early 1886, and an order was issued on January 12 to begin the manufacture of the .45-70 caliber, tin-plated brass Morse Moveable Base Cartridge.

As he had done for the past thirty years, Morse worked to gain the interest of the British government in his new inventions. He provided a special adapter for the use of

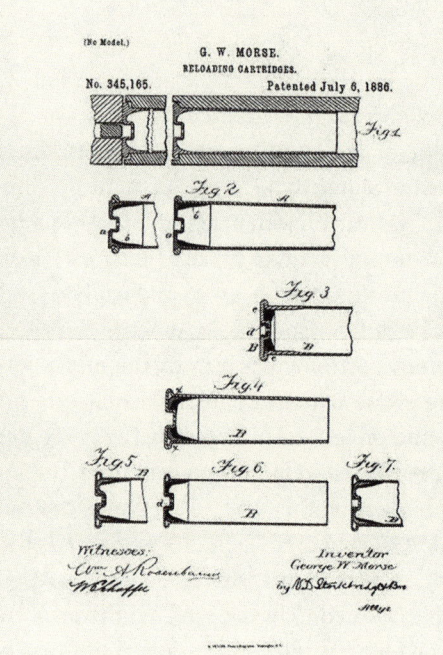

Morse's last cartridge patent, issued July 1886. G. W. Morse, Reloading Cartridge, Letters Patent July 6, 1886, Patent Case File No. 345165; Patented Case Files, 1836–1973; Records of the Patent and Trademark Office, Record Group 241; National Archives at College Park, Md.

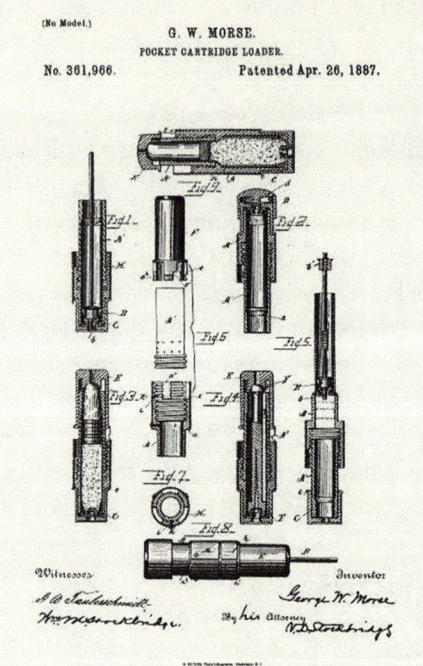

Morse's last patent for an unsuccessful pocket cartridge loader issued April 1887, one year before his death. G. W. Morse, Pocket Cartridge Loader, Letters Patent April 26, 1887, Patent Case File No. 361966; Patented Case Files, 1836–1973; Records of the Patent and Trademark Office, Record Group 241; National Archives at College Park, Md.

his pocket cartridge loader with the British Martini-Henry cartridge. After work had begun at Frankford, Morse obtained permission to alter one thousand .45 caliber Martini-Henry cartridges to his design on December 9, 1886.[4] One was fired and reloaded 278 times. Morse and C. E. Creecy, president of the Pneumatic Gun Carriage and Powder Company, produced a twelve-page booklet in the spring of 1887 for the British military establishment reiterating Morse's claims and projecting an annual savings of twenty-nine thousand dollars. Creecy planned to sail for England on October 1, 1887, to demonstrate Morse's patent to British ordnance officers, but the fate of his effort is unknown.[5]

By April 1887 the Frankford Arsenal had produced one million of Morse's newest cartridges, and they were distributed to various forts, barracks, and posts around the country in June for further testing prior to being approved for general use. Early tests were optimistic. Benet later reported a failure rate of only one in one thousand, each case having been reloaded between 20 and 30 times. But serious problems soon arose with split shell cases and shell separations near the base of the bullet. At some time before manufacture ceased in 1887, Morse made an adjustment to the thickness of the forward end of the cartridge in an effort to correct the problems. Morse died on March 8, 1888, never knowing if his latest invention was a success or a failure. Because the 1887 field trials were inconclusive, new trials were ordered for April 1888, with 140,000 rounds, half of which were the older, thin-walled cartridges and half the newer ones with thicker walls. Troops were instructed to clean the spent cartridges prior to reloading in an attempt to prevent the shell separations. A final report was issued in October 1889, and the existing service cartridge won out easily over both versions of Morse's cartridge in all areas, including failure rate, endurance, cost, complexity of manufacture, and unanimous officer preference.[6] Morse's reloading tool was not adopted by the U.S. military, and the fate of his new cartridge met a similar demise, thus ending the association of George W. Morse with the United States military.

Morse signed his last will and testament on September 4, 1882, in Washington, D.C. He left all of his property, both real and personal, to his "beloved wife Marianne," whom he also named as executrix of his estate.[7] When he modified the will on October 1, 1884, Morse wrote that he left everything to his wife in 1882 because he had entire faith in her wisdom and her desire to do all that may be best for their children. He also wrote that the condition of his affairs required that one mind have the power to act without delay. Further, Morse designated his beloved daughter, Lelia, to take the place of her mother in the event that Marianne died or became incapacitated.[8]

Morse died at his home in Washington, D.C., on March 8, 1888. Had he lived two more months he would have been seventy-six years old. The cause of death was old age complicated for the ten days preceding his death by congestion of the lungs. Morse's occupation on the death certificate was listed as "Inventor." He was buried two days later at Washington's Oak Hill Cemetery in a plot he had purchased four years earlier.[9] His obituary was accurate except for one erroneous statement, repeated in many publications since that time, that he was the nephew of Samuel F. B. Morse, inventor of the telegraph. It appeared in the *Critic-Record* of Washington, D.C., March 8, 1888:

END OF A BUSY LIFE
Death of George W. Morse, the Distinguished Inventor

Colonel George W. Morse, well known as an inventor and engineer, and for many years a prominent citizen of Washington, died at his residence, 1905 F Street northwest, this morning. Colonel Morse was a nephew of Prof. Morse, the inventor of the telegraph, and was the inheritor of his inventive genius.[10]

Born in New Hampshire in 1812, he became as a boy identified with labor-saving inventions and manufactures in Boston, and, in early manhood, became the representative in England of an American firm of manufacturers of printing machinery. During the years spent in England he acquired the knowledge of ordnance manufactures, which led to the invention of the Morse breech-loading gun and Morse cartridge, which have been adopted by the United States Army.

Returning to the United States several years prior to the war, Colonel Morse became State Engineer of Louisiana, and was identified with the engineering works of that and other Southern States. For more than twenty years Colonel Morse has been a resident of Washington. His death was due to inactivity of the heart and the debility of advancing years. His life was one of unceasing activity of the body and brain, and his actions were such as to gain for him the love and esteem of all who knew him. The funeral will occur on Saturday from his late residence and the interment will be at Oak Hill.[11]

Marianne Morse remained in Washington, where she lived with Lelia until at least 1900. Marianne died at the Home for the Incurables in Washington, D.C., on December 10, 1904, at the age of eighty. She was buried beside George on December 12.[12]

The United Daughters of the Confederacy placed a bronze marker at the site of the State Works near the corner of Green Avenue and Nelson Street in Greenville, South

Bronze tablet originally located at the site of the State Works, now at Springwood Park, Main Street, Greenville, S.C. Photograph from the author's file.

Carolina, on December 19, 1937, at a cost of two hundred dollars. The marker was originally mounted on two granite pillars, which came from the foundation of one of the
buildings at the State Works. It was toppled over in the early 1970s and moved to the
entrance of Springwood Cemetery on Main Street.[13] A quotation attributed to General
Wade Hampton was probably borrowed from the minutes of the Greenville Ladies Association in Aid of the Volunteers of the Confederacy: "This carbine was used by General
Wade Hampton and pronounced by him the best he had ever seen."[14] It is correct that
both Jefferson Davis and M. L. Bonham thought Morse's carbine was the best cavalry
breech loader they had seen, but the author has been unable to document such a statement made by Hampton. Hampton commanded mounted troops that were in the service of the Confederate States, and because the carbine was issued only to cavalry and
mounted infantry in the service of the state of South Carolina, Hampton's men should
not have carried Morse's carbine. However, Hampton was surely aware of Morse's carbine
and possibly made the comment, which could have been passed down orally.

The inscription on the bronze plaque reads:

CONFEDERATE ARMORY

1861 1864

ERECTED ON LAND DONATED TO THE STATE

BY VARDRY MCBEE FOR THE MANUFACTURE

OF ARMS FOR THE SOUTH CAROLINA TROOPS

IN THE CONFEDERATE SERVICE

GEORGE W. MORSE, SUPERINTENDENT

OF THE WORK INVENTED AND MANUFACTURED

A BREECH-LOADING CARBINE PRONOUNCED

BY GENERAL WADE HAMPTON THE BEST

THAT HE HAD SEEN

ERECTED BY THE GREENVILLE CHAPTER

UNITED DAUGHTERS OF THE CONFEDERACY

DECEMBER 1937

[reverse side]

THIS STONE IS A PART OF THE ORIGINAL

FOUNDATION OF THE CONFEDERATE ARMORY

Conclusion

This is an account of intelligent men who used their God-given talents to solve problems. David Lopez Jr., a highly regarded carpenter and building contractor, was essentially a very creative and successful businessman, both before and after the war. He knew how to identify a problem, solve it, and move on to the next one, but he lacked military experience, which ultimately hurt him as general superintendent of the South Carolina State Military Works. The State Works never lived up to the expectations of its creators, former governor William H. Gist and Lopez. They had envisioned a facility that could manufacture all the military hardware needs of the state, including the production of small arms, cannon, and ammunition for both. It was not to be. Though it managed to remain in production from mid-1862 to the end of the war, the State Works produced only a moderate amount of shot and shell for cannon, and a large amount of other materiel necessary for conducting a war, but it never manufactured a single cannon. It was most noteworthy for fabricating about 1,032 copies of Morse's breech-loading carbine, a military firearm that was ahead of its time. Novel when it was developed in 1860, both because it was a breech-loader and because it was the first to use a metallic, center-fire cartridge, Morse's carbine was an excellent firearm. As a breech-loading weapon, it could be fired more rapidly than a muzzle loader and with as much accuracy. It was highly portable and weather resistant. Its cartridge was time-consuming to manufacture and used strategic materials, but it was reloadable in the field, thus offsetting those drawbacks.

At a time when large numbers of easily manufactured muzzle loaders were being made, Morse's carbine was a few critical years ahead of its time. By the time the war started, the U.S. government decided to supply most of its infantry with muzzle loaders and to provide smaller numbers of breechloaders to selected troops, but for that need it chose either existing breechloaders such as Burnside's and Sharps's or the newer repeaters such as the Henry and Spencer. The Morse carbine's major drawback was that it was produced in numbers too small to have made a significant difference in the outcome of the war.

George Woodward Morse was born with the mind of an inventor, and he acquired the skills of a machinist, but his insatiable desire for perfection rendered him impatient, often hindering the implementation of his ideas. One supporter summarized Morse's

crowning achievement when he wrote, "in 1856, he invented the 'metallic cartridge case' which made breech-loading small arms a success."[1] Another longstanding advocate, Virgil Stockbridge, agreed when he wrote that "Morse invented all the essential features of the modern breech-loading system of firearms, in use in all parts of the world."[2] Further testimony from the chief of ordnance of the U.S. Army, S. V. Benet, also supported Morse, calling him "undoubtedly a pioneer in this great improvement in arms."[3] Though many men felt that Morse's inventions were world-changing, no one believed it more than Morse himself, a supremely self-confident man who felt that his gun was the best in the world. He was sincere when he wrote in 1876 that his inventions had "vastly benefitted the whole world," placing them on a par with the railroad, telegraph, and the Holy Bible.[4] Morse spent a great deal of his own money on research and development, but his inventions were so far ahead of their time that he had to wait for the U.S. Patent Office, the U.S. Army, and the general public to accept them.

Though his inventions were eventually appreciated by the Ordnance Department, U.S. Patent Office, U.S. Congress, and even Secretary of War Robert Todd Lincoln, Morse's life ended sadly because he never received the full recognition he deserved for his substantial contributions to the modern center-fire cartridge. Morse's primary contribution to the development of firearms technology was to demonstrate the fully functional combination of a successful breech-loading firearm and a metallic center-fire cartridge. He understood very early in his career as a firearms inventor that the essence of his role in their development was the first practical metallic, pre-primed, center-fire, reloadable cartridge. He wrote to the Board of Ordnance in July 1858 that the cartridge "was the best feature of all my plans," and, later, "It is not so much my present peculiar alteration of the arms before you that I value . . . as it is with this cartridge."[5] Sadly, he made two decisions that adversely affected his legacy. The first was borne of ignorance, both on his part and that of his patent attorney. Though Morse patented the firearm and cartridge individually, he never patented the *combination,* most likely because even he did not fully grasp the significance of his ideas in 1856. Had he done so, his name probably would have been generally accepted as the father of the modern cartridge. The second decision was that he sided with the South during the Civil War. His reasons are not fully known, but that choice ultimately resulted in his failure to achieve the recognition he justly deserved. After the war, Morse was *persona non grata* in the North, but his concepts of a breech loader and accompanying metallic, center-fire cartridge lived on in most postwar firearms. His influence in the development of the modern firearm and cartridge was noteworthy and has been largely disregarded.

Lopez was the driving force behind the creation of the State Works. Morse invented a functional breechloader and perfected the metallic, center-fire, pre-primed cartridge. Without the State Works, Morse's brass-frame carbine would likely have never been mass produced. Without the carbine, the State Works would likely be remembered as only one of many small Southern arms factories. Their combined story allows us to appreciate both and to honor the lives of the men involved.

APPENDIX 1

Inventory of Machinery, Tools, and Stock of Tennessee Armory, Atlanta, March 1862

1 hand lathe

1 6-foot engine lathe, complete

1 10.5-foot engine lathe, complete

1 14-foot engine lathe, unfinished and under construction

1 12-foot engine lathe, complete

1 10-foot sat of ways

1 breeching machine

1 milling machine, incomplete

1 compound planer, complete

1 drill press, complete

1 planer without table

1 8-spindle drill press, complete

Parts of profiling machine

1 punch and shear, complete

1 bolt header and dies, incomplete

1 gear cutter, complete

7 arbors or boring bars

10 sets overhead rig, complete

6,584 lbs. shafting and pulleys

1 (old man) drill stand

3 hangers for counters (casting), 103 lbs. to 309 lbs.

17 hangers for main shafting, 1,571 lbs.

5 hangers (without boxes) 51 lbs. to 255 lbs.

1 overhead rig for gear cutter, complete

5 legs for drill lathes, 36 lbs. to 180 lbs.

Parts of old milling machine, 995 lbs.

1 grind stone frame, complete, 374 lbs.

3 polishing heads, complete

2 rifling rolls and stands

1 fixture for holding barrels to mill cone seats

Parts of lathes (some work done), cast iron, 101 lbs.

17 bench vices, wrought iron, 965 lbs.

5 bench vices, cast iron, 261 lbs.

Wire, all sizes, some cast steel, 181 lbs.

Cast steel, 6,600 lbs.

Blister steel, 450 lbs.

Block tin, 80 lbs.

Brass and copper, 170 lbs.

1 gig for lock frames drilling

5 hack saws

12 gig saws

2 hack saw frames

18 dozen gun wipers

1 stand for drilling ramrods

6 sets stocks and dies

134 taps, all sizes, from 1¼ to 1 1/16

1 case drafting tools

1 alphabet

1 set figures

1 2-feet standard scales

1 gig and mill for stirrups

1 gig and mill for main spring pivots

1 ratchet drill

19 dogs

50 bench hammers

40 monkey wrenches

7 draw knives

19 iron braces

34 hand vices (16 with handles)

106 screw drivers

30 pairs pliers

462 sheets sand paper

280 sheets emory paper

43 bits

1 patent brad awl

2 patent brad awls and handles

14 chisel handles

3 smooth planes

1 jack plane

103 cones (small)

300 cones (government)

1 gig for filing tumblers

1 stamp, "Tennessee Amory"

1 hand saw

1 copying press

1 gun gauge, brass

110 drills

18 reamers

25 rose bits

11 cherries

14 drill sockets

2 tumbler punches

2 tumbler mills

10 counter boxes

5 boring arbors

12 cast steel centres for new lathes

6 rests for lathes

2 drill sockets

1 6-inch Fairman chuck

1 set mills for milling tumblers, 5 inch set

1 set mills for milling lock frame, 14 inch set (top and bottom)

1 set mills for milling main springs, 5 inch set (sides)

1 set mills for milling not finished, 11 inch set (top and bottom)

7 plain mills

1 mill for key sets

15 unfinished mills

8 arbors for milling machine

12 sets jaws for milling machine (2 sets cast steel)

5 handles and stands for milling machines

4 sets jaws

2 sets jumpers and dies for forging tumblers

17 hand tools and handles

62 turning tools (engine lathe)

1 index head

11 tap and breeching wrenches

2 knees for planers

6 emory wheels

40½ lbs. sheet brass

2 stands and tools for percussioning old guns

2 paper brads

3 gross screws, ⅝

35 lbs. small nuts

12 bench brushes

15 cone wrenches

75 back action locks

60 common locks (rifle)

4 old musket locks

20 sets triggers for rifles

65 machine oilers

35 tin lamps

74 paint brushes

4 striping brushes

7 parallel stripes for planers

65 machine oilers

56 tumbler pins (turned)

12 lathe wrenches

1 grindstone shaft (turned)

1 lot bolts and nuts

1 chuck for holding lock frame to take cut on sides

2 soldering irons

1 clamp and cutter for cherries

49 main springs for repairing old guns

140 lock frames, milled

33 strips for stirrups (cast steel)

20 bench oil pots

70 lbs. babbit and type metal

35 lbs. lead (old pipe)

400 ferrules for file handles

628 musket flints

1 jumper stand for forging locks

2 sets jumper stands for forging locks

1 set jumper stands for forging triggers

2 sets jumper stands for forging small work

1 roll file card

1 tin box for small articles

3 pieces lace leather

2 gig saws

2 pieces sheet steel

6 pairs strap hinges

9 drawer locks

150 carriage bolts, 2 sets jumper stands for forging 2 to 6 in.

1 counter shaft (turned, 30 lbs.)

6 oil stones

13 anvils, 105 to 150 lbs.

6 blacksmith's sledges

25 pr. blacksmith's tongs

8 pr. blacksmith's set hammers

9 pr. blacksmith's swedges

19 pr. blacksmith's chisels

1 level

7 screw plates

8 steel punches

4 prs. Fullers

3 hand hammers

2,112 ft. new leather belting, assorted sizes, 1 to 12 in.

507½ ft. old leather belting, assorted sizes, 2 to 13½ in.

63 ft. old gum belting, 3¼ to 6 in.

9 doz. gun wipers

7 gross knitting pins, 11s to 14s

10 papers finishing nails, 1 to 2½ in.

4½ lbs. beeswax

78 87–144 gross assorted gimlet screws

348 9–12 oz. assorted files

43 patterns for parts of machines and tools

8 doz. chisel handles and 1 lot core boxes for patterns

500 lbs. emory

APPENDIX 2

Surviving Morse Firearms Listed by Serial Number

Carbines

No legible serial number—Greensboro Historical Museum; brass frame dug from Bentonville battlefield off Goldsboro road where Wade Hampton's cavalry fought the 100th Indiana Regiment; Roman numeral "VII" is damaged.

No serial number. prototype—Greensboro Historical Museum. It was discovered near Greenville, South Carolina, in the 1950s; .54 caliber, rifled with seven lands and grooves.

No serial number. Cased set on display at the South Carolina State Museum. The receiver is a Type III carbine. Finely checkered grip, brass shoulder plate, full set of tools, bullet molds, etc., bullets, cartridges, three barrels—carbine, rifle, and shotgun, each barrel has its own wooden forestock, and the brass forestock is shortened. Governor M. L. Bonham purchased it from the State Works in 1864. It is now owned by the McKissick Museum, University of South Carolina, and is on display at the South Carolina State Museum. (See detailed discussion in chapter 1.)

No serial number. Type III carbine—Atlanta History Center, 30.25-inch, 24 gauge (.56–.57 caliber), smoothbore shotgun barrel. There are no other markings, no cleaning jag, and the breech block is not milled.

#1—referred to by Hill and Anthony, possibly a prototype.

#2—Type I converted to Type II—Greensboro Historical Museum. Barrel is octagonal for the first 2.25 inches, mentioned by Dr. Murphy, .50 caliber, rifled in seven lands and grooves.

#15—Type I.

#16—Type I.

#19—Type I.

#22—Type I, mentioned in a 1976 letter from William Floyd to Wray in Wray-Morse File 002, Atlanta History Center.

#44—Type I.

#57—Type I.

#59—Type I, round barrel, possibly .54 caliber, 2½ milled out circular areas in breech block, possibly leaf sight.

#66—Type I converted to Type II, part octagonal barrel, referred to in Hill and Anthony.

#68—Type I.

#82—Type I.

#85—Type I.

#87—Type I.

#91—Type I, mentioned in Wray-Morse File, 017, Atlanta History Center. Photo in file 047.

#101—Type I.

#103—Type I converted to Type II.

#105—Type I converted to Type II, at South Carolina State Museum via Citadel collection.

#108—Type I, fragment, location unknown. This might be the dug frame at Greensboro Historical Museum.

#110—Type I, at Springfield Armory Museum, .50 caliber.

#113—Type I.

#115—Type I, at West Point Museum. It has a 20¼-inch barrel, .50 caliber. It first appeared in the 1929 catalog at West Point, but no source was listed. Most likely, it was a captured piece that went into U.S. stocks before issue to the museum. The most likely source was Springfield Armory, which officially received the vast bulk of captured Confederate arms.

#116—Type I modified to a Type II, Upcountry History Museum, Greenville, S.C., the two numerals "1" are small die and "6" is large die.

#120—Type I converted to Type II, incorrectly restored Type II breech block.

This appendix is a compilation of the work of several individuals, including Jack A. Meyer. Their efforts are gratefully acknowledged.

#126—Type I, Atlanta History Center; .50 caliber, large die, round barrel, sliding block is brass, 2½ milled-out sides, serial number on underside of breech slide.

#127—Type I.

#129—Type I.

#135—Type I converted to Type II, Museum and Library of Confederate History, Greenville, S.C.

#137—Type I, possibly converted to Type II. Altered to fire a .50 caliber rimfire.

#140—Type I, Museum of the Confederacy, Richmond. It was donated in 1912.

#141—Type I converted to type II, Atlanta History Center; .50 caliber, large die, octagonal barrel for 2.5 inches, sliding block is brass, 2½ milled-out sides, serial number on underside of breech slide.

#142—Type I, Albaugh and Simmons, *Confederate Arms,* 70, plates 97 and 98.

#143—Type I converted to Type II.

#146—Type I.

#148—Probably converted from Type I to Type II, referred to in Hill and Anthony, *Confederate Longarms and Pistols,* round barrel.

#149—Type I.

#150—Type I.

#157—Type I.

#161—Type I, mentioned by Dr. Murphy (*Confederate Carbines and Musketoons*), .50 cal. smooth bore.

#164—Type I, Greensboro Historical Museum. Frame is marked "South Carolina." Barrel is octagonal for the first 2.25 inches. Dr. and Mrs. John M. Murphy Collection. Fore-end is stamped 108.

#166—Type I.

#180—Type I, Virginia Historical Society's Maryland-Steuart Collection, donated by Mr. Richard D. Steuart in the 1940s. It is 40 inches long, 18.5 inch barrel, the first 2 ½ inches of which are octagonal. Barrel is browned, frame, buttplate, fore end tip, trigger guard, front sight, breech block, and hammer are all brass. .52 caliber. "V.I." is stamped on breech block.

#181—Type I.

#186—Type I, sold at auction in 2013. Engraved "Bloomington, Indiana" and "Robert."

#195—Type I, appears Wray-Morse File 017 and File 030, Atlanta History Center.[1]

#196—Type I.

#208—Type I.

#230—Type II, probably originally Type I, Atlanta History Center; .51 caliber, large die, round barrel, sliding block is brass, 2½ milled-out sides, serial number on underside of breech slide, firing pin has been modified to fire rim-fire cartridges. #230 is also mentioned in a handwritten note by Wray in File 030, Atlanta History Center.

#237—Cleaning rod is missing, breech is broken.

#257—Type I.

#274—Type II.

#288—Type II, Greensboro Historical Museum. Marked "J Davis" on the top of the frame.

#289—Type II, .58 caliber.

#298—Type II.

#301—Type II.

#304—Type II, Greensboro Historical Museum. The finger rest is expanded possibly to accept a sling or possibly from use.

#305—Type II.

#308—Type II, Marked "captured at Columbia, So. Car."

#309—Type II, at Atlanta History Center, large die, round barrel, sliding block is brass, no milled-out sides, serial number on underside of breech slide.

#313—Type II.

#325—Type II.

#330—Type II, number hard to read and might be 380.

#333—Type II, large die/font, was on display at the Museum and Library of Confederate History in Greenville, S.C., until recently, when it was sold into a private collection.

#336—Type II, North Carolina Museum of History. This carbine was purchased in 1963 by the Civil War Centennial Commission as a part of a series of Confederate firearms. It has no known North Carolina connections.

#337—Type II, Atlanta History Center; .50 caliber.

#343—Type II.

#347—Type II.

#357—Type II, highest serial number with large die/font according to Dr. John M. Murphy in *Confederate Carbines and Musketoons.*

#368—Type II.

#372—Type II.

#374—Type III. It is a "frontier conversion" with a .32 caliber, rifle-length octagonal barrel.

#375—Type II converted to a Type III according to author's communication with Dr. Jack A. Meyer.

#377—Type III.

#380—Type III, serial number hard to read and could be #330.

#383—Type II.

#384—Type III.

#386—Type II.

#417—Type III, lock has been modified.

#425—Type III, referred to in Hill and Anthony, *Confederate Longarms and Pistols.*

#437—Type III. Dr. Meyer notes that Dr. Sutherland noted that this carbine had a Type II latch.

#440—Type III.

#451—Type III.

#456—Type III.

#462—Type III.

#465—Type III, Milwaukee Public Museum, .54 caliber, purchased in 1960 from collection of William A. Lawrence of New York.

#469—Type III.

#474—Type III, South Carolina Confederate Relic Room.

#475—Type III.

#485—Type III.

#490—Type III.

#499—Type III.

#501—Type III.

#509—Type III.

#514—Type III, Chickamauga Museum, Fuller Gun Collection, carbine, .50 caliber.

#515—Type III.

#518—Type III.

#521—Type III.

#530—Type III.

#536—Type III.

#541—Type III.

#546—Type III.

#548—Type III.

#551—Type III, referred to in the *American Rifleman*, February 1954.

#566—Type III, Frazier Museum of History, Louisville, Ky.

#568—Type III.

#594—Type III.

#601—Type III.

#605—Type III.

#608—Type III.

#614—Type III.

#627—Type III.

#628—Type III.

#631—Type III.

#637—Type III.

#638—Type III.

#643—Type III, on display at the E. Berkley Bowie Firearms Collection at Fort McHenry National Monument and Historic Shrine, .50 caliber, round barrel, "R" on one side of serial number behind the trigger and "X" on the other side, adjustable rear leaf sight; brass breech block has three-milled out circular areas. Dr. Meyer noted that 643 was at Gettysburg at one time.

#645—Type III, at Fort McHenry at one time according to Dr. Sutherland, "Arms Manufactury in Greenville County."

#649—Type III. Barton Cox notes that this carbine was found at a building demolition site possibly in South Carolina about 1986 and is missing the wood stock. The receiver side plate, hammer, and possible internal parts are also missing.

#653—Type III.

#655—Type III.

#660—Type III.

#663—Type III, at Buffalo Bill Center of the West. .50 caliber, three grooves, cleaning rod and jag are present.

#664—Type III.

#666—Type III, Atlanta History Center, .50 caliber, small die, round barrel, sliding block is steel, not milled, serial number on underside of breech slide, flip rear sight.

#676—Type III, at Greensboro Historical Museum. Engraved, "Mar. 1, 1865, Cos. G and K Mass. Inf. by a night march surprised, near St. Stephens, S.C., a squad of Georgia cavalry who had followed Sherman's army from Chattanooga to Savannah. This carbine was captured in the fight by Capt. Chas. C. Soule." Originally from Dr. and Mrs. John M. Murphy Collection.

#682—Type III.

#683—Type III.

#702—Type III.

#705—Type III.

#711—Type III.

#715—Type III, Museum and Library of Confederate History, Greenville, S.C., small die.

#725—Type III.

#727—Type III.

#729—Type III.

#736—Type III.

#737—Type III.

#742—Type III.

#743—Type III, might be a duplication of #745 or #746 or #748.

#751—Type III.

#759—Type III, Texas Civil War Museum, partial octagonal barrel according to Dr. Murphy (*Confederate Carbines and Musketoons*).

#776—Type III, Virginia Historical Society Museum, Richmond, captured from a Confederate trooper in South Carolina in February 1865 by a Fifth Ohio cavalryman. At the same time, the Ohio cavalryman captured a flintlock pistol.

#778—Type III.

#789—Type III.

#790—Type III, mentioned by Dr. Murphy (*Confederate Rifles and Muskets*), altered to fire .56-56 Spencer rimfire cartridge.

#803—Type III, appears in *Civil War Relics from South Carolina* and in Wray-Morse File 017, Atlanta History Center.[2]

#807—Type III, mentioned in a letter in Wray-Morse file 002 at Atlanta History Center. In near-mint condition in 1975.

#830—Type III.

#836—Type III.

#857—Type III.

#865—Type III, Museum and Library of Confederate History, Greenville, S.C., small die.

#867—Type III, mentioned by Dr. Murphy (*Confederate Carbines and Musketoons*), .58 cal., rifled with three lands and grooves.

#874—Type III, South Carolina State Museum, serial number stamped on left side of barrel near the stock. Buttstock has serial number 634.

#875—Type III, mentioned in a letter from William A. Floyd to Wray in Wray-Morse file 002, Atlanta History Center.

#878—Type III.

#879—Type III, Charleston Museum. Also shown as "879R."

#893—Type III, stamped with "R."

#911—Type III, mentioned in Wray-Morse File 015 and File 048, Atlanta History Center. "E.B.L." cut in ⅜-inch letters on left side of frame, and stamped with "LWROE."

#912—Type III, also marked with an "R" and "VII."

#921—Type III.

#938—Type III, marked "MORSE."

#953—Type III, marked "MORSE."

#955—Type III, "MORSE" stamped on the right side of the frame.

#956—Type III, marked "MORSE."

#966—Type III, Atlanta History Center. .50 caliber, stamped "MORSE" on the side of the receiver, small die, round barrel; sliding block is steel, serial number on underside of breech slide.

#978—Type III, marked "MORSE."

#980—Type III.

#985—Type III.

#986—Type III, not stamped with "MORSE."

#988—Type III, marked "MORSE."

#995—Type III.

#998—Type III.

#999—Type III.

#1007—Type III, marked "MORSE."

#1013—Type III, Greensboro Historical Museum, marked "MORSE" on right side of brass side plate. Steel breech block. "R" stamped in front of the hinge on the receiver. The forestock looks maple, and buttstock looks walnut. Appears to be miscast with several deep scratches in the metal behind the trigger and a pitted blemish on the left side of the receiver above the trigger. This gun does not match Murphy and Madaus, *Confederate Carbines and Musketoons,* 189, plate 15. The stamp "MORSE" is not in the same place, and the scratches on the museum's example do not match the photograph. Originally in Dr. Murphy's personal collection now at Greensboro Historical Museum.

#1021 or 1023—Shown in Murphy and Madaus, *Confederate Carbines and Musketoons,* 183, plate 5.

#1025—Type III, marked "MORSE," mentioned in Wray-Morse File 017 and photo in File 047, Atlanta History Center.

#1026—Mentioned in a 1976 letter from William Floyd to Wray in Wray-Morse file 002, Atlanta History Center.

#1032—Type III, marked "MORSE," the highest known serial number according to Dr. Sutherland, "Arms Manufactury in Greenville County."

Muzzy

#4—In 1974 it was privately owned.

#7—Cased set at Atlanta History Center, a 24-gauge shotgun, .45 caliber rifle, and .52 caliber carbine, stamped "Muzzy Worcester."

#9—Cased set. The carbine barrel measures 25-⅝ inches long and is approximately .52 caliber. The octagonal rifle barrel is also .52 caliber, and is approximately 31 inches long, measuring nearly 1 inch across the flats. The 12 gauge shotgun barrel is 30-¾ inches long. The action has scrolled engraving, and the walnut stock has a checkered wrist. The interior of the case shows many loading tools, a brass bullet mold, key, extra front and rear sights, and other small loading tools of the period.

#11—Privately owned in 1961.[3] Rifle is .50 caliber, carbine is .55 caliber, shotgun is 14 gauge.

#20—Buffalo Bill Center of the West, .54 caliber center-fire. It has only the single barrel but was probably originally part of a cased set. The cocking lever is broken off. Also, the left-hand side of the two checkered thumb pieces used to open the action is missing. The barrel is 21¾ inches long. Missing are the other two barrels, the case, and tools. The Buffalo Bill Center of the West also has Morse's world-famous shot shell collection, approximately 10,337 pieces.

#46—A Morse Muzzy cased set with .54, .48, and .69 caliber barrels.

#47—Exists as a receiver only. There are no barrels or stock. It has the unique feature of two sling bars, one on the lower forward end of the receiver and another on the left side of the receiver to which a saddle ring is attached.

#60—Owned by and on display at the Museum and Library of Confederate History in Greenville, S.C. The shotgun is 16-gauge and has a 28½-inch barrel. The carbine is .54 caliber with a 20¾-inch barrel, and the rifle is .54 caliber with a 27-⅞-inch barrel.

#87—Mentioned in a letter in Wray-Morse File 002 at the Atlanta History Center.

#91—On display at the Confederate Memorial Hall in New Orleans. Morse reportedly gave it to Louisiana Representative Sanford as a gift.

#94—Cased set at Atlanta History Center is a 24-gauge shotgun, a .45 caliber rifle, and a .52 caliber carbine, stamped "Morse's patent October 20, 1856" and "Muzzy Worcester."

#103—A privately owned cased set is reportedly serial number 103.

Unknown serial number—At least one additional Morse-Muzzy cased set is privately owned.

Inside Lock

No serial number (unmarked)—At Greensboro Historical Museum, it has no markings and no serial number. It is a short rifle or carbine, and its barrel came from a U.S. model 1842 musket. Originally in Dr. Murphy's collection. Dr. Meyer notes that it is .69 caliber.

No serial number available (unmarked)—Atlanta History Center. This is an unusual inside lock. It has no markings and is possibly a prototype. A note in Wray's File 015 says, "This is the only example of this rifle known. It was probably a prototype made in Nashville in 1861 or Atlanta in 1862." Originally in Dr. Murphy's collection. According to Dr. Gordon Jones it is a .54 caliber.

#5—Recently sold by Brian Akins (rebelrelics .com).

#6—Recently sold by Brian Akins (rebelrelics .com).

#12—South Carolina State Museum. A note in Wray-Morse File 019 at the Atlanta History Center says "L. Sutherland." Dr. Meyer notes that it is .69 caliber.

#19—It is not a complete firearm but includes components of an inside lock discovered on the banks of the Edisto River near Orangeburg, South Carolina, on loan to the South Carolina Confederate Relic Room.

#48—Recently sold by Brian Akins (rebelrelics .com).

#50—At Greensboro Historical Museum. It is a musket marked "Morse's lock, State Works, Greenville, S.C." A note in Wray-Morse File 019 at the Atlanta History Center says "William Floyd → Dr. Murphy."

#70—In Atlanta History Center, marked "Morse's lock, State Works, Greenville, S.C." A note in Wray-Morse File 019 says "Geo. Wray ex-Sutherland."

#93—In Springfield Armory Museum.

#104—In Greensboro Historical Museum, a musket marked "1863, Morse's lock, State Works, Greenville, S.C."

#110—In Virginia Historical Society; behind the trigger guard is stamped "1863," and in front of the trigger guard is stamped "Morse's Lock" and "State Works Greenville, S.C." and "110."

#118—In American History Museum, Smithsonian Institution.

#122—A note in Wray-Morse File 019 at the Atlanta History Center says "Springfield A. Mass."

#129—Stamped "Morse's Lock" and "State Works Greenville, S.C." and "129."[4]

#163—The highest recorded serial number, according to Dr. Sutherland.[5]

....

Patent Models and Prototypes

#15995—In Smithsonian Institution. The maker was probably Daniel Searles. It is on loan to the Buffalo Bill Center of the West, where it is on display. The number "252571" is stamped on the wood above the side tang or faux "lockplate" on the left side. There are no other external markings visible.

#20503—In Atlanta History Center. U.S. Patent Office model for Morse's Breechloading System, patented June 8, 1858. It is a wood model with metal hammer, trigger, and trigger guard.

The Buffalo Bill Center of the West owns a Morse prototype sporting rifle, a different firearm than Muzzy serial number 20, also in the collection. The maker is unknown, probably Daniel Searles and possibly Nathan Muzzy. It is nearly identical to the Smithsonian patent model, which is also on display there, including screws, buttplate, blued finish, and rear sight. It has no markings and no serial number. The Smithsonian patent model has two "ears" used to open the action in the same position as a Muzzy, but of flattened shape and with no checkering. The "ears" are not on the Buffalo Bill prototype sporting rifle but may have been at one time. The barrel is 34½ inches long and measures at .583 caliber. It is a center-fire, not rimfire as is incorrectly listed. The action is nearly identical to a standard Muzzy cased set, but there are several external differences between this piece and a Muzzy. The action measures 5-⅛ inches long whereas a standard Muzzy measures 6½ inches. The breech cover has an elaborate scalloped shape, as seen on the original patent drawing for 15995. The standard Muzzy buttplate is iron and measures 4-⅞ inches, whereas the Buffalo Bill prototype sporting rifle has a brass buttplate which measures 4-⅜ inches. The barrel is held on by two screws, unlike the Muzzy production models.

....

Breechloader Alterations

Serial letter "E" (or "3") model 1841 rifle—In Atlanta History Center. This is one of three survivors from the four sent to Washington.

Model 1816 rifled musket—In Atlanta History Center, stamped "Springfield 1839," originally flint converted to percussion then to breechloader.

Two .69 caliber model 1816 Springfield muskets converted to breechloader—At American History Museum, Smithsonian Institution. Marked "Springfield Armory." Lockplates dated 1839.

Six breechloader conversions of muskets or rifled muskets—At Springfield Armory Museum:

Model 1841 with sliding breech piece (Munck modification of Patent 20503).

Model 1816, no sliding breech piece (Patent 20503).

Model 1816, no sliding breech piece (Patent 20503).

Model 1816 with sliding breech piece (Munck modification of Patent 20503).

Model 1816, no sliding breech piece (Patent 20503).

Model 1816 with sliding breech piece (Munck modification of Patent 20503).

The Springfield Armory Museum website says that there is a Harpers Ferry M1841 conversion privately owned.

#197—Mentioned in Morse Wray File 002 at Atlanta History Center

A Springfield M1839—Privately owned.

M1816/1822—In Buffalo Bill Center of the West; .69 caliber with three-groove rifling with no serial number. Sliding breech block. The label says "alteration number 23."

APPENDIX 3

List of Slave Workers

The following is a list of the owners and the slaves they hired out to the State Works, taken from "Pay Rolls of Negro Mechanics and Laborers."[1]

Charles Alston: Jack—helper, Primus—helper/laborer, Prince—helper, William—laborer, Aleck—laborer, Titus—laborer, George—laborer, Jim—laborer.

Henry Boylston: Dick—carpenter.

A. P. Calhoun: Tom—laborer, Ephraim—laborer, Henry—laborer, Jim—laborer, Mack—laborer, Pinckney—laborer, Chapman—laborer, Jimmy—laborer, Edward—laborer, Simon—laborer. Simon's name also appears in a letter from J. R. Smith to A. R. Calhoun dated October 18, 1864.[2]

W. A. Claire: Tom—carpenter.

T. A. Coffin: Tom—helper.

Thomas M. Cox: Jerry—laborer, Aleck—laborer.

F. Davenport: Dick—laborer.

J. B. Davis: Jim—laborer, Charles—laborer.

J. P. Earle: Joe—brickman, George—brickman, Stephen—brickman, Paul—laborer, William—laborer, Missmiss—laborer.

William S. Elliott: Frank—laborer.

Samuel Fair: Aleck—blacksmith.

T. Filletto:[3] John—carpenter.

E. B. Fuller: Ben—laborer.

E. M. Fuller: Nero—laborer, Sam—laborer, January—laborer.

S. Fuller: Abram—laborer, Mingo—laborer, Prince—laborer, another Prince—laborer, Adam—laborer.

John A. Hamilton: Paul—helper.

J. W. Hayne: Bacchus—laborer.

James B. Heyward: Sam—laborer/blacksmith, Sam—laborer (there are two Sam's), Antony—laborer, Charles—laborer, Waby or Wabby—laborer, Quacco—laborer, June—laborer, Caesar—laborer, Stephney, or Stepney—laborer, Chance—laborer, Charleston—laborer, Guyms—laborer.

Robert Hume: Charles—laborer.

O. B. Irvine: John—helper.

W. J. Jenkins: Prince—laborer.

L. J. Jones: Sam—carpenter, Pompey—carpenter, Noah—laborer, June—laborer, Aleck—laborer.

John Klinck: Elijah—laborer/helper.

G. W. Logan: Joe—carpenter.

David Lopez: Kit—carpenter, Barber—carpenter, George—carpenter, Marcus—carpenter, Jacob—carpenter, Jim—fireman/carpenter, John—carpenter, John—laborer.

John H. Lopez: Thomas—carpenter.

William E. Martin: Robert—carpenter.

John A. Michel: Quash—fireman, George—laborer, William—laborer, Frank—laborer, Peter—laborer, John—laborer, Billy—laborer.

M. C. Mordecai: Tom—laborer.

M. C. Mordecai: Handy—laborer.

George W. Morse: Nat—carpenter.

H. Myers: John—laborer.

M. H. Nalhane: John—helper.

F. P. Pope: Albert—carpenter.

S. E. Porcher: Joseph—carpenter.

William T. Potter: Ferdinand—laborer/helper, George—laborer/helper.

George H. Reynolds: Daniel—helper.

A. G. Rose: Tom—blacksmith, Edward—wagoner.

A. T. Shier: John—blacksmith.

J. Ralph Smith: Charles—fireman.

Thomas P. Smith: Sam—carpenter.

William M. Thomas: James (Jim)—laborer.

Waddy Thompson: Sammy—carpenter.

Thomas B. Thruston: Jim—helper, Tom—helper.

T. P. Thurston: Tom—blacksmith.

E. A. Ward: Jim—helper.

A. Weston: Thomas, or Tom, or Tom

Lewis—blacksmith, John—carpenter,
John—helper, David—carpenter, Sammy—
carpenter.

J. B. Whitridge: Fortune—laborer, Billy—
laborer.

A. Wilkins: Abram—wagoner.

John Wilson (estate of John Wilson): Peter—
laborer.

APPENDIX 4

List of White Employees

Administration

David Lopez—general superintendent.[1]
John Hinton Lopez*—office clerk.
Moses E. Lopez*—office clerk.
James M. Eason—superintendent.
J. Ralph Smith—master mechanic/general
    superintendent.
John A. Michel—draftsman.
Daniel S. Hart—time keeper/store keeper.
Julius Clarence Smith—traveling agent.
William Walton Smith—chief clerk.
E. E. Pritchard*—store keeper.

Carpentry Shop

John W. Sawner—foreman carpentry shop.
Hugh Ferguson*—carpenter.
William McKay—wheelwright.[2]
Thomas Lyons—carpenter.

Morse's Small Arms Shop

George Woodward Morse—superintendent,
    small arms shop.
George T. Brooks—foreman.
Thomas McDonnell—machinist.
E. A. Sinclair—machinist.
E. C. Burden—machinist.
August J. Freitag—machinist.[3]
J. M. Poe—machinist.
John Faulkner—machinist.
John C. C. Turner—machinist.[4]
William Cammer*—machinist.
Henry Young Simpson*—machinist.
Thomas Powell—machinist.
J. B. Elkin—machinist.
Henry Newton Reid*—machinist.
J. F. McGinley—machinist.
E. A. LeBlanc—machinist.
Thomas Gerrard*—machinist.

A. Parkmon, or Parkman—machinist.
William Robert Powell—machinist.[5]
S. Busshardt, or Busshart*—worker.
Robert Charles Williman, or Willimon*—
    apprentice.
W. A. Ferguson—apprentice.
Sanford Vandiver Howard*—apprentice.
T. P. Philips*—apprentice.
O. H. C. Smith*—cartridge maker.
J. G. DuPre—cartridge maker.

Gun (Musket) Repair Shop

James Johnston Mackey—foreman.
H. P. Rives—gunsmith.
John Willis—gunsmith.
G. W. Taylor-gunsmith.
W. R. Buchanan-gunsmith.
C. B. Parks—gunsmith.
Robert Williman—apprentice.
Joseph King—apprentice.
John A. Robertson—apprentice.
W. M. Patton—gunsmith.
William W. Haughton—gunsmith.
James W. Carey—gun stocker.
Francis Davenport—gun stocker.
Benjamin George Happoldt*—gunsmith.[6]
David Thomas Peden*—gunsmith.[7]

Machine Shop

E. T. "Eam" Miller—foreman.
George H. Stein—machinist.
P. Bergin*—machinist.
John T. Guy—machinist.
Joseph Pigott—machinist.
Oliver J. Butts Sr.*—machinist.[8]
Richard Furman Divver*—machinist.[9]
R. K. Gossett*—apprentice.
B. A. Goddard—apprentice machinist.
A. B. Beeco*—apprentice.

Blacksmith Shop

Phillip A. Mullane—foreman until September 19, 1863;[10] apprentice in the gun repair shop after January 15, 1864.
John J. Gossett*—blacksmith.
Daniel Carey—blacksmith.
J. P. Adams—blacksmith.
J. R. Godfrey—blacksmith.
John S. Kilby*—blacksmith.
Patrick Murphy—blacksmith.
J. J. McManus—blacksmith.
W. B. Burdine—blacksmith helper, cupola tender's helper.
James A. Gaines*—blacksmith.
Hampton Turner—apprentice blacksmith.
William H. Mounce*—blacksmith.
Marcus Lafayette Davis—blacksmith.[11]
John Coddington, or Codington—blacksmith.[12]
O. E. Bowen*—blacksmith.

Foundry

Robert McKay—foreman.
J. T. Williams—moulder.
James Doolan—moulder.
John Keenan—cupola tender.
J. R. Willis—apprentice and moulder.
William O'Connor—apprentice.
J. H. Moore—apprentice.
Edward Eugene Tavel*—moulder.
William Henry Halsall*—moulder.

Laborers and Watchmen

James Fraser, or Frazer—foreman.
James Moore—day watchman.

Lemuel Jacobs*—day watchman.
John Love—day watchman.
Patrick Fahey, or Fahy—night watchman.
A. Payne—night watchman.
Joseph Pelley—night watchman.
Henry M. Fliedner*—night watchman.
W. P. T. Satterfield—night watchman's helper.
William Godfrey—night watchman's helper.

Miscellaneous

J. P. Elkins—apprentice.
J. C. Godfrey—helper.
Bernard Lighe, or Fighe—helper.
A. Williman—apprentice.
G. F. Poole—apprentice.
C. D. Gaillard—helper.
George Kruse—pattern maker.
Jon F. Pollard—apprentice.

Men Whose Names Do Not Appear on the Payrolls

J. D. White—employed at State Works prior to July 1863.[13]
H. C. Briggs*—collier.[14]
Thomas Dixon Jr.*—moulder at State Works, June 1862–April 1863.[15]
James F. Barnes*—blacksmith. Detached service at State Works January 21–May 2, 1863.
Eli E. Fleming*—gunsmith.[16]
William Barringer*—December 1862–February 1863.
George Stokes.

* A man who was serving or had served in either State or Confederate service.

APPENDIX 5

Total Production of Morse's Firearms

Manufacturer	Date Made	Applicable Patent #	Notes	Number Made	Fate
Searles	ca. 1855–56	15996	Patent prototype	1	Smithsonian Institution
Searles	ca. 1855–56	15996	Sporting prototype	1	Buffalo Bill Center of the West
Searles	ca. 1856–57	15996	Carbine used in 1857 West Point trials	1	Unknown
Muzzy	1857–58	U.S. 15995, British 1357	Three-barrel cased sets	100	At least 13 survive
Washington Arsenal	December 1857	U.S. 15995, British 1357	Four were ordered, but order was cancelled	none	N/A
Washington Arsenal	ca. March 1858	Probably U.S. 20503	First experimental alterations, .69 caliber rifle, .69 caliber smoothbore, .54 caliber rifle	3	Unknown
Washington Arsenal	May 1858	Probably U.S. 20503	Alteration of a .69 caliber smoothbore	1	Unknown
Muzzy	March–November 1858	?	100 ordered; 1 made	1	Unknown
Muzzy	September 1858	Probably 20503	Pattern firearm	1	Unknown
Munck	late 1858–early 1859	20503, "Munck" hybrid	Alteration of M1816	1	Unknown
Springfield Armory	November 1858	20503, unmodified	M1816/1822 alterations	3	3 survive at Springfield Armory
Springfield Armory	November 1858	20503 and/or 20503, "Munck" hybrid	M1841 alterations	3	Unknown
Springfield Armory	1859	20503, Munck hybrid	M1816/1822 alterations	54	About 5 survive
Springfield Armory	July 1860	20503, Munck hybrid	M1841 alteration	1	Unknown
Springfield Armory	1860	20503, Munck hybrid	New carbines, not alterations	5	Unknown
H. Marshall, Atlanta	1862	20503, Munck hybrid	Brass-frame carbine prototype. No serial number	1	Greensboro Historical Museum
S.C. State Military Works, Greenville	1862	20503, Munck hybrid	Brass-frame carbine. Made for S.C. legislature and governor. No known serial number	2	Unknown
S.C. State Military	1862	20503, Munck hybrid	Brass-frame carbine. Sent to Richmond.	1	Unknown

Manufacturer	Date	Patent	Description	Quantity	Survivors / Location
S.C. State Military Works, Greenville	1864	20503, Munck hybrid	Brass-frame carbine cased set made for Governor Bonham. No serial number	1	McKissick Museum, USC/ S.C. State Museum
S.C. State Military Works, Greenville	1864	20503, Munck hybrid	Brass-frame carbine made for John Hunt Morgan. No known serial number	1	Unknown
S.C. State Military Works, Greenville	1864–65	20503, Munck hybrid	Brass-frame carbine (Type III) with shotgun barrel. No serial number	1	Atlanta History Center
S.C. State Military Works, Greenville	1863–65	20503, Munck hybrid	Brass-frame carbine (types 1, 2, and 3). Serial numbers 1–1,032	1,032	About 180 known survivors
S.C. State Military Works, Greenville	1864	No known patent	Inside lock	140	About 17 known survivors

Total production of Morse firearms   1,355

NOTES

Introduction

1. National Archives, RG 123, Box 449A.

2. "Memorial of George W. Morse to the 44th Congress of the United States in Regard to His Claim as the Inventor of the Modern Metallic Cartridge System of Fire-arms," George W. Wray Jr. File, Atlanta History Center (hereinafter referred to as Wray-Morse File), File 040, p. 19.

3. Ibid., 8.

4. James J. Mackey was born January 26, 1817, in New York City; died on April 13, 1901, at Grove Station in Greenville County; he was buried at Springwood Cemetery in Greenville.

5. "Cash received at State Works 4th Quarter 1864," photocopy, Museum and Library of Confederate History, Greenville, S.C.

6. An intra-office memo states that the carbine barrel is .52 caliber, but most likely it is either .54 caliber as seen in earlier model carbines or .50 as seen in most production models (Ashley Halsey Jr. to E. L. Inabinett, April 30, 1965, Bonham File, University of South Carolina McKissick Museum). In comparison, the two Muzzy cased-sets at the Atlanta History Center, serial numbers 7 and 94, both have a .69 caliber (16-gauge) shotgun, a .48 caliber rifle, and a .52 caliber carbine.

7. Gordon L. Jones, "McKissick Museum's Morse Design Cased Firearm February 16, 1987," unpublished, copy at South Carolina State Museum, HSSI 701.

8. M. L. Bonham to Captain J. J. Mackey, January 14, 1884, Morse File, McKissick Museum, University of South Carolina.

9. Morse to Bonham, August 26, 1885, Morse File, McKissick Museum.

10. http://www.ancestry.com.

11. M. L. Bonham to Captain J. J. Mackey, January 14, 1884, Morse File, McKissick Museum.

12. http://findagrave.com. Some sources show Mackey's middle name as Jonathan and others as Johnston. His son, James Franklin Mackey, started Mackey Mortuary in Greenville.

13. A. S. Salley, Secretary, Historical Commission of South Carolina, to Bonham, September 6, 1932, Morse File, McKissick Museum; Jones, "McKissick Museum's Morse Design Cased Firearm," 11

Chapter 1: George Woodward Morse

1. Haverhill is pronounced "hayvrill." In several U.S. Census records, George W. Morse is shown incorrectly as being from Massachusetts, possibly confusing Haverhill, New Hampshire, with Haverhill, Massachusetts.

2. An extensive family genealogical survey is available in File 025, Wray-Morse File. Marriage contract #2255, George W. Morse and Maryanne Terrett, Natchitoches, Louisiana, Genealogical and Historical Society.

3. *New Hampshire Gazette,* June 18, 1850, 2.

4. "End of a Busy Life," *Critic-Record* (Washington, D.C.), no. 6108 (March 8, 1888): 1. Copyright News Bank and/or the American Antiquarian Society, 2004, from GenealogyBank .com.

5. National Archives, RG 46.

6. Wray-Morse File 026.

7. Ibid., Files 006 and 008.

8. William F. Whitcher, *History of the Town of Haverhill, New Hampshire* (Concord, N.H.: Rumford Press, 1919), 597–99.

9. Wray-Morse File 032.

10. Ibid., File 010.

11. Whitcher, *History of the Town of Haverhill,* 597–99, says 1833. Other sources say he moved to Louisiana in 1831.

12. Wray-Morse File 009.

13. Whitcher says Peabody moved to California in 1843, a date that is clearly incorrect (*History of the Town of Haverhill*), 599.

14. Whitcher, *History of the Town of Haverhill,* 599, writes that P. A. Morse died on November 16, 1878, in Ninock, Louisiana.

15. A different date of birth, September 3, 1812, comes from an affidavit signed by his brother, Isaac, on October 24, 1870. See Records of the U.S. House of Representatives, Forty-second Congress, National Archives, RG 233, Box 21.

16. Wray-Morse File 018.

17. Ibid., Files 019 and 021; "Fertile Plantation Has Been in One Family for 100 Years," *Shreveport Times,* July 25, 1937.

18. National Archives, RG 123, Box 449A.

19. Tufts is listed in Stimpson's directories every year between 1832 and 1840, but Morse appears only in the 1834 and 1835 directories. See Charles Stimpson, *Stimpson's Boston Directory* (Boston: Charles Stimpson Jr., 1834 and 1835).

20. "End of a Busy Life," 1.

21. "A New Printing Press," *Connecticut Gazette* (New London, Conn.), vol. 74, no. 3835 (May 10, 1837): 2. Copyright News Bank and/or the American Antiquarian Society, 2004, from Genealogy Bank.com. National Archives, RG 123, Box 449A.

22. "End of a Busy Life," 1.

23. National Archives, RG 123, Box 449A.

24. "End of a Busy Life," 1.

25. New York, Passenger and Immigration Lists, 1820–1850, Microfilm M237, Roll 35, List 965; http://www.ancestry.com; "Passengers," *Spectator,* New York, December 28, 1837.

26. The sole source of this information is an affidavit signed by Morse's brother Isaac on October 24, 1870. See Records of the U.S. House of Representatives, Forty-second Congress, National Archives, RG 233, Box 21.

27. U.S. Census, 1840.

28. Morse to Louisiana Secretary of State, File 033, Wray-Morse File.

29. www.slodms.doa.la.gov/Historical Document; Morse's papers are preserved under the section "Field Notes." Records of the U.S. House of Representatives Forty-second Congress, National Archives, RG 233, Box 21.

30. Her Christian name appears in other records as Marion and Mary Ann, and it is spelled Maryanne on the marriage contract. Her surname also appears as Bloodworth. Her parents were James Bludworth Jr. (November 25, 1779–1852) and the deceased Aimie Rouquier (November 25, 1789–1835), whose first name is shown as both Josephine and Marie. There were either seven or eight siblings in her generation. For a complete genealogical profile of the Bludworth family, see Wray-Morse File 025. See also Wray-Morse File 021.

31. The 1850 U.S. Census states he was born in Missouri, but both the 1860 and 1870 censuses state he was born in Kansas. Some sources show his birthday as May 30, 1842. An extensive Morse family genealogical review found in File 025 in the Wray-Morse File says he was born at Jefferson Barracks, Missouri. Fort Scott, Kansas,

only four miles from the Missouri border, was where his father was accidentally killed. File 044 in the Wray-Morse File has a daguerreotype of the young Terrett.

32. Harry C. Meyers, ed., "'The Crack Post of the Frontier': Letters of Thomas and Charlotte Sword," *Kansas History* 5 (Autumn 1982): 205–8.

33. Ibid.

34. Marriage contract no. 2255 found in the Natchitoches Genealogical and Historical Society.

35. George's slaves were: Orange—29 years, Victor—27 years, Charles—22 years, Neville—20 years, Elizabeth—37 years, Maria—25 years, William—10 years and child of Maria, Louisa—6 years and child of Maria, Eliza—3 years and child of Maria, Female infant—1 year and child of Maria, Felicity—40 years, Amos—14 years and child of Felicity.

36. The slaves that Marianne owned a half-share in were: Liddy—25 years, James—10 years and child of Liddy, Moses—4 years and child of Liddy, Susan—23 years, Sarah—10 years and child of Susan, Ellen—8 years and child of Susan, Virginia—4 years and child of Susan, Lydia Anne—2 years and child of Susan, Mary—31 years, Adeline—8 years and child of Mary, Louisa—4 years and child of Mary, Jenny—2 years and child of Mary, Jess—45 years, Maria—55 years, Emiline—18 years, Caroline—3 years and child of Emiline, Nat—30 years. The slaves Marianne owned outright were: Ellick—22 years, Sophia—35 years, Emily—13 years.

37. U.S. Census, 1850. This census also says that George was born in Massachusetts. Most likely his wife answered the census taker's questions and, knowing that his father had lived there, assumed that her husband was born there.

38. U.S. Census, 1900.

39. Wray-Morse File 049.

40. Whitcher records, incorrectly, that they had only one child, Peabody Atkinson Morse, born six years before the marriage, on May 12, 1842 in Natchitoches (*History of the Town of Haverhill*), 598–599. Extensive Morse family genealogical research is found in Wray-Morse File 025.

41. Morse's great-granddaughter, Mary M. Kane, referred to Morse as "colonel" in a letter to Dr. H. L. Sutherland dated March 2, 1963. See Wray-Morse File 005.

42. Report of the State Engineer, Baton Rouge, April 7, 1853, Wray-Morse File 028. Records of the U.S. House of Representatives, Forty-second Congress, National Archives, RG 233, Box 21.

In an 1872 deposition, Morse stated that it was Walker who appointed him in 1852. Walker left office on January 18, 1853, following which Morse served under Governor Hebert (National Archives, RG 123, Box 449A).

43. Report of the State Engineer, 1853, Wray-Morse File 028; National Archives, RG 123, Box 449A.

44. "Memorial of George W. Morse to the 44th Congress," 25.

45. Report of the State Engineer, 1855, Wray-Morse File 028.

46. Ibid., 1853, Wray-Morse File 028.

47. Ibid.

48. Report of the State Engineer, 1855, Wray-Morse File 028.

49. Ibid., 1853, dated January 1854, Wray-Morse File 028.

50. Patents 15995 and 20503, U.S. Patent Office.

51. Records of the U.S. House of Representatives, Forty-second Congress, National Archives, RG 233, Box 21.

52. National Archives, RG 123, Box 449A.

....

Chapter 2: Morse's Early Patents

1. In a deposition given in 1887 by V. D. Stockbridge, some names of inventors of either breechloaders or metallic cartridges who preceded Morse were Smith & Wesson (American, 1854); Minessinger (American, 1849); Fay (American, 1827); McCarty, Walter Hunt (American, 1850); Chauden (French, 1847); Pryse & Redman ("the Flobert arm," English, 1853); Beringer (French, 1834, 1840); Bourcier, LeFaucheux (French, 1835); Robert, Pauly (English, 1816); Desnyau (French, 1840); Gastinne (French, 1853); A. E. L. Bellford (English, 1853); Palmer & Bidault (French, 1854); Pottet (French, 1855); Houillier (French, 1846); Loron (French, 1847); and Sharps (American, 1848).

2. National Archives, RG 123, Box 449B.

3. Ibid., Box 449A.

4. Ibid.

5. Ibid.

6. "Gun Locks," *Hampden Whig* (Springfield, Mass.), October 19, 1831. Copyright News Bank and/or the American Antiquarian Society, 2004, from GenealogyBank.com. The author has been unable to locate any records in the U.S. Patent Office that support Morse having patented such an invention. It is probable that he never sought a patent for it.

7. Samuel Blatchford, Reports of Cases Argued and Determined at the Circuit Court of the

United States for the Second Circuit (New York: Baker, Voorhist, and Co., 1888), vol. 24: 510.

8. National Archives, RG 123, Box 449A.

9. In 1870 Morse recalled the date as "about 1853." See National Archives, RG 123, Box 449A.

10. "Memorial of George W. Morse to the 44th Congress," 26.

11. National Archives, RG 123, Box 449A.

12. "Memorial of George W. Morse to the 44th Congress," 26.

13. National Archives, RG 123, Box 449A.

14. Blatchford, Reports of Cases Argued and Determined at the Circuit Court of the United States for the Second Circuit 24: 497.

15. Whitcher, History of the Town of Haverhill, 599.

16. National Archives, RG 123, Box 449B.

17. Ibid.

18. Ibid.

19. Ibid.

20. Ibid.

21. National Archives, RG 123, Box 449A.

22. Ibid.

23. This knife is engraved, "R. P. Bowie to H. W. Fowler, U.S.D." and "Searles, Baton Rouge, La." See William R. Williamson, "Daniel Searles, of Baton Rouge," *Arms Gazette*, June 1977, 15–19, and the *Baton Rouge Advocate,* October 8, 1978, 131.

24. National Archives, RG 123, Box 449A.

25. National Archives, RG 123, Box 449B.

26. "Memorial of George W. Morse to the 44th Congress," 27.

27. Daniel Searles died March 21, 1860, in Baton Rouge. His wife, Jane, died on December 21, 1853, in Baton Rouge. Their burial locations are unknown. For more details of the family's history, see the 1850 Census for Baton Rouge, 174B; Notarial Book N-C, 71; and "Personal Recollections," *Baton Rouge State Times,* December 11, 1914.

28. Richard T. Hill and William E. Anthony, *Confederate Longarms and Pistols: A Pictorial Study* (Charlotte, N.C.: Richard Taylor Hill and William Edward Anthony, 1978), 28.

29. "Memorial of George W. Morse to the 44th Congress," 22.

30. Records of the U.S. House of Representatives, Forty-second Congress, Accompanying Paper Files, Moody to Negley, RG 233, Box 21.

31. National Archives, RG 123, Box 449A.

32. Ibid.

33. Ibid.

34. Patent 15995, U.S. Patent Office.

35. Ibid.

36. National Archives, RG 123, Box 449A.

37. Records of the U.S. House of Representatives, Forty-second Congress, National Archives, RG 233, Box 21.

38. Morse used several terms for the sliding breech block, "sliding block, breech slide, and closer" (British Patent 1357, "Specification of George Woodward Morse, Breech Loading Firearms, London, printed by George E. Eyre and William Spottiswood, Printers to the Queen's Most Excellent Majesty, Published at the Great Seal Patent Office, 25, Southampton Buildings, Holbron, 1857," 2.).

39. Patent 15996, U.S. Patent Office.

40. Ibid.

41. National Archives, RG 123, Box 449A.

42. Patent 15996, U.S. Patent Office.

43. Ibid.

44. Ibid.

45. National Archives, RG 123, Box 449A.

46. National Archives, RG 123, Box 449A, Lawsuit #10270.

47. British Patent 1357, "Specification of George Woodward Morse, Breech Loading Firearms."

48. Ibid., 5.

49. "Memorial of George W. Morse to the 44th Congress," 41.

50. British Patent 1357, "Specification of George Woodward Morse, Breech Loading Firearms," 5.

51. National Archives, RG 123, Box 449A.

52. Morse's 1872 deposition, National Archives, RG 123, Box 449A.

....

Chapter 3: Nathan M. Muzzy

1. Records of the U.S. House of Representatives, Forty-second Congress, National Archives, RG 233, Box 21.

2. George D. Moller, American Military Shoulder Arms, Volume III: Flintlock Alterations and Muzzleloading Percussion Shoulder Arms, 1840–1865 (Albuquerque: University of New Mexico Press, 2011), 177.

3. Report of Brigadier General S. V. Benet, Chief of Ordnance, to Secretary of War, March 6, 1875, Wray-Morse File 030.

4. Ibid.

5. National Archives, RG 123, Box 449A.

6. Ibid.

7. The Morse Arms Manufacturing Company v. the United States, 27 Ct. Cl. 363, U.S. Court of Claims, No. 14340, decided June 20, 1892. Davis rejoined the U.S. Senate on March 4, 1857, and Floyd took office as secretary of state on March 6. Craig wrote to Bell on March 5, probably reflecting the fact that Floyd was acting as secretary earlier than the sixth.

8. Morse's deposition, National Archives, RG 123, Box 449A.

9. Bell's report dated March 6, 1857, National Archives, RG 123, Box 449.

10. Slides 0061–0070, National Archives, RG 123, Box 449B.

11. Bell's report dated March 6, 1857, National Archives, RG 123, Box 449.

12. National Archives, RG 123, Box 449.

13. Morse's deposition, National Archives, RG 123, Box 449A.

14. National Archives, RG 123, Box 449; Report of Brigadier General S. V. Benet, Chief of Ordnance, to Secretary of War, March 6, 1875, Wray-Morse File 030.

15. "Memorial of George W. Morse to the 44th Congress," 20.

16. U.S. House of Representatives, Forty-eighth Congress, First Session, National Archives, Congressional Series Report No. 508.

17. ww3.rediscov.com/spring/spring.htm.

18. Joseph T. Vorisek, The Breechloading Shotgun 1860-1940, Volume II, G through P (Brighton, Mich.: Cornell Publications, 1990), 465.

19. National Archives, RG 123, Box 449A.

20. U.S. Patent 20214.

21. Blatchford, Reports of Cases Argued and Determined at the Circuit Court of the United States for the Second Circuit 24: 503–4.

22. Robert M. Holter, "The Morse-Muzzy Connection," Gun Report, May 2001, 18.

23. Ibid.

24. National Archives, RG 123, Box 449B.

25. James E. Hicks, U.S. Military Firearms 1776–1956 (La Canada, Calif.: James E. Hicks & Son, 1962), 85. One source states that the receiver of a surviving cased set is stamped, "Morse's Patent, Muzzy & Co. Oct. 20, 1858," and the barrel is stamped "Worcester," but the actual stamped date is October 20, 1856, which is also a mistake because the patent was October 28, 1856. See Hill and Anthony, Confederate Longarms and Pistols, 28.

26. National Archives, RG 123, Box 449A.

27. Ibid.

28. Holter, "The Morse-Muzzy Connection," 18.

29. B. R. Lewis states that this is the same type of cartridge that was used in Morse's Springfield alterations and carbines, but offers no documentation for the statement ("Morse Arms and Ammunition." American Rifleman, March 1955, 31–33).

30. Holter, "The Morse-Muzzy Connection," 18; copy of instructions in File Wray-Morse File 030.

31. Personal communication with the author.

32. Hill and Anthony, Confederate Longarms and Pistols, 128.

33. Lewis does not provide references or footnotes in "Morse Arms and Ammunition," 31–33.

....

Chapter 4: War Department Evaluations, 1857–1858

1. Morse Arms Manufacturing Company *v.* the United States, 27 Ct. Cl. 363, U.S. Court of Claims, No. 14340, decided June 20, 1892; Report of Brigadier General S. V. Benet, Chief of Ordnance, to Secretary of War, March 6, 1875, Wray-Morse File 030.

2. National Archives, RG 123, Box 449B.

3. Register of Letters Received by Ordnance Department, vols. 27–31, National Archives, RG 156, entry 20.

4. Morse's deposition, National Archives, RG 123, Box 449A.

5. National Archives, RG 123, Box 449B.

6. "Memorial of George W. Morse to the 44th Congress," 27–34.

7. Letters, Endorsements and Circulars Sent (Miscellaneous Letters), vols. 48–52, National Archives, RG 156, entry 3.

8. Bell's report dated March 6, 1857, National Archives, RG 123, Box 449.

9. John D. McAulay, *Civil War Breech Loading Rifles: A Survey of the Innovative Arms of the American Civil War* (Lincoln, R.I.: Andrew Mowbray Publishers, 1987), 120; Letters, Endorsements and Circulars Sent (Miscellaneous Letters), vols. 48–52, National Archives, RG 156, entry 3.

10. National Archives, RG 123, Box 449B.

11. Ibid.

12. Ibid.

13. Report of Brigadier General S. V. Benet, Chief of Ordnance, to Secretary of War, March 6, 1875, Wray-Morse File 030.

14. Ibid. The reference is repeated in H. L. Sutherland, "Arms Manufactury in Greenville County," *Proceedings and Papers of the Greenville County Historical Society 1968–1971* 4 (1971): 54.

15. National Archives, RG 123, Boxes 449B and 449A.

16. Ibid.

17. Ibid.

18. National Archives, RG 123, Box 449.

19. National Archives, RG 123, Box 449A.

20. McAulay, *Civil War Breech Loading Rifles,* 123. The original reference is #10 in McAulay's book.

21. *Gun Report,* February 1989, 31, 60.

22. Ibid.

23. Morse's letter dated November 12, 1857, National Archives, RG 123, Box 449. Also found in Morse Arms Manufacturing Company *v.* the United States, 27 Ct. Cl. 363, U.S. Court of Claims, No. 14340, decided June 20, 1892.

24. Records of the Office of the Secretary of War, Letters Sent, Military Affairs, vol. 39, March 1–December 31, 1857, National Archives, No. 6, Roll 39. This letter is paraphrased in Morse Arms Manufacturing Company *v.* the United States, 27 Ct. Cl. 363, U.S. Court of Claims, No. 14340, decided June 20, 1892. It is also found in National Archives, RG 123, Box 449.

25. National Archives, RG 123, Box 449A.

26. U.S. Patent 20503.

27. Ibid.

28. Orders and Endorsements Sent by the Secretary of War, 1846–1870, vol. 7, National Archives, Microcopy 444, Roll 4.

29. National Archives, RG 123, Box 449.

30. Ibid.

31. National Archives, RG 123, Box 449A.

32. Letters, Endorsements and Circulars Sent (Miscellaneous Letters), vols. 48–52, National Archives, RG 156, entry 3; RG 123, Box 449.

33. Morse Arms Manufacturing Company *v.* the United States, 27 Ct. Cl. 363, U.S. Court of Claims, No. 14340, decided June 20, 1892.

34. "Memorial of George W. Morse to the 44th Congress," 39.

35. Orders and Endorsements Sent by the Secretary of War, 1846–1870, vol. 7, National Archives, Microcopy 444, Roll 4. This reference is repeated in Morse Arms Manufacturing Company *v.* the United States, 27 Ct. Cl. 363, U.S. Court of Claims, No. 14340, decided June 20, 1892. Also found in National Archives, RG 123, Box 449.

36. Statements of Accounts for Contractors, vol. 3, National Archives, RG 156, entry 152. Letters, Endorsements and Circulars Sent (Miscellaneous Letters), vols. 48–52, National Archives, RG 156, entry 3. This reference is also paraphrased in Moller, *American Military Shoulder Arms, Volume III,* 95.

37. National Archives, RG 123, Box 449.

38. U.S. Patent 20727.

39. Jack W. Gunn, "Ben McCulloch: A Big Captain," *Southwestern Historical Quarterly* 58, no. 1 (July 1954): 1–21.

40. Thomas W. Cutrer, *Ben McCulloch and the Frontier Military Tradition* (Chapel Hill: University of North Carolina Press, 1993).

41. National Archives, RG 123, Box 449.

42. Gunn, "Ben McCulloch," 1–21; Cutrer, Ben McCulloch and the Frontier Military Tradition, 153.

43. "Memorial of George W. Morse to the 44th Congress," 45–46.

44. Ibid., 46.

45. Report of Brigadier General S. V. Benet, Chief of Ordnance, to Secretary of War, March 6, 1875, Wray-Morse File 030.

46. National Archives, RG 123, Box 449A.

47. Ibid.

48. National Archives, RG 123, Box 449.

49. Ibid.

50. National Archives, RG 123, Box 449A.

51. U.S. Patent 20503. The same report is found in "Memorial of George W. Morse to the 44th Congress," 34–37.

52. "Memorial of George W. Morse to the 44th Congress," 20; National Archives, RG 123, Box 449.

53. Sutherland, "Arms Manufactury in Greenville County," 54.

54. National Archives, RG 123, Box 449.

55. U.S. Patent 20214.

56. Ibid.

57. U.S. Patent 20727.

58. Ibid.

59. George A. Hoyem, The History and Development of Small Arms Ammunition, Volume II: Centerfire: Primitive, and Martial Longarms (Tacoma, Wash.: Armory Publications, 1982), 16.

60. Lewis, "Morse Arms and Ammunition," 31–33.

61. National Archives, RG 123, Box 449A.

62. National Archives, RG 123, Box 449; also in Report of Brigadier General S. V. Benet, Chief of Ordnance, to Secretary of War, March 6, 1875, Wray-Morse File 030. Moller, *American Military Shoulder Arms, Volume III,* 95.

63. National Archives, RG 123, Box 449.

64. Lieutenant Colonel J. W. Ripley, Brevet Major George D. Ramsay, and Captain William Maynadier comprised the board to consider altered arms. Major Alfred Mordecai, Captain T. J. Rodman, and Major T. T. S. Laidley comprised the board to consider the best breech-loading carbine. Extensive results from the boards can be found at National Archives, RG 123, Box 449.

65. National Archives, RG 123, Box 449.

66. Ibid.; Report of Brigadier General S. V. Benet, Chief of Ordnance, to Secretary of War, March 6, 1875, Wray-Morse File 030.

67. National Archives, RG 123, Box 449.

68. Letters, Endorsements and Circulars Sent (Miscellaneous Letters), vols. 48–52, National Archives, RG 156, entry 3.

69. Ibid.; National Archives, RG 123, Box 449.

70. National Archives, RG 123, Box 449; "Memorial of George W. Morse to the 44th Congress," 27.

71. National Archives, RG 123, Box 449A.

72. "Memorial of George W. Morse to the 44th Congress," 39.

73. National Archives, RG 123, Box 449.

74. Ibid.

75. National Archives, RG 123, Box 449; "Memorial of George W. Morse to the 44th Congress," 37–45.

76. National Archives, RG 123, Box 449.

77. Ibid.; Report of Brigadier General S. V. Benet, Chief of Ordnance, to Secretary of War, March 6, 1875, Wray-Morse File 030; Morse Arms Manufacturing Company *v.* the United States, 27 Ct. Cl. 363, U.S. Court of Claims, No. 14340, decided June 20, 1892; Records of the U.S. House of Representatives, Forty-second Congress, National Archives, RG 233, Box 21; "Memorial of George W. Morse to the 44th Congress," 4; Records of the U.S. Senate, Forty-second Congress, Ordnance Office, War Department, National Archives, RG 46, Box 83.

78. Moller, American Military Shoulder Arms, Volume III, 95.

··

Chapter 5: Springfield Armory

1. National Archives, RG 123, Box 449.

2. Ibid.

3. Ibid., Box 449A.

4. Ibid., Box 449.

5. Report of Brigadier General S. V. Benet, Chief of Ordnance, to Secretary of War, March 6, 1875, Wray-Morse File 030.

6. National Archives, RG 123, Box 449; Morse Arms Manufacturing Company *v.* the United States, 27 Ct. Cl. 363, U.S. Court of Claims, No. 14340, decided June 20, 1892.

7. Morse Arms Manufacturing Company *v.* the United States, 27 Ct. Cl. 363, U.S. Court of Claims, No. 14340, decided June 20, 1892.

8. Moller, American Military Shoulder Arms, Volume III, 95.

9. Morse Arms Manufacturing Company *v.* the United States, 27 Ct. Cl. 363, U.S. Court of Claims, No. 14340, decided June 20, 1892.

10. National Archives, RG 123, Box 449A.

11. Letters, Endorsements and Circulars Sent (Miscellaneous Letters), vols. 48–52, National Archives, RG 156, entry 3; National Archives, RG 123, Box 449; Report of Brigadier General S. V. Benet, Chief of Ordnance, to Secretary of War, March 6, 1875, Wray-Morse File 030.

12. Orders and Endorsements Sent by the Secretary of War, 1846–1870, vol. 8, National Archives, Microcopy 444, Roll 4; National Archives, RG 123, Box 449.

13. National Archives, RG 123, Box 449; House of Representatives, Forty-eighth Congress, First Session, National Archives, Congressional Series Report No. 508.

14. Hicks, U.S. Military Firearms 1776–1956, 85.

15. National Archives, RG 123, Box 449.

16. Pitman, Statement of the Receipt and Issue of Morse's Breech Loading Arms, 90.

17. Moller, American Military Shoulder Arms, Volume III, 95.

18. ww3.rediscov.com/spring/spring.htm.

19. Morse Arms Manufacturing Company *v.* the United States, 27 Ct. Cl. 363, U.S. Court of Claims, No. 14340, decided June 20, 1892.

20. About four hundred breech-loading rifles designed by William Mont Storm of New York were manufactured at the U.S. Armory at Harpers Ferry in 1859 (McAulay, *Civil War Breech Loading Rifles,* 120).

21. The 1860 U.S. Census shows a Charles H. Munck, a forty-year-old German-born gunsmith living in Washington, D.C. His name appears in several Washington directories between 1862 and 1866. In the 1865 directory his name appears as Christian H. Munck in an advertisement for his business as a gunsmith. In the 1870 Washington directory the name A. Kleinhenn appears as a "successor to C. H. Munck" in the business of gunmaking and locksmithing.

22. National Archives, RG 123, Box 449A.

23. Ibid.

24. National Archives, RG 123, Box 449.

25. Orders and Endorsements Sent by the Secretary of War, 1846–1870, vol. 8, National Archives, Microcopy 444, Roll 4.

26. Letters, Endorsements and Circulars Sent (Miscellaneous Letters), vols. 48–52, National Archives, RG 156, entry 3.

27. Records of the U.S. Senate, Forty-second Congress, Ordnance Office War Department, National Archives, RG 46, Box 83.

28. National Archives, RG 123, Box 449.

29. Ibid.

30. Ibid.

31. Orders and Endorsements Sent by the Secretary of War, 1846–1870, vol. 8, National Archives, Microcopy 444, Roll 4.

32. National Archives, RG 123, Box 449.

33. Orders and Endorsements Sent by the Secretary of War, 1846–1870, vol. 8, National Archives, Microcopy 444, Roll 4; National Archives, RG 123, Box 449.

34. National Archives, RG 123, Box 449.

35. Ibid.

36. Ibid.

37. Ibid.

38. Orders and Endorsements Sent by the Secretary of War, 1846–1870, vol. 8, National Archives, Microcopy 444, Roll 4.

39. Records of the U.S. Senate, Forty-second Congress, Ordnance Office War Department, National Archives, RG 46 Box 83.

40. Cutrer, Ben McCulloch and the Frontier Military Tradition, 164–65.

41. Ibid., 165.

42. Ibid.

43. Ibid.

44. Ibid.

45. Records of the Office of the Secretary of War, Letters Sent, Military Affairs, vol. 39, March 1–December 31, 1857, National Archives, No. 6, Roll 39.

46. National Archives, RG 123, Box 449.

47. Letters, Endorsements and Circulars Sent (Miscellaneous Letters), vols. 48–52, National Archives, RG 156, entry 3; Orders and Endorsements Sent by the Secretary of War, 1846–1870, vol. 9, National Archives, Microcopy 444, Roll 5.

48. Letters, Endorsements and Circulars Sent (Miscellaneous Letters), vols. 48–52, National Archives, RG 156, entry 3.

49. Message of the President of the United States to the Two Houses of Congress at the Commencement of the First Session of the Thirty-sixth Congress, First Session, Ex. Doc. 2, Congressional Series, National Archives.

50. Report of Brigadier General S. V. Benet, Chief of Ordnance, to Secretary of War, March 6, 1875, Wray-Morse File 030.

51. National Archives, RG 123, Box 449.

52. Claud E. Fuller, comp., *Springfield Muzzle-Loading Shoulder Arms* (Glendale, N.Y.: S and S Firearms, 1986), 171. Fifty-four were described as rifled muskets, and one as a percussion rifle.

53. Ibid.

54. Pitman, Statement of the Receipt and Issue of Morse's Breech Loading Arms, 90.

55. National Archives, Waltham, Mass.,

Springfield Armory, Letters Sent, vol. 9 of 26, November 22, 1860, 97–98.

56. ww3.rediscov.com/spring/spring.htm.

57. Ibid.

58. Pitman, Statement of the Receipt and Issue of Morse's Breech Loading Arms, 90.

59. National Archives, RG 123, Box 449A.

60. Extract from Annual Report of Secretary of War, 1859, Mess. & Docs., vol. 2, Thirty-sixth Congress, First Sess. 1859–1860; ww3.rediscov .com/spring/spring.htm.

....

Chapter 6: Morse's New Carbine, 1860

1. Letters, Endorsements and Circulars Sent (Miscellaneous Letters), vols. 48–52, National Archives, RG 156, entry 3.

2. National Archives, RG 123, Box 449.

3. Ibid.

4. Ibid.

5. A. Anderson is a bit of a mystery and cannot be further identified by the author.

6. Report of Brigadier General S. V. Benet, Chief of Ordnance, to Secretary of War, March 6, 1875, Wray-Morse File 030; Records of the U.S. Senate, Forty-second Congress, Ordnance Office, War Department, National Archives, RG 46, Box 83; National Archives, RG 123, Box 449.

7. Cutrer, Ben McCulloch and the Frontier Military Tradition, 168.

8. Ibid.

9. Reprinted as "Long Shots," *Charleston Daily Courier,* November 26, 1860, 1.

10. Cutrer, Ben McCulloch and the Frontier Military Tradition, 177.

11. National Archives, RG 123, Box 449A.

12. Morse recalled that the number was four or five and that one was to be used as a model (National Archives, RG 123, Box 449A).

13. Orders and Endorsements Sent by the Secretary of War, 1846–1870, vol. 9, National Archives, Microcopy 444, Roll 5; Hicks, *U.S. Military Firearms 1776–1956,* 85–86; Orders and Endorsements Sent by the Secretary of War, 1846–1870, vol. 9, National Archives, Microcopy 444, Roll 5.

14. Pitman, Statement of the Receipt and Issue of Morse's Breech Loading Arms, 90.

15. National Archives, RG 123, Box 449A.

16. Ibid., Box 449; Moller, American Military Shoulder Arms, Volume III, 176.

17. ww3.rediscov.com/spring/spring.htm.

18. Letters, Endorsements and Circulars Sent (Miscellaneous Letters), vols. 48–52, National Archives, RG 156, entry 3.

19. Springfield Armory, Waltham, Mass., Letters Sent, vol. 8 of 26, January 16, 1858; March 8, 5, and 6, 1860, National Archives.

20. Ibid.

21. Springfield Armory, Waltham, Mass., Letters Sent, vol. 8 of 26, January 16, 1858; March 8, 5, and 6, 1860, National Archives. National Archives, RG 123, Box 449.

22. Springfield Armory, Waltham, Mass., Letters Sent, vol. 8 of 26, January 16, 1858; March 8, 5, and 6, 1860, National Archives.

23. National Archives, RG 123, Box 449.

24. Ibid.

25. Ibid.

26. Orders and Endorsements Sent by the Secretary of War, 1846–1870, vol. 9, National Archives, Microcopy 444, Roll 5.

27. Report of Brigadier General S. V. Benet, Chief of Ordnance, to Secretary of War, March 6, 1875, Wray-Morse File 030.

28. U.S. Census, 1860.

29. National Archives, RG 123, Box 449A.

30. Records of the U.S. House of Representatives, Forty-second Congress, National Archives, RG 233, Box 21.

31. Letters, Endorsements and Circulars Sent (Miscellaneous Letters), vols. 48–52, National Archives, RG 156, entry 3.

32. Hicks, *U.S. Military Firearms 1776–1956,* 85–86; Orders and Endorsements Sent by the Secretary of War, 1846–1870, vol. 9, National Archives, Microcopy 444, Roll 5. This quote is paraphrased in Hicks, *U.S. Military Firearms 1776–1956,* 85–86.

33. Cutrer, Ben McCulloch and the Frontier Military Tradition, 173.

34. National Archives, RG 123, Box 449.

35. Orders and Endorsements Sent by the Secretary of War, 1846–1870, vol. 9, National Archives, Microcopy 444, Roll 5; National Archives, RG 123, Box 449.

36. Report of Brigadier General S. V. Benet, Chief of Ordnance, to Secretary of War, March 6, 1875, Wray-Morse File 030; National Archives, RG 123, Box 449.

37. Report of Brigadier General S. V. Benet, Chief of Ordnance, to Secretary of War, March 6, 1875, National Archives, RG 123, Box 449.

38. Ibid.

39. Ibid.

40. Ibid.

41. "Memorial of George W. Morse to the 44th Congress," 50.

42. National Archives, RG 123, Box 449.

43. Letters, Endorsements and Circulars Sent

(Miscellaneous Letters), vols. 48–52, National Archives, RG 156, entry 3.

44. National Archives, RG 123, Box 449.

45. Ibid.

46. Letters, Endorsements, and Reports to the Secretary of War, 1854–1864, National Archives, RG 156, entry 5.

47. Register of Letters Received at the Ordnance Department, vols. 27–31, National Archives, RG 156, entry 20; Orders and Endorsements Sent by the Secretary of War, 1846–1870, vol. 10, National Archives, Microcopy 444, Roll 5; National Archives, RG 123, Box 449; Orders and Endorsements Sent by the Secretary of War, 1846–1870, vol. 10, National Archives, Microcopy 444, Roll 5.

48. ww3.rediscov.com/spring/spring.htm.

49. Springfield Armory, Waltham, Mass., Letters Sent, vol. 9 of 26, November 2 and 14, 1860, 78–79 and 93–94, National Archives.

50. National Archives, RG 123, Box 449.

51. Letters, Endorsements and Circulars Sent (Miscellaneous Letters), vols. 48–52, National Archives, RG 156, entry 3.

52. National Archives, RG 123, Box 449.

53. Ibid.

54. Ibid.

55. Letters, Endorsements and Circulars Sent (Miscellaneous Letters), vols. 48–52, National Archives, RG 156, entry 3.

56. Springfield Armory, Waltham, Mass., Letters Sent, vol. 9 of 26, October 18, 1860, 62–63, National Archives.

57. Ibid., 65–66.

58. Ibid.

59. Ibid., 75–76.

60. Springfield Armory, Waltham, Mass., Letters Sent, vol. 9 of 26, January 8 and February 4, 1861, 133 and 149, National Archives.

61. "Memorial of George W. Morse to the 44th Congress," 49.

62. Springfield Armory, Waltham, Mass., Letters Sent, vol. 9 of 26, October 31, 1860, 77, National Archives.

63. George Woodward Morse, "Morse's Breech-Loading Fire-arms," pamphlet, Benjamin McCulloch Papers, Box 3G37, Materials Related to Mexican War, Utah Expedition, and the Civil War 1846–1862, Broadsides 1862, Briscoe Center for American History, Austin, Tex.

64. "Memorial of George W. Morse to the 44th Congress," 48.

65. Morse, "Morse's Breech-Loading Fire-arms"; National Archives, RG 123, Box 449A.

66. John D. McAulay, *Civil War Carbines, Volume II, . . . The Early Years, Appendix I* (Lincoln, R.I.: Andrew Mowbray Inc., 1991), 139. "Report from the Committee on the Subject of the Stores, Machinery, and Property Captured at Harpers Ferry," Document 31, pages 1–11, Greensboro History Museum, Civil War Collection, Mss. Coll. #16 65:1.

67. National Archives, RG 123, Box 449B.

68. Springfield Armory, Waltham, Mass., Letters Sent, vol. 9 of 26, November 6, 1860, 82, National Archives.

69. H. Newton Reid to Thos. D. Belotte, September 3, 1863, South Caroliniana Library; photocopy, Museum and Library of Confederate History, Greenville, S.C.

70. National Archives, RG 123, Box 449A.

71. Springfield Armory, Waltham, Mass., Letters Sent, vol. 9 of 26, November 14, 1860, 93–94, National Archives.

72. Ibid., November 22, 1860, 97–98, National Archives.

73. Moller writes that two of the carbines were made in June 1860 and two in October 1860 and were described as being incomplete on the master armorer's inventory. He writes that the Springfield Armory Historic Site's collections reveal that the carbines were Morse-realtered model 1816 rifles and sighted muskets that had been shortened to carbine length. While it is possible that some of the M1816 muzzle-loader alterations were cut down to carbine length, the carbines made at Springfield in late 1860 were not M1816 alterations, based on the variety of calibers described by Wright in his letter of November 6, 1860. See *American Military Shoulder Arms, Volume III*, 97.

74. *Confederate Veteran* 21 (1913): 173.

75. Report of Brigadier General S. V. Benet, Chief of Ordnance, to Secretary of War, March 6, 1875, Wray-Morse File 030. Benet wrote that this was the last paper generated by the War Office prior to the war relative to Morse.

76. Fuller, comp., *Springfield Muzzle-Loading Shoulder Arms*, 171.

77. National Archives, RG 123, Box 449.

· ·

Chapter 7: Morse in Late 1860 and Early 1861

1. "Morse, "Morse's Breech-Loading Fire-arms."

2. Ibid.

3. Ibid.

4. "Memorial of George W. Morse to the 44th Congress," 47–50.

5. Morse, "Morse's Breech-Loading Fire-arms."

6. National Archives, RG 123, Box 449A.

7. Fuller, comp., Springfield Muzzle-Loading Shoulder Arms, 170–71.

8. National Archives, RG 123, Box 449.

9. Springfield Armory, Waltham, Mass., Letters Sent, vol. 9 of 26, November 14, 1861, 91–92, National Archives.

10. Ibid.

11. Ibid., January 1, 1861, 127, National Archives.

12. Ibid., January 3, 1861, 129, National Archives.

13. Letters and Telegrams Sent to National Armories, 1860–1861, National Archives, RG 156, entry 11.

14. National Archives, RG 123, Box 449A.

....

Chapter 8: Harpers Ferry, Nashville, and Atlanta

1. Records of the U.S. House of Representatives, Forty-second Congress, National Archives, RG 233, Box 21.

2. S165015, General Assembly Petitions, 1863, No. 6, S.C. Department of Archives and History. Morse to the S.C. Legislature on December 1, 1863.

3. *War of the Rebellion: A Compilation of the Official Records of the Union and Confederate Armies* (Washington, D.C.: Government Printing Office, 1899), ser. 4, vol. 1: 131–32. Hereafter cited as *Oreg.*

4. Ibid.

5. Morse Arms Manufacturing Company *v.* the United States, 27 Ct. Cl. 363, U.S. Court of Claims, No. 14340, decided June 20, 1892; Letters Received by the Secretary of War, 1861–1865, 11, National Archives, M437, Roll 1.

6. Cutrer, Ben McCulloch and the Frontier Military Tradition, 188.

7. Gunn, "Ben McCulloch," 18.

8. Cutrer, Ben McCulloch and the Frontier Military Tradition, 191.

9. Ibid., 191.

10. National Archives, RG 123, Box 449.

11. Morse Arms Manufacturing Company *v.* the United States, 27 Ct. Cl. 363, U.S. Court of Claims, No. 14340, decided June 20, 1892; National Archives, RG 123, Box 449.

12. National Archives, RG 123, Box 449.

13. Records of the U.S. House of Representatives, Forty-second Congress, National Archives, RG 233, Box 21.

14. National Archives, RG 123, Box 449A.

15. Jones put the number of small arms at about fifteen thousand, but the figure was subsequently lowered.

16. *Oreg.*, ser. 1, vol. 2: 3-6.

17. Moller, *American Military Shoulder Arms, Volume III,* 176; Hill and Anthony state that about six hundred alterations were in progress when the armory was captured, but they provide no documentation (*Confederate Longarms and Pistols,* 132).

18. Gordon L. Jones, Confederate Odyssey: The George W. Wray Civil War Collection at the Atlanta History Center (Athens: University of Georgia Press, 2014), 109.

19. "Report from the Committee on the Subject of the Stores, Machinery, and Property Captured at Harpers Ferry," Document 31, p. 3.

20. *Oreg.*, ser. 1, vol. 2: 785.

21. Ibid., vol. 51, pt. 2: 20.

22. Ibid., 34.

23. Ibid., ser. 1, vol. 2: 785.

24. Ibid., vol. 51, pt. 2: 53.

25. Jones, Confederate Odyssey,113.

26. *Oreg.*, ser. 1, vol. 2: 6.

27. Ibid., 793–94. An article in the *Confederate Veteran* from 1898 states that John M. Standidge, a prominent figure in prewar Louisiana, was joined in Richmond by his wife and two sons in April 1861, and they worked there for several weeks making center-fire cartridges for Morse rifles. This is the only evidence that supports the production of any Morse-related product in Richmond at any time during the war and is certainly inaccurate because there were no Morse firearms available for use in April 1861 (*Confederate Veteran* 6 [1898]: 572).

28. "Report from the Committee on the Subject of the Stores, Machinery, and Property Captured at Harpers Ferry," Document 31, pp. 3 and 8.

29. *Oreg.*, ser. 1, vol. 51, pt. 2: 98.

30 "Report from the Committee on the Subject of the Stores, Machinery, and Property Captured at Harpers Ferry," Document 31, p. 3.

31. Ibid.

32. U.S. Senate, Forty-fifth Congress, First Session, Congressional Series Report No. 866, National Archives.

33. *Oreg.*, ser. 1, vol. 51, pt. 2: 154.

34. "Report from the Committee on the Subject of the Stores, Machinery, and Property Captured at Harpers Ferry," Document 31, p. 3.

35. *Confederate Veteran* 32 (1924): 166; Confederate Papers Relating to Citizens or Business Firms, Roll 0717, National Archives, NARA M346, RG 109.

36. *Oreg.*, ser. 4, vol. 1: 489–90.

37. John M. Murphy and Howard Michael Madaus, *Confederate Rifles and Muskets: Infantry Small Arms Manufactured in the Southern*

Confederacy, 1861–1865, (Newport Beach, Calif.: Graphic Publishers, 1996), 69. "Inventory of Musket Machinery Taken at Harpers Ferry and Now in the Armory at Richmond," esp. p. 147—"Machinery and Tools Sent to the State of Tennessee, by Order of the Confederate States War Department," Virginia Executive Documents for 1861, Document 40.

38. Letters Sent and Received, Ordnance Department, Richmond, Va., November 1862–January 1864, chap. 4, vol. 90, National Archives, RG 109. "Report from the Committee on the Subject of the Stores, Machinery, and Property Captured at Harpers Ferry," Document 40, p. 147. McAulay, *Civil War Carbines,* 139. "Report from the Committee on the Subject of the Stores, Machinery, and Property Captured at Harpers Ferry," Document 31, pp. 1–11. Murphy and Madaus, *Confederate Rifles and Muskets,* 69. Major William S Downer to Colonel Josiah Gorgas, November 26, 1862, Letters Sent and Received, Ordnance Department, Richmond, Va., November 1862–January 1864, chap. 4, vol. 90: 36, National Archives, RG 109.

39. *Oreg.,* ser. 1, vol. 51, pt. 2: 53.

40. Confederate Papers Relating to Citizens or Business Firms, National Archives, NARA M346, Roll 0717.

41. Ibid.

42. Ibid.

43. Wray-Morse File 030.

44. Samuel Dodd Morgan Papers, Box 1, Folder 6, Tennessee State Library and Archives.

45. Ibid., "Business Correspondence," Tennessee State Library and Archives.

46. *Oreg.,* ser. 4, vol. 1: 991.

47. Murphy and Madaus, Confederate Rifles and Muskets, 65.

48. Wray-Morse File 039.

49. Letters Received by the Secretary of War, 1861–1865, 7–8, National Archives, M437, Roll 1.

·······································

Chapter 9: The South Carolina State Military Works and David Lopez Jr.

1. "Report of W. H. Gist Chief of the Department of Construction and Manufacture, and David Lopez to His Excellency, Governor Pickens, August 29, 1862," 164.

2. Charles Edward Cauthen, *South Carolina Goes to War, 1860–1865* (Chapel Hill: University of North Carolina Press, 1950), 115–16.

3. Ibid., 116.

4. Gist's and Lopez's Report, 164.

5. Cauthen, South Carolina Goes to War, 116.

6. Gist's and Lopez's Report, 165.

7. Ibid.

8. Ibid., 167

9. Wray-Morse File 035.

10. A separate entity, also called the Executive Council, existed in 1861. Its members were appointed by the governor and confirmed by the Secession Convention. Its function was similar to that of the president's cabinet in that it was an advisory board whose members answered to the governor. It ceased to exist in the spring of 1861 (Charles E. Cauthen, ed., *State Records of South Carolina: Journals of the Executive Council* [Columbia: South Carolina Archives Department, 1956], x).

11. Cauthen, South Carolina Goes to War, 142.

12. Ibid., 149.

13. Cauthen, *South Carolina Goes to War,* 149; "Bells for Cannon," *Charleston Mercury,* March 27, 1862.

14. Gist's and Lopez's Report, 166.

15. "First Annual Report of the State Auditor, 1863," p. 157, Records of the General Assembly, Reports and Resolutions, 1863, South Carolina Department of Archives and History, Columbia.

16. Gist's and Lopez's Report, 163.

17. Cauthen, ed., State Records of South Carolina, 95.

18. For an extensive review of Lopez's life, see Barry Stiefel, "David Lopez Jr.: Builder, Industrialist, and Defender of the Confederacy," *American Jewish Archives Journal* 64 (2012): 53–81.

19. Gist's and Lopez's Report, 162.

20. Stiefel, "David Lopez Jr.," 72 and 74.

21. Both are buried at Coming Street Cemetery.

22. Stiefel, "David Lopez Jr.," 56.

23. A physical description of John when he was thirty-three years old was: five feet, five inches tall, dark complexion, dark eyes, and dark hair (Compiled Service Records of Confederate soldiers who served in organizations from the State of South Carolina; the National Archives, National Archives and Records Service General Services Administration, RG 109, Microcopy No. 207, roll 0086).

24. Stiefel, "David Lopez Jr.," 53–57.

25. U.S. Census, 1860.

26. Stiefel, "David Lopez Jr.," 57–58.

27. Rebecca Moise Lopez was born in 1814.

28. Stiefel, "David Lopez Jr.," 62–63.

29. Ibid., 63–65.

30. Ibid., 67.

31. South Carolina Records, p. 14, Canthen, Charles E., South Carolina State Records. . . at Archival Sources.

32. Stiefel, "David Lopez Jr.," 67.

33. Daniel S. Hart (February 1, 1839–April 15, 1877) married Priscilla Lopez Hart (1839–August 9, 1882) on December 17, 1860. They had the following children: Mary, born about 1873; Ida, born about 1869; Isabell, born about 1867; Catherine, born about 1863; Samuel Jr., born about 1865; Daniel, born about 1866; and Israel, born about 1860 (U.S. Census, 1860, 1870, 1880).

34. Moses E. Lopez's Compiled Service Record from http://www/fold3.com CSR; RG 109, M 267, roll 0086.

35. Ellison Capers, *Confederate Military History,* extended ed., vol. 6: *South Carolina* (Wilmington, N.C.: Broadfoot Publishing Co., 1987), 710.

36. Thomas Dixon's CSR M267, Roll 0338.

Chapter 10: 1862

1. Cauthen, ed., State Records of South Carolina, 101.

2. Governor Moore to Governor Pickens, March 8, 1862, Governor Pickens Papers, Letters Received and Sent, S511001, South Carolina Department of Archives and History, Columbia.

3. Cauthen, ed., State Records of South Carolina, 104.

4. Ibid., 117.

5. Ibid.

6. James Tupper to M. L. Bonham, October 27, 1863, Governor Milledge L. Bonham Papers, Letters Received and Sent, Folder 3, 238-L-02, Box 1, South Carolina Department of Archives and History.

7. Return of Ordnance and Ordnance Stores, September 30, 1864, Comptroller General, State Auditor, Records of the Ordnance Bureau, S126175, State Arsenal, Columbia, South Carolina Department of Archives and History.

8. Cauthen, ed., State Records of South Carolina, 120.

9. Report of W. H. Gist, 161.

10. Gist's and Lopez's Report, 161.

11. Cauthen, South Carolina Goes to War, 150.

12. "Notice," advertisement, *Charleston Mercury,* April 3, 1862.

13. Gist did not name the special agent in his report. It is unlikely that it was Lopez since he does name Lopez frequently in the same report (Gist's and Lopez's Report, 162).

14. Gist's and Lopez's Report, 162.

15. http://www.civilwarartillery.com.

16. Gist's and Lopez's Report, 162.

17. Ibid.

18. "The Greenville Railroad," *Southern Patriot,* December 8, 1853, 2.

19. "The Greenville and Columbia Railroad," *Southern Patriot,* February 19, 1852, 1.

20. Stiefel, "David Lopez Jr.," 67–68.

21. Archie V. Huff Jr., *Greenville: The History of the City and County in the South Carolina Piedmont* (Columbia: University of South Carolina Press, 1995), 115 and 152.

22. Ibid., 122–23.

23. Ibid., 113.

24. Cauthen, ed., State Records of South Carolina, 132.

25. Ibid.

26. Ibid.

27. "State Donation," *Carolina Spartan,* April 10, 1862, 1.

28. Ibid.

29. Cauthen, ed., State Records of South Carolina, 181.

30. Vardry McBee's deed, Office of the Register of Deeds for Greenville County, South Carolina. An examination of Vardry McBee's papers at the University of North Carolina at Chapel Hill fails to reveal why McBee waited until August 1862 to sign the deed.

31. J. Tupper opinion, November 7, 1864, Governor Bonham Papers, Letters Received and Sent, S512001, South Carolina Department of Archives and History.

32. John A. Michel, born January 28, 1828, died from chronic hepatitis June 9, 1880, buried at St. Laurence Cemetery, Charleston.

33. U.S. Census, 1860.

34. State Works pay voucher #212, May 3, 1864, South Carolina Department of Archives and History, S126182.

35. Photocopy of original document, "State of South Carolina, State Works, Payment to Greenville & Columbia R.R. Co.", March, 1863, Museum and Library of Confederate History, Greenville, S.C.

36. "Expenses Paid by State Works to Greenville and Columbia Rail Road in August 1864," photocopy, Museum and Library of Confederate History, Greenville, S.C.

37. Committee on the Military, Report to Governor Bonham, December 24, 1864, photocopy, Museum and Library of Confederate History, Greenville, S.C.

38. Cauthen, ed., State Records of South Carolina, 132.

39. Ibid., 156.

40. James Welch Patton, ed., Minutes of the Proceedings of the Greenville Ladies'

Association in Aid of the Volunteers of the Confederate Army (New York: AMS Press, [1970]), 109.

41. Personal communication by author with Gordon Thruston, 2014.

42. *Greenville Magazine,* August 2000, 17–18.

....

Chapter 11: Morse Comes to South Carolina

1. "First Annual Report of the State Auditor, 1863," 156.

2. Governor Isham Harris Papers, Box 2, Folder 6, Tennessee Library and Archives, Nashville, Tennessee.

3. Ibid.

4. "First Annual Report of the State Auditor, 1863," 156; Confederate Papers Relating to Citizens or Business Firms, Roll 0717, National Archives, NARA M346, RG 109.

5. "First Annual Report of the State Auditor, 1863," 156.

6. Letter of George W. Morse to Col. George F. Townes, Records of the General Assembly, Senate Correspondence and Assorted Papers, 1866–1877, S165216.

7. Morse to Garlington, November 7, 1864, in which Morse wrote that he understood that Davis did not like breechloaders, but that he considered Morse's to be the best of them all (photocopy, Museum and Library of Confederate History, Greenville, S.C.).

8. Gist's and Lopez's report, 172.

9. "First Annual Report of the State Auditor, 1863," 156.

10. Letter of George W. Morse to Col. George F. Townes, Records of the General Assembly, Senate Correspondence and Assorted Papers, 1866–1877, S165216.

11. The earliest pay voucher for him is February 15, 1864, but he was personally at the State Works in the early summer of 1862 (State Works voucher #155, February 15, 1864, South Carolina Department of Archives and History, S126182, photocopy, Museum and Library of Confederate History, Greenville, S.C.); State Works voucher #11, May 15, 1864, South Carolina Department of Archives and History, S126182, photocopy, Museum and Library of Confederate History, in Greenville, S.C.; Letters Received by the Secretary of War, 1861–1865, pp. 7–8, National Archives, M437, Roll 1.

12. Auditor James Tupper Papers, South Carolina Department of Archives and History, S126174.

13. Tupper to Bonham, October 27, 1864, Governor Bonham Papers, Letters Received and Sent, S512001, South Carolina Department of Archives and History.

14. Ibid.

15. Stiefel, "David Lopez Jr.," 68.

16. Gist's and Lopez's report, 172.

17. Ibid., 171.

18. Ibid.

19. Ibid.

20. Huff, *Greenville,* 141.

21. Cauthen, ed., State Records of South Carolina, 204.

22. "State Works Voucher #171", Photocopy, Museum and Library of Confederate History, Greenville, S.C.

23. Cauthen, ed., State Records of South Carolina, 132.

24. Ibid., 185 and 220.

25. There were at least three machine shops in Charleston in late March 1862 that produced engines similar to those used in the State Works. Lopez probably used his knowledge of this aspect of Charleston's industry to purchase one or more of these engines. See "Charleston Business Directory," *Charleston Mercury,* March 31, 1862.

26. Gist's and Lopez's report, 172.

27. Ibid.

28. Cauthen, ed., State Records of South Carolina, 236.

29. Ibid., 262.

30. Ibid., 244, 257, and 263.

31. Ibid., 244.

32. Ibid., 251.

33. Ibid.

34. Gist's and Lopez's Report.

35. Cauthen, ed., State Records of South Carolina, 262.

36. Gist's and Lopez's Report, 172.

37. Ordnance and Ordnance Stores Received, Purchased and Issued, August 5, 1862, to September 30, 1862, S126175, State Auditor, Ordnance Department, South Carolina Department of Archives and History.

38. Gist's and Lopez's Report, 172.

39. Ibid., 173.

40. Ibid.

41. "State Works Voucher #195," Photocopy, Museum and Library of Confederate History, Greenville, S.C.

42. Gist's and Lopez's Report, 165.

43. Ibid., 167.

44. It is not clear why Gist bought so much unnecessary bacon, but Lopez sold all 8,730 pounds of it in September 1862 for $1,632.51 (South Carolina Records, pp. 264, 271).

45. Gist's and Lopez's Report, 166.

46. Ibid.

47. Ibid.

48. Ibid.

49. Ibid.

50. South Carolina Records, p. 271.

51. Ibid., 275.

52. Ibid., 279.

53. Thomas Dixon's CSR M267, Roll 0338.

54. Ibid.

55. South Carolina Records, p. 287.

56. *Journal of the Senate of the State of South Carolina,* archive.org/stream/journalsenate1862 1863/journalsenate18621863_djvu.txt.

57. Cauthen, South Carolina Goes to War, 152–63.

58. The actual weight of most surviving carbines is six pounds, six ounces.

59. "Mr. Morse's Patent Improved Breech-Loading Carbine," *Intelligencer* (Atlanta), December 13, 1862. Most researchers show the date as December 19, 1862, but the article clearly says "Saturday" in its title. Though hard to read, the date appears to be December 13, and the Saturdays in December 1862 were the thirteenth and the twentieth.

60. http://www/findagrave.com; U.S. Census, 1850 and 1870.

61. Records for H. Marshall and Company can be found under "Marshall H & Co." and "Marshall and Co., H," Confederate Citizens Records, Fold3.com.

62. David Lopez to M. L. Bonham, December 22, 1862, Governor Bonham Papers, Letters Received and Sent, S512001, South Carolina Department of Archives and History.

63. Reports and Resolutions of the South Carolina House of Representatives, December 18, 1862, Records of the General Assembly, pp. 317–18, South Carolina Department of Archives and History, Columbia; also in Wray-Morse File 035.

64. General Assembly Petitions, 1863, No. 6, S165015, South Carolina Department of Archives and History.

65. House Journal, December 9, 1862, South Carolina Department of Archives and History.

66. Reports and Resolutions of the South Carolina House of Representatives, December 18, 1862, Records of the General Assembly, pp. 317–18. South Carolina Department of Archives and History, Columbia; also in Wray-Morse File 035.

67. Ibid.

68. David Lopez to M. L. Bonham, December 22, 1862, Governor Bonham Papers, Letters Received and Sent, S512001, South Carolina Department of Archives and History.

69. Ibid.

70. Ibid.

71. Reports and Resolutions of the South Carolina House of Representatives, December 18, 1862, Records of the General Assembly, pp. 317–18, South Carolina Department of Archives and History, Columbia.

72. Letters Sent—Ordnance Department at Richmond, November 1862–January 1864, War Department Collection of Captured Confederate Records, RG 109, National Archives; Chapter 4, vol. 90: 137–38; Major William S. Downer to Colonel Josiah Gorgas, February 2 or 3, 1863, in John M. Murphy and Howard Michael Madaus, *Confederate Carbines and Musketoons: Cavalry Small Arms Manufactured in and for the Southern Confederacy, 1861–1865* (Santa Ana, Calif.: Graphic Publishers, 2002), 176–98.

73. David Lopez to M. L. Bonham, January 20, 1863, Governor Bonham Papers, Letters Received and Sent, S512001, South Carolina Department of Archives and History.

74. Ibid.

75. Ibid.

76. Ibid.

······································

Chapter 12: Labor

1. J. F. Barnes, CSR M267, Roll 0039.

2. State Auditor, Payrolls, 1863–1864 (Oversize), S126182, South Carolina Department of Archives and History.

3. Ibid.

4. South Carolina Records, p. 113.

5. Ibid., 273.

6. Edward Tavel's CSR, http://www.fold3.com

7. Cauthen, *South Carolina Goes to War,* 178.

8. *Recollections and Reminiscences, 1861–1865 through World War I* (South Carolina Division of the United Daughters of the Confederacy, 2002), vol. 12: 608.

9. From letter dated February 8, 1954, Museum and Library of the Confederacy, Greenville, S.C.

10. Cauthen, South Carolina Goes to War, 178–87.

11. Ibid.

12. Ibid.

13. State Works pay vouchers #199, May 20, 1864; #230, June 20, 1864; #396, March 1, 1865; and #410, March 31, 1865, South Carolina Department of Archives and History, S126182.

14. Smith to Calhoun, October 18, 1864, #2744,

copy of original, Museum and Library of Confederate History, Greenville, S.C.

15. Undated letter, J. Ralph Smith most likely to A. P. Calhoun, copy of the original, Museum and Library of Confederate History, Greenville, S.C.

16. State Auditor, Payrolls, 1863–1864 (Oversize), South Carolina Department of Archives and History.

17. Payments to Julius C. Smith, #407. Copy of the original is available at the Museum and Library of Confederate History, Greenville, S.C., Museum and Library of Confederate History, Greenville, S.C.

18. Julius Clarence Smith, born in Charleston on December 27, 1830; died in Greenville on July 12, 1903; buried at Springwood Cemetery in Greenville, http://www.findagrave.com, http://www.ancestry.com. State Works pay vouchers #217, June 10, 1864; #218, June 10, 1864, copies in Museum and Library of Confederate History, Greenville, S.C. Payments to Julius C. Smith, #407, copy in Museum and Library of Confederate History, Greenville, S.C.

19. http://www.findagrave.com; http://www.ancestry.com.

20. http://www.findagrave.com; http://www.ancestry.com.

21. "May 1st, 1865, and May 14th, 1884," *Enterprise and Mountaineer,* May 14, 1884, 2.

....

Chapter 13: 1863

1. "The relatives and friends of Mr. David Lopez," funeral announcement, *Charleston Mercury,* September 13, 1862. The notice refers to the deceased as "David Lopez, Jr.," and to his father as "David Lopez."

2. John H. Lopez's Compiled Service Record, http://www.fold3.com. Research by staff at the Citadel Archives and Museum shows no record of John H. Lopez having attended that school.

3. John H. Lopez's Compiled Service Record, http://www.fold3.com.

4. Ibid.

5. Second Annual Report, ca. November 1864, Comptroller General–State Auditor, Miscellaneous Papers of the State Auditor, S126174, South Carolina Department of Archives and History, Columbia.

6. South Carolina Department of Archives and History, S126182. A similar abstract of articles purchased at the State Works from August 15, 1862, to April 1, 1863 also exists in the same File.

7. *Oreg.,* ser. 3, vol. 1: 4.

8. The reference to pine lumber actually shows up in a July 1863 voucher. See South Carolina Department of Archives and History, S126182, photocopy, Museum and Library of Confederate History, Greenville, S.C.

9. Photocopy of "Abstract of Articles fabricated at "State Works' during the second quarter of 1863", Museum at Library of Confederate History, Greenville, S.C.

10. Abstract of Articles fabricated at State Works during the second quarter of 1863, photocopy, Museum and Library of Confederate History, Greenville, S.C.

11. Morse to Garlington, November 7, 1864, photocopy, Museum and Library of Confederate History, Greenville, S.C.

12. "First Annual Report of the State Auditor, 1863," p. 155. These were probably the thirty-two artillery projectiles known as Brooks Rifle bolts later documented to have been made at the State Works.

13. "Abstract of Disbursement of 'State Works' by D. Lopez, Gen'l. sup't. in the month ending July 31, 1863.", Photocopy, Museum and Library of Confederate History, Greenville, S.C.

14. "Abstract of Receipts and of Cash by D. Lopez, Gen. Supt. State Military Works at Greenville from 1st Jany. 1863 to 19th August 1863," photocopy, Museum and Library of Confederate History, Greenville, S.C.

....

Chapter 14: Lopez's Resignation and Successor

1. David Lopez to James Tupper, with enclosure, April 8, 1863, Comptroller General–State Auditor, Miscellaneous Papers of the State Auditor, S120174.

2. Ibid.

3. Ibid.

4. Ibid.

5. David Lopez to James Tupper, April 23, 1863, Comptroller General–State Auditor, Miscellaneous Papers of the State Auditor, S126174.

6. Ibid., April 27, 1863, Comptroller General–State Auditor, Miscellaneous Papers of the State Auditor, S126174.

7. Ibid., May 6, 1863, Comptroller General–State Auditor, Miscellaneous Papers of the State Auditor, S126174.

8. Ibid., May 8, 1863, Comptroller General–State Auditor, Miscellaneous Papers of the State Auditor, S126174.

9. G. W. Morse to James Tupper, July 6, 1863, Comptroller General–State Auditor, Miscellaneous Papers of the State Auditor, S126174.

10. Lopez was paid through August 19, 1863, and Smith signed a document as general superintendent on August 19, 1863 (J. R. Smith to James Tupper, with enclosures, August 21, 1863, Comptroller General–State Auditor, Miscellaneous Papers of the State Auditor, S126174). H. C. Briggs CSR M267, Roll 0076.

11. J. R. Smith to James Tupper, August 21, 1863, Comptroller General–State Auditor, Miscellaneous Papers of the State Auditor, S126174.

12. J. R. Smith to James Tupper, August 21, 1863, LeBlanc's account, Comptroller General–State Auditor, Miscellaneous Papers of the State Auditor, S126174.

13. Ibid., W. W. Smith's account, Comptroller General–State Auditor, Miscellaneous Papers of the State Auditor, S126174.

14. Ibid., Mullane's account, Comptroller General–State Auditor, Miscellaneous Papers of the State Auditor, S126174.

15. Stiefel, "David Lopez Jr.," 69.

16. J. R. Smith to James Tupper, with enclosures, August 21, 1863, Comptroller General–State Auditor, Miscellaneous Papers of the State Auditor, S126174.

17. Stiefel, "David Lopez Jr.," 70.

18. Moses E. Lopez's Compiled Service Record, http://www/fold3.com.

19. "Abstract of Receipts and of Cash by D. Lopez, Gen. Supt. State Military Works at Greenville from 1st Jany. 1863 to 19th August 1863," and State Works voucher #154, February 15, 1864. Photocopy, Museum and Library of Confederate History, Greenville, S.C.

20. Payroll of mechanics and laborers at the State Works for the month ending July 15, 1863, photocopy, Museum and Library of Confederate History, Greenville, S.C.

21. For James Ralph Smith's genealogy, see the 1860 U.S. Census, 1870 U.S. Census, http://www.findagrave.com, and http://www.ancestry.com.

22. G. W. Morse to James Tupper, August 20, 1863, Comptroller General–State Auditor, Miscellaneous Papers of the State Auditor, S126174.

23. "Statement of Articles Repaired at the State Military Works Greenville, So. Car., from the Establishment of the Works to 1st October 1863," photocopy, Museum and Library of Confederate History, Greenville, S.C.

24. http://www.wikipedia.com.

25. H. Newton Reid to Thos. D. Belotte, September 3, 1863, original at South Caroliniana Library; photocopy , Museum and Library of Confederate History, Greenville, S.C. The actual weight of most surviving carbines is six pounds, six ounces.

26. Ordnance and Ordnance Stores Received, Purchased and Issued from August 5, 1862, to September 30, 1862, State Auditor Ordnance Department, South Carolina Department of Archives and History, S126175.

27. "First Annual Report of the State Auditor, 1863," p. 155.

....

Chapter 15: Morse's Brass-frame Carbine

1. Precise measurements for the various carbines, both the prototype and Types I, II, and III, are in Murphy and Madaus, *Confederate Carbines and Musketoons,* 176–98. The actual weight of most surviving carbines is six pounds, six ounces.

2. Morse wrote in 1866 that the state had never paid him the agreed-upon five hundred dollars for the use of his patent on the "tenth hundred guns," meaning serial numbers 901–1,000. He did not mention being unpaid for the production of any guns above serial number 1,000. He was paid per 100-gun batch, supporting the likelihood that he never completed the "11th hundred" batch. See George W. Morse to Col. George F. Townes, Records of the General Assembly, Senate Correspondence and Assorted Papers 1866–1877, S165216.

3. Morse to Bonham, March 9, 1864, Wray-Morse File 035.

4. Ibid.

5. J. R. Smith to M. L. Bonham, June 1, 1864, Governor Bonham Papers, Letters Received and Sent, South Carolina Department of Archives and History, S512001.

6. Murphy and Madaus, Confederate Carbines and Musketoons, 185–91.

7. The following statement came from Wray-Morse File 043. Its author and source are unknown: "Primitive Centerfire, Morse Guns and Cartridges: The Morse carbine was not a success in its original form because the soft-metal bolt tended to set back under firing. Punctured primers, undoubtedly caused by excessive headspace, eroded the bolt face. This defect was dealt with by covering the brass bolt face with a flanged iron firing pin. A latch was added to keep the operating lever closed. Finally, Morse replaced the brass bolt with one of iron and provided a projection on its upper forward end to deflect escaping gases. Thus it is evident that the cartridge used was not an entirely satisfactory device for obturating the bolt/barrel joint."

8. Murphy and Madaus, Confederate Carbines and Musketoons, 176–98.

9. Ibid.

10. Ibid.

11. Ibid.

12. Ibid.

13. Ibid., 179.

14. Ibid., 176–98.

15. Sutherland, "Arms Manufactury in Greenville County."

16. Murphy and Madaus, Confederate Carbines and Musketoons, 180.

....

Chapter 16: Sale of the State Works

1. J. Tupper to Bonham, October 27, 1864, Governor Bonham Papers, Letter Received and Sent, South Carolina Department of Archives and History, S512001.

2. Wray-Morse File 035.

3. "First Annual Report of the State Auditor, 1863," p. 156.

4. Ibid.

5. Bonham to I. G. Harris, September 21, 1863, Governor Bonham Papers, Letters Received and Sent, Folder 3, 238-L-02, Box 1, South Carolina Department of Archives and History.

6. Sutherland, "Arms Manufactury in Greenville County," 52–53.

7. J. Ralph Smith to James Tupper, October 13, 1863, Comptroller General–State Auditor, Miscellaneous Papers of the State Auditor, S126174.

8. Ibid., October 16, 1863, Comptroller General–State Auditor, Miscellaneous Papers of the State Auditor, S126174.

9. M. L. Bonham to J. T. Trezevant, October 19, 1863, Governor Bonham Papers, Letter Received and Sent, South Carolina Department of Archives and History, S512001.

10. M. L. Bonham to J. A. Seddon, October 15, 1863, Governor Bonham Papers, Letter Received and Sent, South Carolina Department of Archives and History, S512001.

11. Ibid.

12. M. L. Bonham to J. T. Trezevant, October 19, 1863, Governor Bonham Papers, Letter Received and Sent, South Carolina Department of Archives and History, S512001.

13. "First Annual Report of the State Auditor, 1863," p. 154.

14. Ibid., 155.

15. Ibid., 156.

16. John LeConte to James Tupper, November 4, 1863, Comptroller General–State Auditor, Miscellaneous Papers of the State Auditor, S126174.

17. Committee Report on the Second Annual Report, December 20, 1864, Comptroller General–State Auditor, Miscellaneous Papers of the State Auditor, S126174.

18. *Journal of the Senate of South Carolina,* 1863, 46–47, docsouth.unc.edu/imls/scsess63/scsess63 .html.

19. Second Annual Report, ca. November 1864, Comptroller General–State Auditor, Miscellaneous Papers of the State Auditor, S126174.

20. M. L. Bonham to J. A. Seddon, December 17, 1863, Records of the Comptroller General, State Auditor, Papers of the State Works, 209-D-04, South Carolina Department of Archives and History.

21. Second Annual Report, ca. November 1864, Comptroller General–State Auditor, Miscellaneous Papers of the State Auditor, S126174.

22. Benjamin Evans to Tupper, October 26, 1864, Comptroller General–State Auditor, Miscellaneous Papers of the State Auditor, S126174.

23. James Tupper to M. L. Bonham, October 27, 1863, Governor Bonham Papers, Letters Received and Sent, Folder 3, 238-L-02, Box 1, South Carolina Department of Archives and History.

24. Ibid.

25. Tupper to Bonham (addendum by Bonham, October 29, 1864), October 27, 1864, Comptroller General–State Auditor, Miscellaneous Papers of the State Auditor, S126174.

26. Second Annual Report, ca. November 1864, Comptroller General–State Auditor, Miscellaneous Papers of the State Auditor, S126174.

27. *Daily South Carolinian,* November 2, 1864.

28. Second Annual Report, ca. November 1864, Comptroller General–State Auditor, Miscellaneous Papers of the State Auditor, S126174.

29. J. Tupper opinion, November 7, 1864, Governor Bonham Papers, Letters Received and Sent, South Carolina Department of Archives and History, S512001.

30. G. A. Trenholm, November 5, 1864, Executive Council Papers, Miscellaneous Folder 2, 1st Floor Processing, South Carolina Department of Archives and History.

31. Second Annual Report, ca. November 1864, Comptroller General–State Auditor, Miscellaneous Papers of the State Auditor, S126174.

32. G. A. Trenholm, November 5, 1864, Executive Council Papers, Miscellaneous Folder 2, 1st Floor Processing, South Carolina Department of Archives and History.

33. George W. Morse letter to A. C. Garlington, November 7, 1864, South Carolina Department of Archives and History, S126182.

34. J. R. Smith to M. L. Bonham, November 12, 1864, Governor Bonham Papers, Letters Received and Sent, South Carolina Department of Archives and History, S512001.

35. Second Annual Report, ca. November 1864, Comptroller General–State Auditor, Miscellaneous Papers of the State Auditor, S126174.

36. Ibid.

37. Ibid.

38. "Gov. Bonham's Message," *Edgefield Advertiser,* November 30, 1864.

39. Ibid.

40. Committee on the Military, Report to Bonham, December 24, 1864. South Carolina Department of Archives and History, S126182.

41. Ibid. The state originally approved the manufacture of 1,000 carbines, which were completed. Under this new approval of an additional 1,500 carbines, production continued, but not for very long. The highest known survivor is 1032, indicating that only a few examples were actually manufactured under the "1,500 allotment."

42. Records of the General Assembly, Codes and Session Laws, vol. 13 (1861–66): 391–92, South Carolina Department of Archives and History, Columbia, S165004.

....

Chapter 17: Late 1863 and 1864

1. Journal of the Senate of the State of South Carolina for the Annual Session of 1863, 43–44, docsouth.unc.edu/imls/scsess63/scsess63.html.

2. Personal communication by author with L. F. Knudsen.

3. R. S. Seigler, South Carolina's Military Organizations during the War Between the States (Charleston: History Press, 2008), 309–11.

4. General Assembly Petitions, 1863, No. 6, South Carolina Department of Archives and History, S165015.

5. Ibid., No. 152, South Carolina Department of Archives and History, S165015.

6. Wray-Morse File 035.

7. George W. Morse to Col. George F. Townes, Records of the General Assembly, Senate Correspondence and Assorted Papers, 1866–1877, S165216.

8. Ibid.

9. State Works voucher #158, December 28, 1863; January 29, 1864; and February 5, 1864, photocopy, Museum and Library of Confederate History, Greenville, S.C.

10. Greensboro Historical Museum, Greensboro, North Carolina. The State Works probably also conducted business with several other firms

in Greenville District. See Alfred Ward Grayson Davis File, South Caroliniana Library.

11. *Oreg.,* ser. 1, vol. 28, pt. 2: 448–59.

12. Ibid., vol. 32: 746–49.

13. Abstract of items purchased at the State Works, 2nd Quarter 1864, photocopy, Museum and Library of Confederate History, Greenville, S.C.

14. J. R. Smith to Tupper, March 2, 1864, Comptroller General–State Auditor, Miscellaneous Papers of the State Auditor, S126174.

15. Ibid., March 15, 1864, Comptroller General–State Auditor, Miscellaneous Papers of the State Auditor, S126174.

16. Ibid., April 14, 1864, Comptroller General–State Auditor, Miscellaneous Papers of the State Auditor, S126174.

17. State Works voucher #229, June 8, 1864, South Carolina Department of Archives and History, S126182, photocopy, Museum and Library of Confederate History, Greenville, S.C.

18. Precise measurements for the various carbines, both the prototype and Types I, II, and III, are in Murphy and Madaus, *Confederate Carbines and Musketoons,* 176–98.

19. Lewis, "Morse Arms and Ammunition," 31–33.

20. William B. Edwards, Civil War Guns: The Complete Story of Federal and Confederate Small Arms (Secaucus, N.J.: Castle Books, 1978), 379.

21. Sutherland, "Arms Manufactury in Greenville County," 56.

22. Smith's expense report from July 1 to September 30, 1864, photocopy, Museum and Library of Confederate History, Greenville, S.C.

23. Abstract of materials expended and consumed at the State Works 3rd Quarter 1864, photocopy, Museum and Library of Confederate History, Greenville, S.C.

24. Expense reports #321 and 322, photocopies, Museum and Library of Confederate History, Greenville, S.C.

25. Morse to A. C. Garlington, November 7, 1864, photocopy, Museum and Library of Confederate History, Greenville, S.C.

26. "Proclamation, State of South Carolina, Executive Department," *Charleston Daily Courier,* September 7, 1864, 2.

27. Percival was born in Columbia, South Carolina, about 1814. He moved to Aiken about 1856 and was a respected physician there. He led two companies during the war, both from the Aiken area. Percival died in Aiken after a lingering and painful illness on May 16, 1874. He was buried

at St. Thaddeus Episcopal Church in Aiken (*Charleston News and Courier*, May 20, 1873, 4).

28. W. F. Percival Papers, 1864–1865, South Carolina Historical Society, Charleston. Percival had apparently requested that his company be placed in Gregg's Battalion, but was denied. This "Gregg" is not further identified, but was possibly William Gregg.

29. Adjutant and Inspector General of South Carolina, Order and Letterbook, 1862–1864, in Lewis F. Knudsen Jr., compiler, *Captain William F. Percival's Co. Mounted Infantry, S.C. State Troops, (Aiken Mounted Infantry) 1864–1865,* 2009, 3, South Carolina Department of Archives and History.

30. "For the Advertiser," letter to the editor from member of Aiken Mounted Infantry, *Edgefield Advertiser,* January 25, 1865. The actual weight of most surviving carbines is six pounds, six ounces.

31. Ibid.

32. Personal communication with author L. F. Knudsen Jr.

33. W. F. Percival Papers, 1864–1865, South Carolina Historical Society, Charleston.

34. "State of South Carolina, Adjt. and Inspector General's Office," *Charleston Daily Courier,* November 21, 1864, 2.

35. Paul F. Hammond to Governor A. G. Magrath, January 23, 1865, Papers of Andrew G. Magrath, Box 1, Folder 10, S513004, South Carolina Department of Archives and History, Columbia.

36. Adjutant and Inspector General of South Carolina, Order and Letterbook, 1862–1864, in Knudsen, comp, *Captain William F. Percival's Co. Mounted Infantry, S.C. State Troops,* 2009, South Carolina Department of Archives and History.

37. "For the Advertiser," letter to the editor from member of Aiken Mounted Infantry, *Edgefield Advertiser,* January 25, 1865.

38. Alford to Zimmerman, September 30, 1964, Wray-Morse File 001.

39. Second Annual Report, ca. November 1864, Comptroller General–State Auditor, Miscellaneous Papers of the State Auditor, S126174.

40. Ibid.

41. Return of Ordnance and Ordnance Stores, September 30, 1864, Comptroller General, State Auditor, Records of the Ordnance Bureau, S126175, State Arsenal, Columbia, South Carolina Department of Archives and History.

42. Smith's expense report for the fourth quarter 1864, photocopy, Museum and Library of Confederate History, Greenville, S.C.

Chapter 18: 1865

1. James M. Eason should not be confused with W. G. Eason, who was state ordnance officer of South Carolina.

2. http://www/findagrave.com; http://www.ancestry.com; http://www.fold3.com.]

3. Papers of Andrew G. Magrath, Box 1, Folder 21, S513004.

4. *Oreg.,* ser. 1, vol. 53: 53; vol. 47, pt. 2: 503.

5. Ibid., vol. 48, pt. 1: 84; vol. 47, pt. 1: 181.

6. We know that Federal raiders captured some guns at the State Works on May 2, 1865, but cannot be certain they were Morse carbines.

7. Papers of Andrew G. Magrath, Box 1, Folder 24, S513004.

8. Ibid.

9. Ibid.

10. Papers of Andrew G. Magrath, Box 1, Folder 25, S513004.

11. Smith's expense report for the first quarter 1865, photocopy, Museum and Library of Confederate History, Greenville, S.C.

12. Abstract of disbursements at State Works in the quarter ending March 31, 1865, photocopy, Museum and Library of Confederate History, Greenville, S.C.

13. State Works voucher #242, South Carolina Department of Archives and History, S126182.

14. Payments to Julius C. Smith, #407, copy of the original, Museum and Library of Confederate History, Greenville, S.C.

15. Abstract of materials expended or consumed at the State Works during the first quarter of 1865, South Carolina Department of Archives and History, S126182.

16. State Works voucher #394, February 24, 1865, South Carolina Department of Archives and History, S126182.

17. Taff was born in 1829 or 1830, and in 1860 she was the wife of a Greenville carpenter, William Thomas Taff. She lived in the Upcountry—Greenville, Williamston, and Anderson—and died July 14, 1917. She was buried at Graceland West Cemetery West in Greenville (http://www.findagrave.com; http://www.ancestry.com).

18. Papers of Andrew G. Magrath, Box 1, Folder 31, S513004.

19. Ibid., Folder 40, S513004.

20. Ibid., Folder 44, S513004.

21. Ibid., Folder 43, S513004.

22. *Oreg.,* ser. 1, vol. 49, pt. 2: 407.

23. Ibid., pt. 1: 546–48.

24. Chris J. Hartley, *Stoneman's Raid, 1865* (Winston-Salem, N.C.: John F. Blair, 2010), 362.

25. *Oreg.*, ser. 1, vol. 49, pt. 2: 555. Taylor's recollection ("May 1st, 1865, and May 14th, 1884," *Enterprise and Mountaineer*, May 14, 1884, 2), places the number of Federal troopers at two hundred.

26. Thomas Bland Keys, "The Federal Pillage of Anderson, South Carolina: Brown's Raid," *South Carolina Historical Magazine* 76 (1975): 80–86.

27. Louise Ayer Vandiver, *Traditions and History of Anderson County* (1928; n.p.: McNaughton and Gunn, 1991), 244–45.

28. William Pierce Price was born January 29, 1835, in Dahlonega, Lumpkin County, Georgia. He moved to Greenville in 1851 and attended Furman but left before graduating to take charge of the editorial department of the *Greenville Enterprise*. Admitted to the bar in 1856, he established a law practice in Greenville. He mustered in as first sergeant in Hoke's Company B, Second Regiment, South Carolina Volunteers, on April 15, 1861, for twelve months. He was absent sick since December 8, 1861, and was discharged by the secretary of war on February 28, 1862, for disability. Price was a member of the South Carolina House of Representatives from 1864 to 1866. He moved to Dahlonega, Georgia, in 1866 and was a member of the Georgia House of Representatives in 1868–70. He was elected to the U.S. House of Representatives and served from December 1870 to March 1873. He then served in both the Georgia House (1877–79 and 1894–95) and Senate (1880–81). He practiced law and was president of North Georgia Agricultural College 1870–1908. Price founded North Georgia and State University in Dahlonega in 1873. He died November 4, 1908, and was buried in Dahlonega.

29. Vandiver, *Traditions and History of Anderson County*, 244–45; Thomas Simmons Arthur (November 10, 1819–June 20, 1904) was captain of a company of Home Guards in Greenville at the end of the war. He wrote in 1894 that his company participated in diverting Palmer's force from its intended route. An extant roster of Arthur's Company shows no one named Hoke or Price. It is possible that there were two home-guard companies in Greenville in May 1865 or that Vandiver confused the names of the company officers. See *Confederate Historian: Rolls of South Carolina Volunteers in the Confederate States Provisional Army*, Volume 5: *Miscellaneous Troops*, South Carolina Department of Archives and History.

30. Gary R. Baker, *Cadets in Gray* (Columbia, S.C.: Palmetto Bookworks, 1989), 180.

31. "May 1st, 1865, and May 14th, 1884," *Enterprise and Mountaineer*, May 14, 1884, 2.

32. John Henry Spangler, "Arming the Militia: South Carolina Longarms 1808–1903," Mass. thesis, University of Florida, 1977.

33. "By the Commissioners Appointed by the Legislature of South Carolina to Sell the State Works, Extensive Sale of Machinery, Materials, Tools, etc.," *Daily Phoenix* (Columbia), October 5, 1866, 1.

34. CSR M397, Roll 0109.

35. Nancy Ashmore Cooper, "When the Yankees Sacked Greenville: Stoneman's Raid, May 2, 1865." *Carologue* 10, no. 1 (Spring 1994): 10.

36. Huff, *Greenville*, 144.

37. Mrs. Arthur M. Huger to Mrs. William Mason Smith, May 8, 1865, in Daniel E. Huger, ed., *Mason Smith Family Letters, 1860–1868*, 206–207 (Columbia: University of South Carolina Press, 1950).

38. Caroline Howard Gilman, "Letters of a Confederate Mother: Charleston in the Sixties," *Atlantic Monthly* 137, no. 4 (April 1926): 503–15.

39. The date is shown as May 1, 1865. See Patton, ed., Minutes of the Proceedings of the Greenville Ladies' Association, 70.

40. Gilman, "Letters of a Confederate Mother," 512.

41. John A. Broadus, *Memoir of James Petigru Boyce, D.D., L.L.D.* (New York: A. C. Armstrong and Son, 1893), 189–90.

42. Ibid.

43. Gilman, "Letters of a Confederate Mother," 503–15.

44. "May 1st, 1865, and May 14th, 1884," *Enterprise and Mountaineer*, May 14, 1884, 2.

45. Mrs. Arthur M. Huger to Mrs. William Mason Smith, May 8, 1865, in Huger, ed., *Mason Smith Family Letters*, 206–207.

46. West cannot be further identified.

47. Gilman, "Letters of a Confederate Mother," 512.

48. Ibid., 513.

49. Cooper, "When The Yankees Sacked Greenville," 10.

50. Choice was born on March 22, 1808. He was a private in Company B, Hampton Legion Cavalry Battalion, in 1861 and 1862 and was discharged in August 1862.

51. "May 1st, 1865, and May 14th, 1884," *Enterprise and Mountaineer*, May 14, 1884, 2.

52. Ibid.

Chapter 19: State Works and David Lopez, Postwar

1. Report of the Committee on the Military on a Resolution from the Senate in Relation to the Sale of the State Works at Greenville, No. 32, Wray-Morse File 035.

2. John Wickliffe Stokes (1818–1889) was a forty-seven-year-old Greenville lawyer in 1865. He was buried at Christ Church Episcopal, Greenville. Charles James Elford (November 11, 1821–May 25, 1867) was a forty-four-year-old Greenville lawyer in 1865. He had been colonel of the Sixteenth Regiment, South Carolina Volunteers, and the Third Regiment, South Carolina Reserves. He was buried at Springwood Cemetery, Greenville. George Franklin Townes (June 9, 1809–April 11, 1891) was a fifty-six-year-old Greenville lawyer and state senator in 1865. He was buried at Springwood Cemetery, Greenville. Henry Pinckney Hammett (December 31, 1822–May 8, 1891) was a forty-three-year-old Greenville manufacturer in 1865 and later president of a cotton factory. He was buried at Christ Church, Greenville.

3. General Assembly Reports and Resolutions, 1865, 175–76, St 749, South Carolina Department of Archives and History.

4. "The State Works at Greenville," *Charleston Daily Courier,* February 16, 1866, 2.

5. "The State Works," *Southern Enterprise,* March 29, 1866, 2.

6. Letters, Endorsements, and Reports to the Secretary of War, 1854–1864, National Archives, RG 156, Entry 5.

7. "By the Commissioners Appointed by the Legislature of South Carolina to Sell the State Works, Extensive Sale of Machinery, Materials, Tools, etc.," *Daily Phoenix* (Columbia), October 5, 1866, 1.

8. "Message of Governor James L. Orr to the Senate and House of Representatives of South Carolina," *Charleston Mercury,* November 28, 1866.

9. Records of the General Assembly, Codes and Session Laws, vol. 13 (1861–66): 480, South Carolina Department of Archives and History, Columbia, S165004.

10. "Sale of the State Works," *Mountaineer* (Greenville), November 22, 1866, 2.

11. Office of the Register of Deeds for Greenville County, South Carolina, recorded October 9, 1866, pp. 285–86.

12. Ibid., March 4, 1872, pp. 232–33.

13. George W. Morse to Col. George F. Townes, November 27, 1866, Records of the General Assembly, Senate Correspondence and Assorted Papers, 1866–1877, S165216.

14. *Charles Emerson's Greenville Directory, 1876–77* (Greenville, S.C.: Daily Job Office, 1876).

15. Huff, *Greenville,* 159.

16. Ibid., 159–60.

17. E. D. Sloan Jr., "Gunsmithing in Greenville," paper presented at Club of Thirty-nine, December 15, 1994, and at South Carolina Department of Archives and History, January 9, 1995. A March 19, 2006, revision is in the author's file.

18. "Rules, Regulations, and Schedule of Premiums of the Fourth Grand Annual Fair of the Greenville Agricultural and Mechanical Association of South Carolina Commencing Tuesday, October 17th, 1876," Greenville County Library System, South Carolina Pamphlets, Box 59.

19. Sloan, "Gunsmithing in Greenville."

20. "A Greenville Landmark, Old Warehouse That Served Confederacy as Armory," *Keowee Courier* (Pickens, S.C.), June 1, 1910, 2.

21. "Advertisement for Wm. Montgomery, Machinist," *Charleston Daily News,* November 9, 1865, 3.

22. "South Carolina Lime and Cement," *Charleston Daily News,* Sept 24, 1866, 8; "Dissolutions of Copartnership," *Charleston Daily News,* November 22, 1869, 2.

23. U.S. Census, 1870.

24. Ibid., and 1880; Charleston Business Directory, 1869, 1872, 1875, 1878, 1879, 1881, 1883, and 1884, Sholes' Directory of the City of Charleston, A. E. Sholes, publisher at Charleston, S.C., City Directory at U.S. City Directories, 1822–1995 at http://www.ancestry.com.

25. U.S. Census, 1880.

26. http://www.ancestry.com.

Chapter 20: Morse's Postwar Patent Petitions and Lawsuits

1. Records of the U.S. Senate, Forty-second Congress, National Archives, RG 46.

2. Patton, ed., Minutes of the Proceedings of the Greenville Ladies' Association, 118.

3. Wray-Morse File 012.

4. National Archives, RG 123, Box 449B.

5. U.S. Census, 1870.

6. National Archives, RG 123, Box 449A. The 1870 U.S. Census also shows that Morse was born in New Hampshire.

7. Certificate of Death, Wray-Morse File 025.

8. National Archives, RG 123, Box 449A.

9. Records of the U.S. House of Representatives, Forty-second Congress, National Archives, RG 233, Box 21, p. 3.

10. Memorial of Colonel George Woodward Morse on the subject of the construction of the Mississippi levees, Forty-third Congress, First Session, House of Representatives, p. 11, Wray-Morse File 041.

11. U.S. Patent 20503 letter of disclaimer, June 7, 1872; U.S. Patent 20727 letter of disclaimer, June 29, 1872; U.S. Patent 15995, U.S. Patent Office.

12. George Morse and his family were not listed in *Charles Emerson's Greenville Directory, 1876–77.*

13. National Archives, RG 123, Box 449A.

14. Ibid.

15. The Morse Arms Manufacturing Company *v.* the United States, 16 Ct. Cl. 296, U.S. Court of Claims, December term, 1880.

16. National Archives, RG 123, Box 449A.

17. U.S. Senate, Forty-fifth Congress, Third Session, Congressional Series Report No. 866, National Archives.

18. National Archives, RG 123, Box 449A.

19. Blatchford, Reports of Cases Argued and Determined at the Circuit Court of the United States for the Second Circuit 24: 496ff.

20. National Archives, RG 123, Box 449A.

21. Records of the U.S. House of Representatives, Forty-second Congress, National Archives, RG 233, Box 21.

22. Ibid.

23. "Memorial of George W. Morse to the 44th Congress," 8.

24. Records of the U.S. House of Representatives, Forty-second Congress, National Archives, RG 233, Box 21.

25. Ibid.

26. Ibid.

27. National Archives, RG 123, Box 449A.

28. U.S. Senate, Forty-fifth Congress, First Session, Congressional Series Report No. 866, National Archives.

29. Records of the U.S. House of Representatives, Forty-second Congress, National Archives, RG 233, Box 21.

30. Report of Brigadier General S. V. Benet, Chief of Ordnance, to Secretary of War, March 6, 1875, Wray-Morse File 030.

31. National Archives, RG 123, Box 449A. There are no known Confederate patent records for George Morse.

32. Moody to Negley, Records of the U.S.

House of Representatives, Forty-second Congress, Accompanying Paper Files, National Archives, RG 233, Box 21.

33. Ibid.

34. Records of the U.S. House of Representatives, Forty-second Congress, National Archives, RG 233, Box 21.

35. National Archives, RG 123, Box 449A.

36. Ibid.

37. Records of the U.S. House of Representatives, Forty-second Congress, National Archives, RG 233, Box 21.

38. Records of the U.S. Senate, Forty-second Congress, National Archives, RG 46.

39. Records of the U.S. House of Representatives, Forty-second Congress, National Archives, RG 233, Box 21.

40. H.R. 1108, Records of the U.S. Senate, Forty-second Congress, National Archives, RG 46, Box 145; National Archives, RG 46.

41. Ibid.

42. National Archives, RG 123, Box 449A.

43. Handwritten note, U.S. Patent 15995, p. 1.

44. Patent 15995, U.S. Patent Office.

45. Ibid.

46. "Memorial of George W. Morse to the 44th Congress," 8.

47. Morse Arms Manufacturing Company *v.* the United States, 27 Ct. Cl. 363, U.S. Court of Claims, No. 14340, decided June 20, 1892; "Memorial of George W. Morse to the 44th Congress," 8.

48. U.S. Patent 20503.

49. "Memorial of George W. Morse to the 44th Congress," 8.

50. Morse's great-granddaughter, Mary M. Kane, referred to Morse as "colonel" in a letter to Dr. H. L. Sutherland dated March 2, 1963 (Wray-Morse File 005).

51. "Memorial of Colonel George Woodward Morse on the Subject of the Construction of the Mississippi Levees", U.S. House of Representatives, Forty-third Congress, First Session, Wray-Morse File 041.

52. "Memorial of George W. Morse to the 44th Congress."

53. Ibid., 18.

54. Ibid.

55. Ibid., 9.

56. Ibid.

57. Ibid., 13.

58. National Archives, RG 123, Box 449B.

59. "Memorial of George W. Morse to the 44th Congress," 14–15; National Archives, RG 123, Box 449B.

60. "Memorial of George W. Morse to the 44th Congress," 16.

61. Ibid.

62. Ibid.

63. Ibid., 21.

64. Ibid.

65. Ibid., 22.

66. Ibid., 24.

67. U.S. House of Representatives, Forty-fifth Congress, Third Session, Congressional Series Report No. 1, National Archives.

68. Ibid.

69. Forty-fifth Congress, Third Session, Congressional Series Report No. 866, National Archives.

70. U.S. Senate, Forty-fifth Congress, First Session, Congressional Series Report No. 866, National Archives.

71. Ibid.

72. U.S. House of Representatives, Forty-eighth Congress, First Session, Congressional Series Report No. 508, National Archives.

73. Ibid.

74. Ibid.

75. Kane to Sutherland, March 2, 1963, Wray-Morse File 005.

76. Blatchford, Reports of Cases Argued and Determined at the Circuit Court of the United States for the Second Circuit 24: 504.

77. Report of Brigadier General S. V. Benet, Chief of Ordnance, to Secretary of War, March 6, 1875, Wray-Morse File 030.

78. Blatchford, Reports of Cases Argued and Determined at the Circuit Court of the United States for the Second Circuit 24: 496ff.

79. National Archives, RG 123, Box 449A.

80. Ibid.

81. Ibid.

82. Ibid.

83. Ibid.

84. U.S. Senate, Forty-fifth Congress, First Session, Congressional Series Report No. 866, National Archives.

85. Report of Brigadier General S. V. Benet, Chief of Ordnance, to Secretary of War, March 6, 1875, Wray-Morse File 030.

86. Ibid.

87. Morse Arms Manufacturing Company *v.* the United States, 16 Ct. Cl. 296, U.S. Court of Claims, December term, 1880.

88. Ibid.

89. Morse Arms Manufacturing Company *v.* the United States, 27 Ct. Cl. 363, U.S. Court of Claims, No. 14340, decided June 20, 1892. Another date, February 12, 1884, is given in a court

document found in National Archives, RG 123, Box 449A.

90. George W. Morse to Milledge L. Bonham, August 26, 1885, Bonham Papers, South Caroliniana Library, Columbia. A copy is in the Morse File at the McKissick Museum.

91. Ibid., November 12, 1885, Bonham Papers, South Caroliniana Library, Columbia.

92. National Archives, RG 123, Box 449A.

93. Morse Arms Manufacturing Company *v.* the United States, 27 Ct. Cl. 363, U.S. Court of Claims, No. 14340, decided June 20, 1892.

94. National Archives, RG 123, Box 449A; dates are shown as both April 27, 1896, and May 25, 1896.

95. National Archives, RG 123, Box 449A.

Chapter 21: Morse's Final Productivity

1. Samuel L. Maxwell Sr., Lever Action Magazine Rifles Derived from the Patents of Andrew Burgess (published by the author, 1976).

2. The Buffalo Bill Center of the West lists this firearm as having both serial number 170 and 70.

3. James Zuppan, "The Morse Model 1886 Moveable Base Cartridge, Parts I and 2," *Gun Report,* September 1986, 56–62, and October 1986, 30–33.

4. Ibid.

5. Ibid.

6. Ibid.

7. Wray-Morse File 025.

8. Ibid.

9. Certificate of death, Wray-Morse File 025.

10. The statement as it appears in Morse's obituary is inaccurate. George Woodward Morse was very distantly related to Professor Samuel F. B. Morse. They likely shared a great-great-great-great-grandfather.

11. "End of a Busy Life," 1.

12. The following members of the Morse family are buried in lot 2372 at Oak Hill Cemetery, Georgetown, Washington, D.C.: George W. Morse and his wife, Marianne; their spinster daughter, Lelia E. Morse; their son, Bryan H. Morse, and his wife, Elizabeth Ringwalt Morse (the wife of Peabody, who was George and Marianne's son); Hattie H. Morse and her son, George E. Morse; and Bryan and Elizabeth's son, Bryan Woodward Morse, and his wife, Eloise (Wray-Morse File 022).

13. Robert S. Seigler, *A Guide to Confederate Monuments on South Carolina* (Columbia: South Carolina Department of Archives and History, 1997).

14. Patton, ed., Minutes of the Proceedings of the Greenville Ladies' Association, 118.

....

Conclusion

1. Whitcher, History of the Town of Haverhill, 599.
2. "Memorial of George W. Morse to the 44th Congress," 21.
3. Ibid.
4. Ibid., 9.
5. Ibid., 39.

....

Appendix 2

1. Celeste C. Topper and David L. Topper, *Civil War Relics from South Carolina* (privately printed, 1988), 91.
2. Ibid.
3. Graham Burnside, "Amazing Morse Rifle," *Shooting Times,* March 1961, 1.
4. James B. Whisker et al., *Arming the Glorious Cause* (Livonia, N.Y.: R and R Books, 1998), 57.
5. Sutherland, "Arms Manufactury in Greenville County," 56.

....

Appendix 3

1. Available payrolls and the number of slaves working for each payroll are: March 15, 1863, 73 total; July 15, 1863, 53 total; August 15, 1863, 37 total; November 15, 1863, 49 total; January 15, 1864, 49 total; February 15, 1864, 40 total; March 15, 1864, 39 total; July 15, 1864, 39 total; August 15, 1864, 37 total; October 31, 1864, 40 total; December 31, 1864, 44 total; January 31, 1865, 48 total (Comptroller General–State Auditor, Records of the State Works, Payroll [Negro], South Carolina Department of Archives and History, S126182).
2. J. Ralph Smith most likely to A. P. Calhoun, undated. Copy in Museum and Library of Confederate History, Greenville, S.C.

3. Probably Theodore Fillette, a thirty-two-year-old Charleston druggist in 1860. He was born in February 1828. He lived in New York in 1870 and in Washington, D.C., in 1880. He died in May 1887 and was buried in Washington, D.C.

....

Appendix 4

1. Eleven payrolls are available: July 15, 1863; January 15, 1864; February 15, 1864; April 15, 1864; June 15, 1864; July 15, 1864; August 15, 1864; September 30, 1864; December 31, 1864; January 31, 1865; and February 28, 1865 (South Carolina Department of Archives and History, State Auditor, Payrolls, 1863–1864 [Oversize]).
2. *Charles Emerson's Greenville Directory*; http://www.findagrave.com; http://www.ancestry.com.
3. Freitag, born in Germany about 1823 or 1824; gunsmith on Camden and Columbia; died about 1903.
4. Turner's obituary appears in *The State,* September 17, 1903, 3.
5. W. R. Powell (April 7, 1831–June 11, 1895).
6. Sloan, "Gunsmithing in Greenville."
7. Sutherland, "Arms Manufactury in Greenville County."
8. B. September 6, 1828–D. April 19, 1900.
9. Richard Furman Divver, born October 21, 1840, physician in Anderson, S.C., died September 14, 1930 (Vandiver, *Traditions and History of Anderson County,* 261).
10. J. R. Smith to James Tupper, September 24, 1863, Comptroller General–State Auditor, Miscellaneous Papers of the State Auditor, S126174.
11. Sloan, "Gunsmithing in Greenville."
12. A payroll note says, "gone to the enemy."
13. State Reports of South Carolina, 251. Cawthen, Charles E., South Carolina State Records. . . @Archival Sources.
14. Briggs' CSR, M267, Roll 0076.
15. Thomas Dixon's CSR M267, Roll 0338.
16. Flemming's CSR M267, Roll 345.

BIBLIOGRAPHY

Archival Sources

Adjutant and Inspector General of South Carolina, Order and Letterbook, 1862–1864, in Lewis F. Knudsen Jr., compiler, *Captain William F. Percival's Co. Mounted Infantry, S.C. State Troops, (Aiken Mounted Infantry) 1864–1865,* 2009, 3, South Carolina Department of Archives and History.

Governor Milledge L. Bonham Papers, Letters Received and Sent. South Carolina Department of Archives and History, Columbia. S512001.

British Patent 1357. "Specification of George Woodward Morse, Breech Loading Firearms, London, printed by George E. Eyre and William Spottiswood, Printers to the Queen's Most Excellent Majesty, Published at the Great Seal Patent Office, 25, Southampton Buildings, Holbron, 1857."

Cauthen, Charles E., ed. *The State Records of South Carolina: The Journals of the South Executive Councils of 1861 and 1862.* Columbia: South Carolina Archives Department, 1956.

Confederate Historian: Rolls of South Carolina Volunteers in the Confederate States Provisional Army, Volume 5: *Miscellaneous Troops,* South Carolina Department of Archives and History.

Compiled Service Records of Confederate Soldiers Who Served in Organizations from the State of South Carolina. National Archives Microfilms Publications, Microcopy No. 267.

Comptroller General–State Auditor. Miscellaneous Papers of the State Auditor. South Carolina Department of Archives and History, Columbia.

———. Records of the Ordnance Bureau. Return of Ordnance and Ordnance Stores. September 30, 1864, State Arsenal, Columbia. Department of Archives and History, Columbia.

"First Annual Report of the State Auditor, 1863." Records of the General Assembly, Reports and Resolutions, 1863. South Carolina Department of Archives and History, Columbia.

Papers of Andrew G. Magrath. South Carolina Department of Archives and History, Columbia. Box 1, S513004.

"Memorial of George W. Morse to the 44th Congress of the United States in Regard to His Claim as the Inventor of the Modern Metallic Cartridge System of Fire-arms." Wray-Morse File 040, Atlanta History Center.

Morse, George Woodward. "Morse's Breech-Loading Fire-arms." Pamphlet. Benjamin McCulloch Papers, Box 3G37, Materials Related to Mexican War, Utah Expedition, and the Civil War 1846–1862, Broadsides 1862. Briscoe Center for American History, Austin, Tex.

National Archives, Record Groups 46, 123, 156, 233.

Patented Case Files, 1836–1973. Records of the Patent and Trademark Office, Record Group 241; National Archives at College Park, Md.

Governor Pickens Papers, Letters Received and Sent. South Carolina Department of Archives, Columbia. S511001.

"Report from the Committee on the Subject of the Stores, Machinery, and Property Captured at Harper's Ferry." Document 31. Greensboro History Museum, Civil War Collection, Mss. Coll. #16 65:1.

S.C. Dept. Archives & History, Columbia. 160–181.

"Report of W. H. Gist Chief of the Department of Construction and Manufacture, and David Lopez to His Excellency, Governor Pickens, August 29, 1862." Constitutional Papers, Convention of 1860–1862, South Carolina Department of Archives and History.

"Reports and Resolutions of the South Carolina House of Representatives, 18 December 1862, Records of the General Assembly, Reports and Resolutions." South Carolina Department of Archives and History, Columbia.

Sloan, E. D., Jr. "Gunsmithing in Greenville." Paper presented at Club of Thirty-nine, December 15, 1994, and at South Carolina Department of Archives and History, January 9, 1995. March 19, 2006, revision in the author's file.

"State Works Pay Vouchers." South Carolina Department of Archives and History, Columbia, S126182.

"George W. Wray, Jr., File." Atlanta History Center.

....

Published and Internet Sources

Albaugh, William A., III, and Edward N. Simmons. *Confederate Arms.* New York: Bonanza Books, 1958.

Blatchford, Samuel. *Reports of Cases Argued and Determined at the Circuit Court of the United States for the Second Circuit.* New York: Baker, Voorhist, and Co., 1888.

Capers, Ellison. *Confederate Military History.* Extended ed., vol. 6: *South Carolina.* Wilmington, N.C.: Broadfoot Publishing Co., 1987.

Cauthen, Charles Edward. *South Carolina Goes to War, 1860–1865.* Chapel Hill: University of North Carolina Press, 1950.

Charles Emerson's Greenville Directory, 1876–77. Greenville, S.C.: Daily Job Office, 1876.

Cooper, Nancy Ashmore. "When The Yankees Sacked Greenville: Stoneman's Raid, May 2, 1865." *Carologue* 10, no. 1 (Spring 1994): 8–13.

Cutrer, Thomas W. *Ben McCulloch and the Frontier Military Tradition.* Chapel Hill: University of North Carolina Press, 1993.

Edwards, William B. *Civil War Guns: The Complete Story of Federal and Confederate Small Arms: Design, Manufacture, Identification, Procurement, Issue, Employment, Effectiveness, and Postwar Disposal.* Secaucus, N.J.: Castle Books, 1978.

"End of a Busy Life." *Critic-Record* (Washington, D.C.), no. 6108 (March 8, 1888).

Fuller, Claud E., comp. *Springfield Muzzle-Loading Shoulder Arms. A Description of the Flint Lock Muskets, Musketoons and Carbines and the Muskets, Musketoons, Rifles, Carbines and Special Models from 1795 to 1865 with Ordnance Office Reports, Tables and Correspondence and a Sketch of Springfield Armory.* Glendale, N.Y.: S and S Firearms, 1986.

Gilman, Caroline Howard. "Letters of a Confederate Mother: Charleston in the Sixties." *Atlantic Monthly* 137, no. 4 (April 1926): 503–15.

Gunn, Jack W. "Ben McCulloch: A Big Captain." *Southwestern Historical Quarterly* 58, no. 1 (July 1954): 1–21.

Haverhill Historical Society. www.Haverhill HistoricalSociety.blogspot.com. Last modified December 15, 2014.

Hicks, Major James E. *U.S. Military Firearms 1776–1956.* La Canada, Calif.: James E. Hicks & Son, 1962.

Hill, Richard T., and William E. Anthony. *Confederate Longarms and Pistols: A Pictorial Study.* Charlotte, N.C.: Richard Taylor Hill and William Edward Anthony, 1978.

Holter, Robert M. "The Morse-Muzzy Connection." *Gun Report,* May 2001, 18–23.

Hoyem, George A. *The History and Development of Small Arms Ammunition, Volume II: Centerfire: Primitive, and Martial Longarms.* Tacoma, Wash.: Armory Publications, 1982.

Huff, Archie V., Jr. *Greenville: The History of the City and County in the South Carolina Piedmont.* Columbia: University of South Carolina Press, 1995.

Huger, Daniel E., ed. *Mason Smith Family Letters, 1860–1868.* Columbia: University of South Carolina Press, 1950.

Jones, Gordon L. *Confederate Odyssey: The George W. Wray Civil War Collection at the Atlanta History Center.* Athens: University of Georgia Press, 2014.

———. "McKissick Museum's Morse Design Cased Firearm February 16, 1987." Unpublished. Copy at South Carolina State Museum, HSSI 701.

Lewis, Colonel B. R. "Morse Arms and Ammunition." *American Rifleman,* March 1955, 31–33.

Lustyik, Andrew F. "Civil War Carbines." *Worldwide Gun Report,* 1962.

McAulay, John D. *Civil War Breech Loading Rifles: A Survey of the Innovative Arms of the American Civil War.* Lincoln, R.I.: Andrew Mowbray Publishers, 1987.

———. *Civil War Carbines, Volume II, . . . The Early Years, Appendix I.* Lincoln, R.I.: Andrew Mowbray Inc., 1991.

Moller, George D. *American Military Shoulder Arms, Volume III: Flintlock Alterations and Muzzleloading Percussion Shoulder Arms, 1840–1865.* Albuquerque: University of New Mexico Press, 2011.

Murphy, John M., and Howard Michael Madaus. *Confederate Carbines and Musketoons: Cavalry Small Arms Manufactured in and for the Southern Confederacy, 1861–1865.* Santa Ana, Calif.: Graphic Publishers, 2002.

———. *Confederate Rifles and Muskets: Infantry Small Arms Manufactured in the Southern Confederacy, 1861–1865.* Newport Beach, Calif.: Graphic Publishers, 1996.

Murphy, John M. *Confederate Carbines and Musketoons.* Dallas, Texas: Taylor Publishing Co., 1986

Myers, Craig A. *Greenville Railroad History since 1853: An Overview of Greenville's Railroads, Streetcars, Buildings and Businesses.* 2nd ed. Greenville: By the author, 2002.

Patton, James Welch, ed. *Minutes of the Proceedings of the Greenville Ladies' Association in Aid of the Volunteers of the Confederate Army.* New York: AMS Press, [1970].

Pitman, John. *The Pitman Notes on U.S. Martial Small Arms and Ammunition 1776–1933, Volume One, Breech-loading Carbines of the United States Civil War Period, from Original Drawings made by Brigadier General John Pitman.* Tacoma, Wash.: Armory Publications, 1987.

"Rules, Regulations, and Schedule of Premiums of the Fourth Grand Annual Fair of the Greenville Agricultural and Mechanical Association of South Carolina Commencing Tuesday, October 17th, 1876." Greenville County Library System, South Carolina Pamphlets, Box 59.

Smith, Roy McBee. *Vardry McBee, 1775–1864: Man of Reason in an Age of Extremes.* Columbia, S.C.: R. L. Bryan Co., 1992.

Stiefel, Barry. "David Lopez Jr.: Builder, Industrialist, and Defender of the Confederacy." *American Jewish Archives Journal* 64 (2012): 53–81.

Sutherland, H. L. "Arms Manufactury in Greenville County." *Proceedings and Papers of the Greenville County Historical Society 1968–1971* 4 (1971): 46–61.

Topper, Celeste C. and David L. Topper, *Civil War Relics from South Carolina.* (privately printed, 1998), 91.

Vandiver, Louise Ayer. *Traditions and History of Anderson County.* 1928. N.p.: McNaughton and Gunn, 1991.

Vorisek, Joseph T. *The Breechloading Shotgun, 1860–1940, Volume II, G Through P.* Brighton, Mich.: Cornell Publications, 1990.

War of the Rebellion: A Compilation of the Official Records of the Union and Confederate Armies. Washington, D.C.: Government Printing Office, 1893.

Whitcher, William F. *History of the Town of Haverhill, New Hampshire.* Concord, N.H.: Rumford Press, 1919.

INDEX

Page numbers in italic type indicate illustrations.